PROTEIN MUTATIONS

Consequences on Structure, Functions, and Diseases

PROTEIN MUTATIONS

Consequences on Structure, Functions, and Diseases

Editor

M Michael Gromiha

Indian Institute of Technology Madras, India

World Scientific

EW JERSEY • LONDON • SINGAPORE • BEIJING • SHANGHAI • HONG KONG • TAIPEI • CHENNAI • TOKYO

Published by

World Scientific Publishing Co. Pte. Ltd.
5 Toh Tuck Link, Singapore 596224
USA office: 27 Warren Street, Suite 401-402, Hackensack, NJ 07601
UK office: 57 Shelton Street, Covent Garden, London WC2H 9HE

British Library Cataloguing-in-Publication Data
A catalogue record for this book is available from the British Library.

PROTEIN MUTATIONS
Consequences on Structure, Functions, and Diseases

ISBN 978-981-12-9325-2 (hardcover)
ISBN 978-981-12-9326-9 (ebook for institutions)
ISBN 978-981-12-9327-6 (ebook for individuals)

For any available supplementary material, please visit
https://www.worldscientific.com/worldscibooks/10.1142/13841#t=suppl

Typeset by Stallion Press
Email: enquiries@stallionpress.com

Dedicated to the memory of my beloved Father.

https://doi.org/10.1142/9789811293269_fmatter

Preface

Proteins perform various functions in living organisms, including enzymes, receptors, transporters, defense proteins, nutrients, and regulators. These functions are mainly dictated by their structures. The mutation of amino acid residues in a protein affects its structure and function, and some of them lead to diseases. The influence of amino acid mutations in a protein on its folding, stability, binding affinity, function, and disease has been extensively studied using experimental and computational approaches. The availability of high-speed computers with extreme storage capacity prompted computational biologists to develop efficient and reliable databases and algorithms. Further, recent advances in big data and artificial intelligence have narrowed down the gap between the performance of computational methods and experimental observations.

This book is dedicated to providing a forum for both fundamental aspects and applications of computational methods to understand the effects of mutations in proteins from different perspectives: change in aggregation, stability, and folding rate, alterations in the binding affinity of protein–protein, protein–nucleic acid, and protein–carbohydrate complexes, influence on protein functions and diseases; as well as next-generation sequence analysis.

The highlight of the book is the coverage of the consequences of protein mutations from different perspectives, such as aggregation, folding, stability, binding affinity, function, and diseases, as well as a wide range of investigations ranging from database development to drug design. The web addresses of available databases and online tools are provided in the appropriate chapters. In addition, the literature has been thoroughly surveyed, and important methods have been highlighted in detail. This book will be of immense use and a valuable guide to students and researchers working on mutational effects in proteins, as well as to those who have an interest to study protein mutations using computational approaches.

The book is broadly classified into three categories: (i) protein structure, folding, and stability, (ii) protein function and binding affinity, and (iii) disease-causing mutations. Part I deals with structural aspects of proteins with a systematic analysis of mutations in protein aggregation through *in silico* methods, understanding and predicting the stability of proteins upon mutations, important factors influencing the folding rates of proteins and their mutants, as well as protein disorder upon mutation with details on databases, algorithms, and applications.

Part II is devoted to protein function and the binding affinity of protein–protein, protein–nucleic acid, and protein–carbohydrate complexes. This part includes computational resources such as databases and tools for predicting the binding affinity change upon mutation in protein–protein, protein–DNA, protein–RNA, and protein–carbohydrate complexes. These chapters also include factors for understanding the binding affinity and comparison of different computational algorithms. In addition, the effects of mutations on protein functions have been discussed.

Part III is focused on disease-causing mutations, including the influence of mutations in protein sequences and structures, predicting hotspot residues and disease-causing mutations, and next-generation sequence analysis. It covers diverse aspects such as understanding disease-causing mutations in proteins with applications to HIV, identifying cancer hotspot residues and mutations in proteins, deciphering disease-causing mutations in membrane proteins, decoding the evolution of COVID-19 through mutational studies on SARS-CoV-2, and transcriptome-based analysis for understanding the effects of mutations in neurodegenerative diseases.

In essence, this book would be a valuable resource for students and researchers to strengthen their knowledge about studies on the consequences of mutations in proteins.

M. Michael Gromiha

https://doi.org/10.1142/9789811293269_fmatter

Acknowledgments

I am deeply indebted to Professor P.K. Ponnuswamy, who introduced me to the field of proteins, and encouraged to strengthen the knowledge in protein research.

I am grateful to Prof. S. Pongor, Prof. Y. Akiyama, Prof. Y-h. Taguchi and Prof. I. Simon for their continuous support and encouragements.

I also extend my thanks to my collaborators Dr. S. Kumar, Prof. D. Frishman, Prof. Y-h. Taguchi and Dr. L-T. Huang, and students, K. Yugandhar, S Jemimah, R. Prabakaran, P. Rawat, Y. Dhanusha, A. Srivastava, S. Akila, A. Kulandaisamy, N.R. Siva Shanmugam, M. Pandey, R. Nikam, G. Sudha, N.P. Sneha, K. Harini, D. Sharma, F. Ridha, S.R. Krishnan, P.R. Reddy, S. Lekshmi, S.K. Shah, A. Phogat and V. Sankaran for their efforts to prepare necessary chapters for the book.

It is my pleasure to acknowledge my wife A. Mary Thangakani and children Michael Mozim, Angela Shalom and Michael Abejo for their support to edit the book. I also immensely acknowledge my mother for her valuable guidance.

I warmly thank Ms. Sandhya Devi for efficiently managing the production of the book and World Scientific for publishing the book.

Finally, I thank all my well wishers and friends who encouraged me to edit this book.

https://doi.org/10.1142/9789811293269_fmatter

Contents

Part I

Protein structure, folding, and stability

https://doi.org/10.1142/9789811293269_0001

Chapter 1

Deciphering the modulatory role of mutations in protein aggregation through *in silico* methods

R. Prabakaran[1,2,#,*], Puneet Rawat[2,3,#], Sandeep Kumar[4], and M. Michael Gromiha[2,5,*]

[1]*Department of Biology, Emory University, Atlanta, GA 30322, USA*
[2]*Protein Bioinformatics Lab, Department of Biotechnology, Indian Institute of Technology Madras, Chennai, Tamil Nadu 600036, India*
[3]*University of Oslo and Oslo University Hospital, Oslo, Norway*
[4]*Biotherapeutics Discovery, Boehringer-Ingelheim Inc. 5571 R & D Building, 175 Briar Ridge Road, Ridgefield, CT 06877 USA*
[5]*International Research Frontiers Initiative, School of Computing, Tokyo Institute of Technology, Yokohama 226-8501, Japan*

Abstract

A substantial portion of the cellular homeostasis network is dedicated to managing the consequences of protein aggregation, a process that

*Corresponding authors
RP: prabakaran@emory.edu
MMG: gromiha@iitm.ac.in
[#]These authors contributed equally to this work.

cannot be avoided. In recent decades, research into human diseases, including neurodegenerative disorders, diabetes, and various forms of amyloidosis, has highlighted the profound impact of protein aggregation and its variants on human health. While numerous experimental techniques exist for studying protein aggregation, *in silico* methods have become invaluable due to their speed, cost-effectiveness, and adaptability. This chapter provides a concise overview of the decades-long development of computational tools aimed at identifying, quantifying, and exploring the aggregation propensity (AP) and kinetics of proteins and their variants. With a particular emphasis on the influence of protein mutations, we delve into the prediction of protein aggregation kinetics and the comprehension of the underlying mechanisms through molecular dynamic simulations.

Keywords: Protein aggregation; prediction; kinetics; databases; molecular dynamics; review; amyloid

1.1 Introduction

1.1.1 *Protein aggregation*

Life as we know it is dependent on the molecular activities of biomolecules such as nucleic acids, lipids, carbohydrates, and notably proteins. Every life-sustaining biological process is a result of evolutionarily screened biomolecular interactions. Proteins, being the molecular machinery of the cell, play a central role in carrying out and regulating almost all essential cellular functions. However, like any colloidal molecule, proteins undergo concentration-dependent agglomeration. Agglomeration of protein molecules within or outside the cell membrane could disrupt the cellular process by depleting structurally active conformations and compromise the cellular structural integrity, if not limited.

Agglomeration of protein molecules could lead to the formation of ordered structures with detectable periodic arrangements or amorphous structures. In either scenario, protein molecules tend to phase out of the bulk cellular solution, leading to loss of function and sometimes causing undesirable effects. One such process is amyloid fibrilization, which is a result of the assembly of protein molecules into long thread-like fibrils, often observed in brain tissues associated with neurodegenerative diseases (Yerbury *et al.*, 2016; Chiti & Dobson, 2017). The amyloid term was taken from the Latin word "Amylum," which means sugar. In the mid-nineteenth century, Rudolph Virchow introduced the term "Amyloid" for protein aggregates, when a test to detect plant starch called the iodine-sulfuric acid test also showed positive results for the deposits in the nervous system.

As evident from the increasing number of research publications in the last couple of decades, a considerable amount of effort has been invested in understanding protein assembly, especially protein aggregation. Fibrillization, crystallization, phase separation, condensate formation, or unstructured amorphous clumps have been important topics of scientific curiosity in the past two decades. One of the foremost reasons is the association of protein aggregates with a wide range of human pathologies, including neurodegenerative diseases such as Alzheimer's, Parkinson's, dementia, and Huntington's disease (Dogan, 2017; Nguyen *et al.*, 2021; Louros *et al.*, 2023; Panda *et al.*, 2023). The other equally driving motivation is to understand protein aggregation is to mitigate aggregate formation during the manufacturing and storage of protein- and peptide-based biotechnological products such as antibodies, vaccines, and other therapeutic molecules. In addition, several research groups are focused on developing techniques to mimic the biological phenomenon of protein assembly to produce nanostructures and novel materials with desired characteristics such as shape, stereospecificity, tensile strength, elasticity, and hydrophobicity.

1.1.2 *Proteostasis*

To understand protein aggregation, one must comprehend the complicated network of proteostasis. Protein molecules are synthesized by ribosomes in the cytoplasm, folded to their native conformation, trafficked to the location of function, and over time degraded and recycled. Like every biological process, protein synthesis, folding, and trafficking are prone to error. Naturally, organisms, especially multicellular organisms, have evolved to regulate and mitigate these errors through regulatory networks.

Protein homeostasis or proteostasis refers to the processes that synthesize, transport, regulate, and maintain the functional proteome of the cell. The proteostasis network consists of ~2,000 proteins that mitigate errors and maintain the balance of the human proteome (Yerbury *et al.*, 2016; Klaips *et al.*, 2018; Kurtishi *et al.*, 2019).

Chaperones are an integral part of proteostasis. Several proteins exclusively depend on molecular chaperones to fold to their native state. Molecular chaperones consist of proteins such as GroEL, GroES, DnaK, DnaJ, and GrpE that act as foldase or disaggregase on proteins (Kerner *et al.*, 2005; Tyedmers *et al.*, 2010; Finka & Goloubinoff, 2013; Miller *et al.*, 2015). In humans, chaperones consist of ~330 genes with distinct functions and diverse substrates (Klaips *et al.*, 2018). Misfolded proteins and protein aggregates are degraded through a selective ubiquitin-proteasome system (UPS) or nonspecific autophagy (Ebrahimi-Fakhari *et al.*, 2011; Miller *et al.*, 2015; Kurtishi *et al.*, 2019). When refolding or degradation fails, aggregates are isolated into cytosolic inclusion bodies: the JUxta Nuclear Quality (JUNQ) control compartment and insoluble protein deposit (IPOD) (Kaganovich *et al.*, 2008).

Several human pathological conditions, collectively called protein conformational diseases or proteinopathies have been associated with the misfolding and aggregation of proteins (Chiti & Dobson, 2017). In particular, amyloidosis is the formation and deposition of insoluble fibrillar amyloids within a tissue. Though the constituents of these amyloid deposits vary with the pathology, they are primarily made up of proteins specific to the tissue. Structurally, these deposits share a common architecture of cross-β zipper motifs at their core (Sunde & Blake, 1997). Several proteins of interest are associated with diseases, such as α-synuclein with Parkinson's, amyloid β with Alzheimer's', immunoglobulin light chains with Atrial amyloidosis, and amylin with diabetes, have been extensively studied (Li & Liu, 2020).

1.1.3 *Amyloidogenic proteins*

One classic example of an amyloidogenic protein is islet amyloid polypeptide (IAPP). IAPP, also called Amylin, is a 37-residue intrinsically disordered hormone known to form amyloid *in vitro* and *in vivo*.

Interestingly, the amyloidogenicity of IAPP varies drastically across species. Researchers have carried out several experiments to understand the aggregation propensity of amylin. For example, Palato *et al.* (2019) studied the amyloidogenicity of 12 different naturally occurring full-length animal variants of IAPP by measuring the fibrillization using Thioflavin T binding and atomic force microscopy (AFM). Among 12 homologous protein sequences, six orthologs showed significantly higher AP and formed β-rich aggregation. Further analysis of these sequences revealed the influence of a focused mutation in the sequence, which enhances or suppresses the AP. **Figure 1.1** shows the alignment of aggregation-prone human IAPP against orthologs and sequence variations (colored red) occurring around a short region "FGAIL." "FGAIL" has been identified as an aggregation-prone region (APR) in IAPP and observed to form amyloid-like fibrils in several independent *in vitro* experiments. Removal of this region depletes or attenuates the ability of Amylin to aggregate.

Similar studies on other amyloid-forming proteins lead to identifying segments prone to aggregation. Aggregation is driven by the interaction of short APRs in proteins. Serrano and Lopez de la Paz studied the effect of the insertion of APRs in globular proteins without disrupting protein stability (López de la Paz & Serrano, 2004; Esteras-Chopo *et al.*, 2005; Maurer-Stroh *et al.*, 2010). The study involved the insertion of

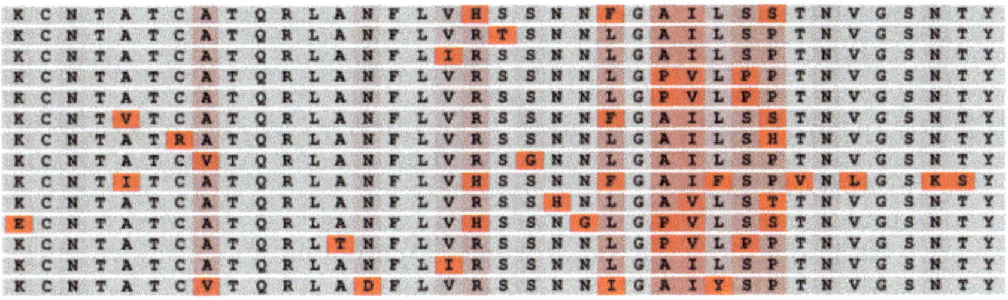

Figure 1.1 Human IAPP and its orthologs. The multiple sequence alignment shown highlights conserved and variable residues in the IAPP sequence as compared across species (in order: *Homo sapiens*, *Canis lupus familiaris*, *Felis catus*, *Mus musculus*, *Rattus norvegicus*, *Gorilla gorilla*, *Odobenus rosmarus divergens*, *Dasypus novemcinctus*, *Ochotona princeps*, *Ictidomys tridecemlineatus*, *Microtus ochrogaster*, *Peromyscus maniculatus*, *Panthera tigris*, and *Calypte anna*). The sequences were obtained by blasting human IAPP against reference proteomes

amyloidogenic peptides such as "STVIIE" in proteins at various locations in sequence without disturbing the native structure. Studies showed that insertion can accelerate or trigger amyloid aggregation in proteins that do not form fibrils prior to destabilizing conditions. Through extensive mutagenesis experiments, researchers have identified that a few mutations were sufficient to trigger amyloid aggregation or suppress aggregation. The ability of a few short stretches to increase AP was studied extensively using synthetic peptides and led to the development of several computational models.

1.2 Investigations on protein aggregation using experimental methods

Numerous researchers have conducted in-depth investigations into the mechanisms, underlying biophysical characteristics, and significance of protein aggregation utilizing various experimental methodologies (Maurer-Stroh *et al.*, 2010; Engelsman *et al.*, 2011; Buck *et al.*, 2012; Lorber *et al.*, 2012; Malmos *et al.*, 2017; Sahoo *et al.*, 2020; Vu *et al.*, 2022). At present, a multitude of experimental methodologies exists to facilitate the quantitative and qualitative analysis of protein aggregate abundance, localization, structural characteristics, and kinetics (Mahler *et al.*, 2009; Gregoire *et al.*, 2012). Some of the widely used experimental techniques are discussed in detail below.

1.2.1 *Circular dichroism and Fourier transform infrared spectroscopy*

The two techniques, circular dichroism (CD) and Fourier transform infrared spectroscopy (FTIR) are frequently employed as qualitative assays for the characterization of protein aggregates (Miller *et al.*, 2013; Micsonai *et al.*, 2015). These methods analyze the secondary structure of the protein sample, as amyloid aggregates predominantly consist of β-sheet arrangements. Generally, these experiments are conducted to correlate changes or overall secondary structure content with a control or reference. FTIR is readily used to determine antiparallel or parallel β-sheet structures. CD, on the other hand, harnesses circularly polarized light, which traverses through an optically active medium. The disparity in absorption between right- and left-polarized light yields the CD spectrum. The absorption of light is mathematically represented by the Beer–Lambert law.

$$\Delta A = (\epsilon_L - \epsilon_R)Cl \qquad \textbf{(Eq. 1.1)}$$

Where, ϵ_L and ϵ_R are molar extinction coefficients for left and right circularly polarized light, respectively. C is the molar concentration of protein and l is the path length in centimeters.

1.2.2 *Dynamic light scattering*

Dynamic light scattering (DLS) operates on the principles of Rayleigh scattering, wherein incident light interacts with diminutive particles, scattering in multiple directions. This methodology involves the transmission of monochromatic light through a polarizer into the protein specimen, enabling the determination of protein aggregate dimensions in terms of hydrodynamic radius. This method is a quantitative analysis of protein aggregates and provides information on size distribution, rate of aggregation, polydispersity (variation in aggregate size), and aggregation state (oligomers or fibrils) (Lorber *et al.*, 2012). However, it does not discriminate between amyloid and amorphous aggregates and frequently complements the fluorescent dye-based assays.

1.2.3 *Fluorescent dye-based binding assay*

Fluorescent dye-based assays (such as Thioflavin T (ThT) and Congo red) stand as the preeminent and extensively adopted methodologies for the detection and characterization of amyloid fibrils. These dyes serve the dual purpose of visualizing and quantifying the presence of amyloid fibrils. In fluorescent dye-based assays, proteins are subjected to incubation alongside the dye within controlled experimental parameters encompassing factors such as temperature, pH, protein concentration, ionic levels, buffer composition, and additives, among others, and measurements are collected at designated time intervals. Crucially, the dyes exhibit an affinity solely toward fibrillar structures. Consequently, a surge in fluorescent intensity is observed after binding of the dye with nascent amyloid fibrils (Biancalana & Koide, 2010).

1.2.4 *ANS fluorescence assay*

The 1-anilinonaphthalene-8-sulfonate (ANS) dye has emerged as a pivotal tool for the appraisal of protein surface hydrophobicity. This dye showcases an elevation in fluorescence intensity when it interacts with exposed hydrophobic domains on the protein surface (Hawe *et al.*, 2008). Notably, hydrophobic interactions wield a substantive influence within the realm of protein aggregation mechanisms. Therefore, the evaluation of protein surface hydrophobicity assumes significance in unraveling the intricate facets of aggregation pathways. For example, a study on Aβ42 fibrils underscores the observation that prefibrillar oligomers distinctly exhibit the most pronounced augmentation in ANS fluorescence (Ladiwala *et al.*, 2011). These prefibrillar oligomers, in turn, have been associated with cellular toxicity, thereby emphasizing their relevance within the context of cellular dynamics (Kayed *et al.*, 2004; Butterfield & Lashuel, 2010).

1.2.5 *Antibody dot blot assay*

Numerous investigators have devoted their efforts to devising conformation-specific antibodies, instrumental in the identification and surveillance of protein aggregation states. A noteworthy example encompasses three distinct antibodies tailored for the identification of aggregates, where (i) "A11" exhibits an affinity toward prefibrillar oligomers (Kayed *et al.*, 2003); (ii) "OC" targets the cross β-sheet structures inherent in oligomers, protofibrils, and fibrils (Kayed *et al.*, 2007); and (iii) "αAPF" exhibits specificity for annular protofibrils (Kayed *et al.*, 2009).

1.2.6 *Microscopy techniques*

Transmission electron microscopy (TEM) and AFM stand as the predominant techniques employed for the comprehensive delineation of protein aggregate size and morphology. These methodologies provide critical insights on structural parameters encompassing length, width, and surface topography of aggregates within the nanometer scale (Gras *et al.*, 2011).

The sample preparation in AFM is relatively more complex than TEM techniques. The former necessitates a more prolonged duration and entails complex procedures, such as the crystallization of the protein sample. Notably, AFM emerges as the method of choice for the scrutiny of smaller, prefibrillar aggregates (~20 nm), surpassing the capabilities of the TEM approach. If a series of images depicting the progression of aggregate formation is captured across time intervals, it becomes possible to analyze the dimensions of these aggregates and ascertain their expansion. Nevertheless, it is important to note that this methodology, while informative, may not yield highly accurate measurements of the rates of aggregation.

1.3 Computational resources for handling experimental protein aggregation-related information

The computational resources play a crucial role in facilitating investigations into the aggregation behavior of proteins and peptides. While primary protein databases like UniProt (UniProt Consortium, 2023) and Protein Data Bank (PDB; Berman *et al.*, 2000) offer sequence and structural data, curated databases have emerged to empower the scientific community further (**Table 1.1**). These specialized databases focus on protein aggregation and offer a comprehensive and organized repository of knowledge.

Lopez de la Paz and Serrano (2004) curated the first collection of amyloidogenic peptides by systematically mutating the residues of amyloidogenic STVIIE peptides. The dataset was soon extended by including the peptides from insulin, β_2-microglobulin, amylin, tau protein, and so on (Thompson *et al.*, 2006). Fibril_one (Siepen & Westhead, 2002) was the first amyloidogenic protein database, which contains 250 mutations and 50 experimental conditions associated with 22 proteins. WALTZ-DB (Beerten *et al.*, 2015) comprised experimentally verified amyloidogenic

Table 1.1 List of databases for protein aggregation-related information

Database	Link	References
Fibril_one*	http://www.bioinformatics.leeds.ac.uk/group/online/fibril_one	Siepen and Westhead (2002)
WALTZ-DB 2.0	http://waltzdb.switchlab.org/	Louros *et al.* (2020)
AmyLoad	http://compreclin.iiar.pwr.edu.pl/amyload/	Wozniak and Kotulska (2015)
AmyPro	https://amypro.net/#/	Varadi *et al.* (2018)
PDB_Amyloid	https://pitgroup.org/amyloid/	Takács *et al.* (2019)
A3D	http://biocomp.chem.uw.edu.pl/A3D2/yeast	Badaczewska-Dawid *et al.* (2022)
AL-Base	http://albase.bumc.bu.edu/aldb	Bodi *et al.* (2009)
AmyloGraph	https://amylograph.com/	Burdukiewicz *et al.* (2023)
AmyloBase	http://150.217.63.173/biochimica/bioinfo/amylobase/pages/view.html	Belli *et al.* (2011)
SNPeffect	https://snpeffect.switchlab.org/	De Baets *et al.* (2012), Janssen *et al.* (2023)
CPAD 2.0	https://web.iitm.ac.in/bioinfo2/cpad2/	Rawat *et al.* (2020)

*Not available online (last accessed on October 2, 2023).

and non-amyloidogenic hexapeptides, which was recently expanded to WALTZ-DB 2.0 (Louros *et al.*, 2020), incorporating new sequences and structural information. The AmyLoad database (Wozniak & Kotulska, 2015) compiled amyloidogenic and non-amyloidogenic sequence fragments from diverse sources and literature. AmyPro (Varadi *et al.*, 2018) is another comprehensive resource on precursor proteins and their APRs. The PDB_Amyloid (Takács *et al.*, 2019) database focuses on amyloid structures and globular structures with amyloid-like substructures. The A3D database (Badaczewska-Dawid *et al.*, 2022) utilizes AlphaFold (AF)-generated structures to predict the surface-exposed APRs and AP of the human proteome. Recently, predicted APRs from the yeast proteome were also added to the A3D database (Garcia-Pardo *et al.*, 2023). Moreover, it also facilitates the exploration of the effects resulting from user-selected mutations on both protein solubility and stability. SNPeffect is available as a database and pipeline, which provides the impact of missense mutations in the human genome (De Baets *et al.*, 2012; Janssen *et al.*, 2023). The AL-Base (Bodi *et al.*, 2009) database curates the light chain sequence of antibodies derived from patients with light chain (AL) amyloidosis. AmyloGraph (Burdukiewicz *et al.*, 2023) is another unique database that contains information related to amyloid-amyloid interactions.

AmyloBase (Belli *et al.*, 2011) was the first resource to curate the aggregation kinetics experiments of amyloidogenic proteins and peptides. Thangakani *et al.* (2016) developed a comprehensive database, "CPAD" on aggregating proteins and APRs of different lengths. CPAD also included mutational data from aggregation kinetics experiments collected from various literature. The database was updated to CPAD 2.0 (Rawat *et al.*, 2020) by significantly expanding the amyloidogenic proteins and peptides, APR, and aggregation kinetics information.

1.4 Computational tools to study protein aggregation

Several computational techniques and tools have been developed to study protein aggregation and amyloidogenicity in the last few decades. A detailed review of various prediction tools and techniques and computational studies on protein β-aggregation is provided in subsequent sections. **Table 1.2** lists various computational methods to identify the APRs in proteins and peptides and predict protein AP (Morris *et al.*, 2009; Redler *et al.*, 2014; Meric *et al.*, 2017; Prabakaran *et al.*, 2021a; Prabakaran *et al.*, 2021b).

Protein aggregation prediction methods can be broadly categorized into three main groups: sequence-based, structure-based, and molecular dynamics (MD)-based approaches.

Sequence-based approaches focus on predicting protein aggregation tendencies, rates, or amyloidogenic peptide regions (APRs) using information derived solely from the protein or peptide sequence. These methods utilize various features such as amino acid physicochemical properties, sequence patterns, statistically derived propensity values, knowledge-based

Table 1.2 Computational resources to predict the aggregation propensity and aggregation-prone regions in a protein

Methods	Link	References
Amyloidogenic pattern	—	López de la Paz and Serrano (2004)
AGGRESCAN	http://bioinf.uab.es/aggrescan/	Conchillo-Solé *et al.* (2007)
Zyggregator	—	Tartaglia and Vendruscolo (2008)
Pafig	http://www.mobioinfor.cn/pafig/	Tian *et al.* (2009)
PAGE	—	Tartaglia *et al.* (2005)
WALTZ	https://waltz.switchlab.org/	Maurer-Stroh *et al.* (2010)
AbAmyloid	http://iclab.life.nctu.edu.tw/abamyloid	Liaw *et al.* (2013)
FoldAmyloid	http://bioinfo.protres.ru/fold-amyloid/	Garbuzynskiy *et al.* (2010)
iAMY-SCM	http://camt.pythonanywhere.com/iAMY-SCM	Charoenkwan *et al.* (2021)
SALSA β-Strand Contiguity (β-SC)	—	Zibaee *et al.* (2007)
APPNN	http://cran.r-project.org/web/packages/appnn/index.html	Família *et al.* (2015)
Amylogram	http://www.smorfland.uni.wroc.pl/shiny/AmyloGram/; http://github.com/michbur/AmyloGram Analysis	Burdukiewicz *et al.* (2017)
ANuPP	https://web.iitm.ac.in/bioinfo2/ANuPP/	Prabakaran *et al.* (2021a)
TANGO	http://tango.crg.es/	Fernandez-Escamilla *et al.* (2004)
SecStr	http://biophysics.biol.uoa.gr/SecStr/	Hamodrakas *et al.* (2007)
NetCSSP	http://cssp2.sookmyung.ac.kr/	Kim *et al.* (2009)
Archcandy	NA	Ahmed *et al.* (2015)

Table 1.2 (*Continued*)

Methods	**Link**	**References**
BetaSerpentine	https://bioinfo.crbm.cnrs.fr/index.php?route=tools&tool=25	Bondarev *et al.* (2018)
BETASCAN	http://betascan.csail.mit.edu	Bryan *et al.* (2009)
AmyloidMutants	http://amyloid.csail.mit.edu/	O'Donnell *et al.* (2011)
PASTA 2	http://old.protein.bio.unipd.it/pasta2/	Walsh *et al.* (2014)
GAP	http://www.iitm.ac.in/bioinfo/GAP/	Thangakani *et al.* (2014)
FISH Amyloid	http://comprec-lin.iiar.pwr.edu.pl/fishInput/	Gasior and Kotulska (2014)
AgMata	https://bitbucket.org/bio2byte/agmata	Orlando *et al.* (2020)
3D PROFILE	https://services.mbi.ucla.edu/zipperdb/submit	Thompson *et al.* (2006)
CORDAX	https://cordax.switchlab.org/	Louros *et al.* (2020)
PATH	https://github.com/KubaWojciechowski/PATH	Wojciechowski and Kotulska (2020)
AMYLPRED2	http://aias.biol.uoa.gr/AMYLPRED2/	Tsolis *et al.* (2013)
MetAmyl	http://metamyl.genouest.org/	Emily *et al.* (2013)
SAP	NA	Chennamsetty *et al.* (2009)
Developability index	NA	Lauer *et al.* (2012)
Aggscore	NA	Sankar *et al.* (2018)
AGGRESCAN3D 2.0	http://biocomp.chem.uw.edu.pl/A3D2/; https://bitbucket.org/lcbio/aggrescan3d	Kuriata *et al.* (2019)

NA: Not available; Last accessed on October 2, 2023.

scoring functions, secondary structure propensities, residue–residue contact potentials, and threading. One of the simplest sequence-based techniques involves pattern matching. For instance, López de la Paz and Serrano (2004) developed sequence patterns based on in vitro positional scanning mutagenesis experiments on the STVIIE peptide to represent amyloid-forming peptides.

Structure-based methods, on the other hand, require protein structure as input. Approaches like SAP, developability index, AGGRESCAN 3D, and Aggscore (Chennamsetty *et al.*, 2009; Lauer *et al.*, 2012; Zambrano *et al.*, 2015; Sankar *et al.*, 2018) predominantly consider the solvent accessibility of protein residues and atoms to estimate surface hydrophobicity. These methods not only account for the static protein structure but also incorporate short MD simulations to calculate ensemble statistics over time. Unlike sequence-based methods, structure-based approaches take into consideration a protein's folding and native state. However, the limitation of MD simulations on timescales may lead to bias when predicting the native state, particularly in highly dynamic proteins with multiple metastable states and disordered regions.

In addition to sequence- and structure-based methods, more computationally intensive simulation techniques are also employed to study protein AP and kinetics. These techniques encompass a wide range of modeling and simulation approaches, from coarse-grained to all-atom models, implicit to explicit solvation models, and Monte Carlo to molecular dynamic simulations (Morriss-Andrews & Shea, 2015; Carballo-Pacheco & Strodel, 2016). These approaches allow for a deeper understanding of the intricate dynamics and behavior of proteins with respect to aggregation.

1.4.1 *Predicting effect of mutation in protein aggregation propensity*

A variety of amino acid properties, including hydrophobicity, size, surface area, charge, aromaticity, contact frequency, β-sheet propensity, and numerous other physicochemical traits, are employed to detect regions in protein sequences that are prone to aggregation. For instance, AGGRESCAN utilizes an AP scale for amino acid residues obtained from in vivo experiments on amyloidogenic proteins (Conchillo-Solé *et al.*, 2007). Similarly, WALTZ combines physicochemical properties with a position-specific scoring matrix (PSSM) derived from amyloidogenic peptides and pseudo-energy estimates from modeled structures (Maurer-Stroh *et al.*, 2010).

The propensity to form β-sheets is a crucial characteristic of peptides and proteins that form amyloid fibrils, and it has been extensively harnessed in the development of prediction algorithms. TANGO employs various empirically and statistically derived potential functions to estimate the likelihood of a segment forming β-strand-mediated aggregates (Fernandez-Escamilla *et al.*, 2004). In essence, TANGO assesses the likelihood of a segment adopting various secondary structural states such as α-helix, β-sheet, coil, and turn. The transition from α-helix to β-sheet conformation is also the underlying principle behind SecStr and NetCSSP (Hamodrakas *et al.*, 2007; Kim *et al.*, 2009).

As previously discussed, the cross-β zipper motif is a common feature in all amyloid fibrils. The intertwining of side chains and the establishment of backbone hydrogen bonds along the fibril axis fortify the supramolecular structure. Furthermore, the stacking of aromatic residues and the formation of hydrogen bond ladders involving Asn, Gln, Thr, and Ser residues contribute additional stability to the structure. These residue-to-residue interactions are pivotal for an Amyloidogenic Peptide Region (APR) and are thus leveraged in various prediction methodologies. PASTA2 and BETASCAN employ residue-to-residue probabilities and scoring functions for β-sheet hydrogen bond formation and contacts, drawing from protein structure databases (Bryan *et al.*, 2009; Walsh *et al.*, 2014). GAP utilizes a position-specific residue-pair energy potential derived from amyloid and amorphous β-aggregating hexapeptide sequences to identify amyloidogenic peptides (Thangakani *et al.*, 2014). Beyond forecasting the APR stretch within the protein sequence, these approaches shed light on β-strand orientation and residue pairing.

Thompson *et al.* (2006) predicted amyloidogenicity based on the crystal structure of the cross-β spine of the NNQQNY peptide (Thompson *et al.*, 2006). Each hexapeptide from a given protein sequence was mapped onto an ensemble of steric zipper templates and subsequently scored. The underlying assumption of the 3D profile method is that APRs constitute the conserved cross-β motif in amyloid fibrils. In addition to identifying amyloidogenic peptides and regions within protein sequences, this approach can forecast the orientation between strands forming the zipper. Regardless of the different techniques employed, the overall impact of mutations on AP is typically estimated by equation 1, assuming that the implicit assumption of the prediction method holds true for both the wild type and mutant proteins.

$$\Delta AP = AP_{Mutant} - AP_{Wild\ type} \qquad \textbf{(Eq. 1.2)}$$

1.4.2 *Computational methods to study protein aggregation kinetics upon point mutation*

Aggregation kinetics pertains to quantifying the pace at which amyloid fibrils are formed under specific experimental parameters from amyloidogenic proteins and peptides. The mechanisms of aggregation, sigmoidal curve fitting, and associated experimental methodologies have been reviewed in existing literature (Morris *et al.*, 2009; Hirota *et al.*, 2019). In this section, we have focused on the role of computational methodologies dedicated to forecasting the aggregation kinetics of proteins and peptides, especially concerning the influence of point mutations and their underlying biophysical features and algorithms.

The aggregation kinetics assays have shown that aggregation kinetics is sensitive to even minute variations in experimental parameters such as concentration of protein/buffer/ions, pH, temperature, seeding with fibrils, or agitation (Hortschansky *et al.*, 2005; Morel *et al.*, 2010; Ow & Dunstan, 2013; Brudar & Hribar-Lee, 2019; Hao *et al.*, 2019). Consequently, the precise forecasting of absolute aggregation rates becomes an intricate endeavor, given the multifaceted interplay among diverse experimental factors. Therefore, most of the computational approaches are developed to predict the change in aggregation rate upon point mutation. This preference arises because mutations are studied in identical experimental settings, highlighting the impact of inherent sequence and structural characteristics.

The methods for predicting change in aggregation rate due to point mutations are outlined in the following section, along with a brief outlook at tools for predicting absolute aggregation rates. While absolute aggregation rate prediction tools can compute change in aggregation rate by subtracting mutant and wild-type absolute aggregation rates, they often lack sensitivity toward subtle sequence or structural variations. On the bright side, these absolute aggregation rate

Table 1.3 Computational resources for the prediction of aggregation kinetics of proteins and mutants

Methods	Link	References
Prediction of change in aggregation rate upon mutation		
Chiti *et al.*	NA	Chiti *et al.* (2003)
AggreRATE-Disc	https://www.iitm.ac.in/bioinfo/aggrerate-disc/	Rawat *et al.* (2018)
AggreRATE-Pred	https://www.iitm.ac.in/bioinfo/aggrerate-pred/	Rawat *et al.* (2019)
Prediction of absolute aggregation rate*		
DuBay *et al.*	NA	DuBay *et al.* (2004)
Tartaglia *et al.*	NA	Tartaglia *et al.* (2005)
AbsoluRATE-Pred	https://web.iitm.ac.in/bioinfo2/absolurate-pred/	Rawat *et al.* (2021)
Yang *et al.*	NA	Yang *et al.* (2019)

NA: Not available; Last accessed on October 2, 2023.

prediction methods have the advantage of handling multiple mutations in sequences, a capability not yet achievable with change in aggregation rate prediction methods.

1.4.2.1 *Methods for aggregation kinetics prediction upon point mutation*

Point mutation experiments to study changes in aggregation rate are widely used to unveil the aggregation mechanism of different amyloidogenic proteins. Chiti *et al.* (2003) first proposed a mathematical equation to predict the change in aggregation rate using intrinsic protein sequence features, which includes a change in the hydrophobicity of the polypeptide chain ($\Delta Hydr$), propensity to convert from α-helical to β-sheet structure ($\Delta\Delta G_{coil-\alpha} + \Delta\Delta G_{\beta-coil}$) and change in overall charge ($\Delta Charge$).

$$ln\left(\frac{v_{mut}}{v_{wt}}\right) = A\,\Delta\,Hydr + B(\Delta\Delta G_{coil-\alpha} + \Delta\Delta G_{\beta-coil}) + C\,\Delta\,Charge \quad \textbf{(Eq. 1.3)}$$

A, B, and C in the above equation are constants, which are estimated by fitting the equation to an experimental change in aggregation rate. The model achieved a correlation of 0.85 on a set of 27 mutations found in short peptides or natively unfolded proteins, including amylin, amyloid β-peptide, tau, and α-synuclein.

Rawat *et al.* (2018) introduced a machine learning-based method "AggreRATE-Disc" (Discrimination of Aggregation Rate Change Upon Mutation), which predicts the aggregation rate enhancer or mitigator mutations using sequence-based features. AggreRATE-Disc discerned distinct sequence-based characteristics influencing aggregation rate across various mutation site conformational classes. For instance, changes in protein stability and flexibility in the helical region influence the rate of aggregation. In β-strand regions, charge, polarity, and β-strand propensity exerted notable influence on the change in aggregation rate. For mutation sites located in coil regions (including bends, turns, and disordered regions), aggregation rates are affected by both helical tendency and AP.

The AggreRATE-Pred model (Rawat *et al.*, 2019) was an improvement over AggreRATE-Disc, introducing structure-based features to predict the quantitative change in aggregation rate. The dataset encompassing 183-point mutations within 23 amyloidogenic proteins underwent primary segmentation into two categories: (i) short peptides (length <40 residues) and (ii) long polypeptides and proteins (length ≥ 40 residues). The long polypeptides and proteins dataset was further stratified into helix, strand, and coil categories, mirroring the methodology of the preceding study (Rawat *et al.*, 2018). Further, statistical models were developed by fitting the sequence and structure-based features into the regression equations for each category, such as short peptides, helices, strands, and coils resulting in four regression equations, **Eqs. (1.4–1.7)**, respectively:

$$\Delta ln\ K_{app(SP)} = -0.95 + 0.13(NA_4VSS) - 2.2(H_t) + 2.7(P_{\varphi-\psi}) \quad \textbf{(Eq. 1.4)}$$

$$\Delta ln\ K_{app(H)} = 0.59 + 10.5(\Delta H) + 10.0(-T\Delta S) - 440(PEII) - 1.2(CP_M) \quad \textbf{(Eq. 1.5)}$$

$$\Delta ln\ K_{app(S)} = 0.07 - 0.73(R_d) + 1.53(P_t) + 1.19(N_m) - 4.01(CP_V) \quad \textbf{(Eq. 1.6)}$$

$$\Delta ln\ K_{app(C)} = -0.29 + 0.67(H_b) - 2.9(-T\Delta S_c) - 114.6(\Delta G) - 7.0(CP_S) + 0.22(CP_T) \quad \textbf{(Eq. 1.7)}$$

The features used in these equations are NA_4VSS (AP from AGGRESCAN (Conchillo-Solé *et al.*, 2007), H_t (hydrophobicity), $P_{\varphi-\psi}$ (residual backbone dihedral angle probability), ΔH (change in enthalpy), $-T\Delta S$ (change in entropy), *PEII* (electron-ion potential), CP_M (contact potential), R_d (residue depth upon mutation) (Tan *et al.*, 2013), P_t (local tendency to form turns), N_m (change in average number of residual medium-range contacts), CP_V (statistical contact potential), H_b (change in backbone H-bonds), $-T\Delta S_c$ (unfolding entropy change), ΔG (normalized Gibbs free energy change), CP_S and CP_T (statistical contact potentials). The numerical values of each property are obtained from Gromiha *et al.* (1999) and Kawashima *et al.* (2008).

Regression analysis highlighted the significance of local structural context, thermodynamic stability fluctuations, and the impact of neighboring residues at the mutation site.

1.4.2.2 *Method for absolute aggregation rate prediction*

The methods described in this section predict the absolute rate of aggregation for amyloidogenic proteins and peptides and usually consider the environmental condition. The change in aggregation rate (ΔK_{app}) upon mutation (whether single or multiple) is calculated by independently predicting the absolute aggregation rate for mutant and wild-type protein using identical experimental parameters and subtracting the aggregation rates:

$$\Delta Aggregation\ Rate(\Delta K_{app}) = K_{app}\ (mutant) - K_{app}\ (wild\ type) \quad \textbf{(Eq. 1.8)}$$

DuBay *et al.* (2004) developed the first method to predict the absolute aggregation rate of polypeptides using intrinsic features, such as hydrophobicity (I^{hydr}), alternating hydrophobic–hydrophilic residue pattern (I^{pat}) and the absolute value of net charge (I^{ch}), and extrinsic features, pH (E^{pH}), ionic strength (E^{ionic}), and polypeptide concentration (E^{conc}). The mathematical formula for prediction of absolute rate of aggregation is given below:

$$log(k) = \alpha_0 + \alpha_{hydr} I^{hydr} + \alpha_{pat} I^{pat} + \alpha_{ch} I^{ch} + \alpha_{pH} E^{Ph} + \alpha_{ionic} E^{ionic} + \alpha_{conc} E^{conc} \quad \textbf{(Eq. 1.9)}$$

Tartaglia *et al.* (2005) proposed a sequence-based algorithm to predict the aggregation rate and APRs in protein/polypeptide sequences. The AP (π_{il}) of the sequence was calculated using position-dependent factors (Φ_{il}) and composition-dependent factors (φ_{il}).

$$\pi_{il} = \Phi_{il}\varphi_{il} \quad \textbf{(Eq. 1.10)}$$

Where l is the length of the segment starting at the position i in the sequence. The position-dependent factors include aromaticity (A_{il}), β-propensity (B_{il}), and charge (C_{il}).

$$\Phi_{il} = e^{A_{il}+B_{il}+C_{il}} \quad \textbf{(Eq. 1.11)}$$

The amino acid composition-dependent factors include the side-chain accessible surface area of apolar (S_j^a), polar (S_j^p) and all residues (S_j^t); solubility (σ_j); parallel ($\theta\uparrow\uparrow$); and antiparallel ($\theta\uparrow\downarrow$) tendency to aggregate. The hatted values are average over 20 standard amino acids.

$$\varphi_{il} = \left[\prod_{j=1}^{i+l-1}\left(\frac{S_j^a}{\hat{S}^a}\theta\uparrow\uparrow + \frac{S_j^p}{\hat{S}^p}\theta\uparrow\downarrow\right)\frac{\hat{S}^t}{S_j^t}\frac{\hat{\sigma}}{\sigma_j}\right]^{\frac{1}{l}} \quad \textbf{(Eq. 1.12)}$$

The aggregation rates can be calculated for the respective protein from the predicted AP using the following formula (**Eq 1.13**):

$$v_{il} = \alpha(c,T)\pi_{il} \quad \textbf{(Eq. 1.13)}$$

Where $\alpha(c,T)$ is included in the equation to take account of the protein concentration and temperature used in the aggregation experiment.

AbsoluRATE (Rawat *et al.*, 2021) is a support vector machine (SVM)-based regression model to predict absolute rates of protein and peptide aggregation. The model takes into account two key aspects: the experimental conditions and the inherent sequence-structural traits of the protein. The experimental conditions were consolidated into a single feature, "Extrinsic PCA," using principal component analysis (PCA) applied to major experimental conditions, including buffer, pH, ionic strength, and protein concentration. Among the intrinsic features, protein disorderness predicted from PrDOS (Ishida & Kinoshita, 2007) and FoldUnfold (Galzitskaya *et al.*, 2006) was among the top correlating features. Hence, another "disorderness PCA" was calculated using the aforementioned features. Two additional features, "AP_{nGK2}" (number of APRs without gatekeeper residues on both sides) and "AP_{SSL}" (lowest AP for hexapeptides) were derived from the ANuPP method (Prabakaran *et al.* 2021a). The P_{β} features, representing the fraction of residues favoring β-strand conformation, was adopted from a previous study (Tartaglia *et al.*, 2005). Furthermore, for the APR regions predicted by ANuPP, the "Ra" (maximum reduction in solvent accessibility) and "P" (minimum polarity) features were calculated based on predefined values for each amino acid.

Yang *et al.* (2019) employed a machine learning model built upon a feedforward fully connected neural (FCN) network architecture to predict the absolute aggregation rates of amyloidogenic proteins. It integrates 16 intrinsic sequence-based features, including sequence length, molecular mass, isoelectric point, hydrophobicity, secondary structure contents, HP pattern, proline residue fraction, radius of gyration, and folding energy. Additionally, it considers four experimental conditions, namely temperature, pH, ionic strength, and protein concentration.

1.5 Inferring the effect of mutations on aggregation through molecular simulations

Computational simulations have been widely used to understand the mechanism of protein aggregation and the associated molecular dynamics (MD) and kinetics of self-association. The first step in *in silico* simulation is modeling, that is, abstract description of biological systems in terms of parameters that can be fed to the computers. For example, a protein can be described as a chain of spheres of radius "r," where each sphere represents an atom. In the simplest scenario, the entire protein can be abstracted to a solid sphere of radius "r." While both models can be simulated, the former provides more information on atomic interactions, while the latter facilitates faster simulation due to the simplistic abstraction of the system. Since protein aggregation is a process that occurs at the meso-macroscale, it requires a 10 to 100 times larger system than a typical protein simulation and is to be studied at least to a microsecond. In other words, the simulation of protein aggregation is much more computationally intensive.

To overcome these challenges, researchers often vary the model and techniques: from coarse-grained to all-atom models, Monte Carlo to molecular dynamic simulations, implicit to explicit solvation models, and dimers to bulk simulations. For example, Bellesia and Shea employed off-lattice simulation of peptide aggregation using a coarse-grained model, consisting of 2 and 1 beads representing the backbone and Cβ, respectively (Bellesia & Shea, 2007, 2009). They analyzed the kinetics, thermodynamics, and aggregate structure through simulations of different peptide sequences. Interestingly, their work highlighted the role of charge residues in stabilizing and changing the preference of orientation of peptides during aggregation. Frederix *et al.* explored the entire sequence space of dipeptides (400) and tripeptides (8000) self-assembly through coarse grain simulations using the MARTINI force field (Frederix *et al.*, 2011, 2015). They simulated each peptide individually in a system with 300 peptides to measure the AP of the peptide and the nature of the nanostructure formed.

Quite often, APRs from the protein are studied extensively as peptides instead of the whole protein to understand the aggregation mechanistic and residue–residue interaction at a limited computational cost. Similarly, the effect of mutations on the proteins is also extensively studied by mutating the simulated peptide. Lopez de la Paz *et al.* studied the amyloidogenic peptide STVIIE and its five variants using MD to understand the sequence dependence for amyloid formation (López de la Paz *et al.*, 2005). They simulated four conformational isomers of six-stranded β-sheets with six different mutated sequences and analyzed the stability of the four conformations with respect to the net charge and the relative disposition and orientation of hydrophobic side-chains. Guo *et al.* (2015)

investigated the effect of the R18 mutation on the Human IAPP 1–19 peptide in a peptide-membrane system using Replica Exchange Molecular Dynamics (REMD) simulations in the GROMACS package and demonstrated the effect of the mutation in altering the secondary structure propensity and membrane affinity of the protein. Similar studies have been carried out on polyglutamine and Amyloid β 1–40 peptides (Peng *et al.*, 2004; Marchut & Hall, 2007).

Computer simulations not only provide an estimate of the effect of mutations on protein aggregation but are also immensely useful in understanding the underlying mechanistic details. For example, in a recent study, the effect of a Valine to Lysine mutation on the aggregation of a monoclonal antibody was extensively studied (Prabakaran *et al.*, 2022). The effect of the mutations on agglomeration was studied through MD simulation of a large system of 125 Å containing a varying number of VLVIY and KLVIY peptides (**Figure 1.2**). The change in solvent accessible surface area (SASA) of the peptides, oligomer size, and formation rate were monitored over 100 ns. The researchers showed a significant

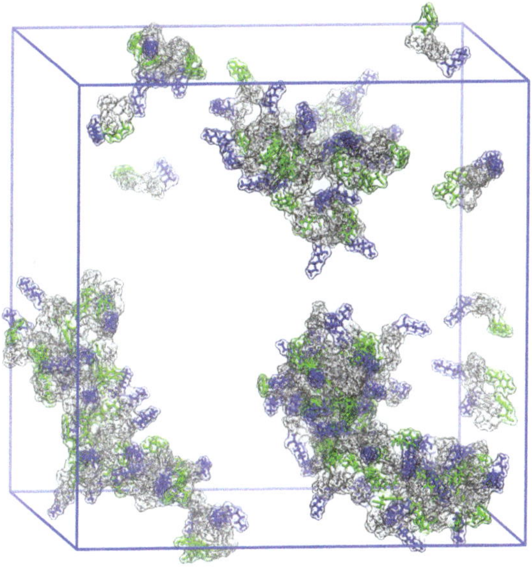

Figure 1.2 Illustration of an all-atom system containing oligomers formed by 105 "KLVIY" peptides

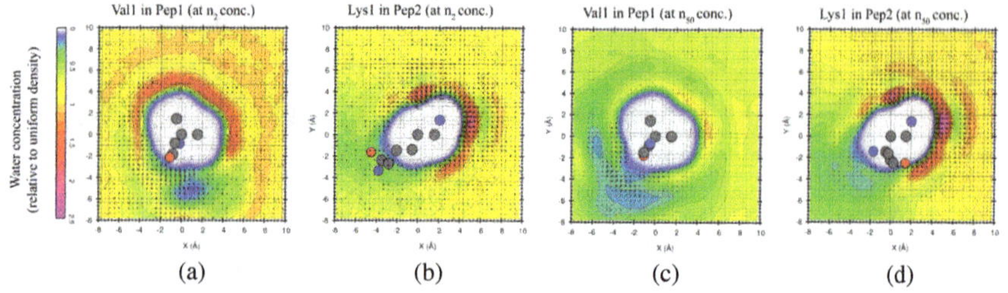

Figure 1.3 Orientation and density of water molecules around amino acid residues computed through molecular dynamic simulation: (a) Wild-type peptide with VAL at position 1 and concentration of two peptides, (b) mutant peptide with LYS at position 1 and concentration of two peptides, (c) wild-type peptide with VAL at position 1 and concentration of 50 peptides, and (d) mutant peptide with LYS at position 1 and concentration of 50 peptides. Colored spheres represent atoms of the amino acid residue of interest.

increase in SASA at the mutation site by 102 $Å^2$ in comparison to the wild type. While the rest of the residues were buried within the oligomers, the mutated residue positions containing Lysine were exclusively positioned on the oligomer surface (**Figure 1.3**). The preference of Lys residues to be on the oligomer surface led to an increase in SASA and disrupted the agglomeration of peptides into larger aggregates due to charge repulsion, and thus limiting the agglomeration.

In addition to peptide–peptide interactions, all atom MD simulations involving explicit solvent models are also used to understand the solvent–solute interactions, that is, the interaction between the peptide/protein with water molecules, such as the residence time of water molecules, hydrogen bonding between amino acid residues and water, clathrate formation, and orientation of water molecules. The arrangement of water molecules around a solute depends on the chemical natures of its constituent residues, the presence of multiple hydrogen bond donors and acceptors, and the local geometry of the solute surface (Makarov *et al.*, 2000). Prabakaran *et al.* (2022) showed that the introduction of lysine residue disrupted the surrounding water clathrate around the hydrophobic residues and the concentrated shell around Val residue. The introduction of Lys residue also led to the orientation of water molecules as represented by the arrows denoting the (OH) vectors directed away from the Lys1 residue, indicating the polar interactions between Lys1 and water molecules.

The disruption of the solvation shell upon the Lys mutation can reduce the entropic contribution to the free energy change of peptide aggregation and enhance protein solubility. The study highlights the usefulness of computer simulations in illustrating the underlying molecular process.

1.6 Conclusion

In summary, this chapter underscores the extensive research conducted in recent decades to unravel the profound effects of protein aggregation on human health and biotechnology. It provides a comprehensive overview of the key concepts and advancements in computational methodologies used for the study of protein aggregation. These computational tools are categorized into three main groups: sequence-based, structure-based, and MD-based approaches. They utilize a wide range of physicochemical properties, sequence patterns, and structural information to predict both the propensity and kinetics of protein aggregation. These tools have significantly deepened our understanding of the factors influencing this complex phenomenon. Additionally, the significance of computational resources, particularly curated databases, cannot be overstated. These databases serve as invaluable repositories of knowledge pertaining to amyloidogenic proteins, regions susceptible to aggregation, and experimental findings.

Furthermore, computational methods have been instrumental in predicting how point mutations impact protein aggregation likelihood and kinetics. These techniques take into account changes in physicochemical properties, shifts in secondary structure tendencies, and altered amino acid interactions caused by mutations. MD simulations are frequently employed in conjunction with these tools to gain deeper insights into the underlying mechanisms and the role of mutations in either promoting or inhibiting protein aggregation.

Acknowledgments

PR has received funding from the European Union's Horizon 2020 research and innovation programme under the Marie Skłodowska-Curie grant agreement No. 801133. We thank the Department of Biotechnology and Indian Institute of Technology Madras for computational facilities.

References

Ahmed, A. B., Znassi, N., Château, M.-T., & Kajava, A. V. (2015). A structure-based approach to predict predisposition to amyloidosis. *Alzheimer's & Dementia: The Journal of the Alzheimer's Association*, *11*(6), 681–690.

Badaczewska-Dawid, A. E., Garcia-Pardo, J., Kuriata, A., Pujols, J., Ventura, S., & Kmiecik, S. (2022). A3D database: Structure-based predictions of protein aggregation for the human proteome. *Bioinformatics*, *38*(11), 3121–3123.

Beerten, J., Van Durme, J., Gallardo, R., Capriotti, E., Serpell, L., Rousseau, F., & Schymkowitz, J. (2015). WALTZ-DB: A benchmark database of amyloidogenic hexapeptides. *Bioinformatics*, *31*(10), 1698–1700.

Bellesia, G., & Shea, J.-E. (2007). Self-assembly of beta-sheet forming peptides into chiral fibrillar aggregates. *The Journal of Chemical Physics*, *126*(24), 245104.

Bellesia, G., & Shea, J.-E. (2009). Effect of beta-sheet propensity on peptide aggregation. *The Journal of Chemical Physics*, *130*(14), 145103.

Belli, M., Ramazzotti, M., & Chiti, F. (2011). Prediction of amyloid aggregation in vivo. *EMBO Reports*, *12*(7), 657–663.

Berman, H. M., Westbrook, J., Feng, Z., Gilliland, G., Bhat, T. N., Weissig, H., Shindyalov, I. N., & Bourne, P. E. (2000). The Protein Data Bank. *Nucleic Acids Research*, *28*(1), 235–242.

Biancalana, M., & Koide, S. (2010). Molecular mechanism of Thioflavin-T binding to amyloid fibrils. *Biochimica Et Biophysica Acta*, *1804*(7), 1405–1412.

Bodi, K., Prokaeva, T., Spencer, B., Eberhard, M., Connors, L. H., & Seldin, D. C. (2009). AL-Base: A visual platform analysis tool for the study of amyloidogenic immunoglobulin light chain sequences. *Amyloid: The International Journal of Experimental and Clinical Investigation: The Official Journal of the International Society of Amyloidosis*, *16*(1), 1–8.

Bondarev, S. A., Bondareva, O. V., Zhouravleva, G. A., & Kajava, A. V. (2018). BetaSerpentine: A bioinformatics tool for reconstruction of amyloid structures. *Bioinformatics*, *34*(4), 599–608.

Brudar, S., & Hribar-Lee, B. (2019). The role of buffers in wild-type Hewl amyloid fibril formation mechanism. *Biomolecules*, *9*(2), 65. https://doi.org/10.3390/biom9020065

Bryan, A. W., Jr., Menke, M., Cowen, L. J., Lindquist, S. L., & Berger, B. (2009). BETASCAN: Probable beta-amyloids identified by pairwise probabilistic analysis. *PLoS Computational Biology*, *5*(3), e1000333.

Buck, P. M., Kumar, S., Wang, X., Agrawal, N. J., Trout, B. L., & Singh, S. K. (2012). Computational methods to predict therapeutic protein aggregation. *Methods in Molecular Biology*, *899*, 425–451.

Burdukiewicz, M., Rafacz, D., Barbach, A., Hubicka, K., Bąkała, L., Lassota, A., Stecko, J., Szymańska, N., Wojciechowski, J. W., Kozakiewicz, D., Szulc, N., Chilimoniuk, J., Jęśkowiak, I., Gąsior-Głogowska, M., & Kotulska, M. (2023). AmyloGraph: A comprehensive database of amyloid-amyloid interactions. *Nucleic Acids Research*, *51*(D1), D352–D357.

Burdukiewicz, M., Sobczyk, P., Rödiger, S., Duda-Madej, A., Mackiewicz, P., & Kotulska, M. (2017). Amyloidogenic motifs revealed by n-gram analysis. *Scientific Reports*, *7*(1), 12961.

Butterfield, S. M., & Lashuel, H. A. (2010). Amyloidogenic protein-membrane interactions: Mechanistic insight from model systems. *Angewandte Chemie*, *49*(33), 5628–5654.

Carballo-Pacheco, M., & Strodel, B. (2016). Advances in the simulation of protein aggregation at the atomistic scale. *The Journal of Physical Chemistry B*, *120*(12), 2991–2999.

Charoenkwan, P., Kanthawong, S., Nantasenamat, C., Hasan, Md.-M., & Shoombuatong, W. (2021). iAMY-SCM: Improved prediction and analysis of amyloid proteins using a scoring card method with propensity scores of dipeptides. *Genomics*, *113*(1 Pt 2), 689–698.

Chennamsetty, N., Voynov, V., Kayser, V., Helk, B., & Trout, B. L. (2009). Design of therapeutic proteins with enhanced stability. *Proceedings of the National Academy of Sciences of the United States of America*, *106*(29), 11937–11942.

Chiti, F., & Dobson, C. M. (2017). Protein misfolding, amyloid formation, and human disease: A summary of progress over the last decade. *Annual Review of Biochemistry*, *86*, 27–68.

Chiti, F., Stefani, M., Taddei, N., Ramponi, G., & Dobson, C. M. (2003). Rationalization of the effects of mutations on peptide and protein aggregation rates. *Nature*. *424*(6950), 805–808.

Conchillo-Solé, O., de Groot, N. S., Avilés, F. X., Vendrell, J., Daura, X., & Ventura, S. (2007). AGGRESCAN: A server for the prediction and evaluation of "hot spots" of aggregation in polypeptides. *BMC Bioinformatics*, *8*, 65.

De Baets, G., Van Durme, J., Reumers, J., Maurer-Stroh, S., Vanhee, P., Dopazo, J., Schymkowitz, J., & Rousseau, F. (2012). SNPeffect 4.0: On-line prediction of molecular and structural effects of protein-coding variants. *Nucleic Acids Research. 40*(Database issue), D935–D939.

Dogan, A. (2017). Amyloidosis: Insights from proteomics. *Annual Review of Pathology, 12*, 277–304.

DuBay, K. F., Pawar, A. P., Chiti, F., Zurdo, J., Dobson, C. M., & Vendruscolo, M. (2004). Prediction of the absolute aggregation rates of amyloidogenic polypeptide chains. *Journal of Molecular Biology, 341*(5), 1317–1326.

Ebrahimi-Fakhari, D., Wahlster, L., & McLean, P. J. (2011). Molecular chaperones in Parkinson's disease–present and future. *Journal of Parkinson's Disease, 1*(4), 299–320.

Emily, M., Talvas, A., & Delamarche, C. (2013). MetAmyl: A METa-predictor for AMYLoid proteins. *PLoS One, 8*(11), e79722.

Engelsman, J., Garidel, P., Smulders, R., Koll, H., Smith, B., Bassarab, S., Seidl, A., Hainzl, O., & Jiskoot, W. (2011). Strategies for the assessment of protein aggregates in pharmaceutical biotech product development. *Pharmaceutical Research, 28*(4), 920–933.

Esteras-Chopo, A., Serrano, L., & López de la Paz, M. (2005). The amyloid stretch hypothesis: Recruiting proteins toward the dark side. *Proceedings of the National Academy of Sciences of the United States of America, 102*(46), 16672–16677.

Família, C., Dennison, S. R., Quintas, A., & Phoenix, D. A. (2015). Prediction of peptide and protein propensity for amyloid formation. *PLoS One, 10*(8), e0134679.

Fernandez-Escamilla, A.-M., Rousseau, F., Schymkowitz, J., & Serrano, L. (2004). Prediction of sequence-dependent and mutational effects on the aggregation of peptides and proteins. *Nature Biotechnology, 22*(10), 1302–1306.

Finka, A., & Goloubinoff, P. (2013). Proteomic data from human cell cultures refine mechanisms of chaperone-mediated protein homeostasis. *Cell Stress & Chaperones, 18*(5), 591–605.

Frederix, P. W. J. M., Scott, G. G., Abul-Haija, Y. M., Kalafatovic, D., Pappas, C. G., Javid, N., Hunt, N. T., Ulijn, R. V., & Tuttle, T. (2015). Exploring the sequence space for (tri-)peptide self-assembly to design and discover new hydrogels. *Nature Chemistry, 7*(1), 30–37.

Frederix, P. W. J. M., Ulijn, R. V., Hunt, N. T., & Tuttle, T. (2011). Virtual screening for dipeptide aggregation: Toward predictive tools for peptide self-assembly. *Journal of Physical Chemistry Letters, 2*(19), 2380–2384.

Galzitskaya, O. V., Garbuzynskiy, S. O., & Lobanov, M. Y. (2006). FoldUnfold: Web server for the prediction of disordered regions in protein chain. *Bioinformatics*, *22*(23), 2948–2949.

Garbuzynskiy, S. O., Lobanov, M. Y., & Galzitskaya, O. V. (2010). FoldAmyloid: A method of prediction of amyloidogenic regions from protein sequence. *Bioinformatics*, *26*(3), 326–332.

Garcia-Pardo, J., Badaczewska-Dawid, A. E., Pintado-Grima, C., Iglesias, V., Kuriata, A., Kmiecik, S., & Ventura, S. (2023). A3DyDB: Exploring structural aggregation propensities in the yeast proteome. *Microbial Cell Factories*, *22*(1), 186.

Gasior, P., & Kotulska, M. (2014). FISH Amyloid—a new method for finding amyloidogenic segments in proteins based on site specific co-occurrence of aminoacids. *BMC Bioinformatics*, *15*, 54.

Gras, S. L., Waddington, L. J., & Goldie, K. N. (2011). Transmission electron microscopy of amyloid fibrils. *Methods in Molecular Biology*, *752*, 197–214.

Gregoire, S., Irwin, J., & Kwon, I. (2012). Techniques for monitoring protein misfolding and aggregation in vitro and in living cells. *The Korean Journal of Chemical Engineering*, *29*(6), 693–702.

Gromiha, M. M., Oobatake, M., & Sarai, A. (1999). Important amino acid properties for enhanced thermostability from mesophilic to thermophilic proteins. *Biophysical Chemistry*, *82*(1), 51–67.

Guo, C., Côté, S., Mousseau, N., & Wei, G. (2015). Distinct helix propensities and membrane interactions of human and rat $IAPP_{1-19}$ monomers in anionic lipid bilayers. *The Journal of Physical Chemistry B*, *119*(8), 3366–3376.

Hamodrakas, S. J., Liappa, C., & Iconomidou, V. A. (2007). Consensus prediction of amyloidogenic determinants in amyloid fibril-forming proteins. *International Journal of Biological Macromolecules*, *41*(3), 295–300.

Hao, X., Zheng, J., Sun, Y., & Dong, X. (2019). Seeding and cross-seeding aggregations of Aβ40 and its N-terminal-truncated peptide $A\beta_{11-40}$. *Langmuir: The ACS Journal of Surfaces and Colloids*, *35*(7), 2821–2831.

Hawe, A., Sutter, M., & Jiskoot, W. (2008). Extrinsic fluorescent dyes as tools for protein characterization. *Pharmaceutical Research*, *25*(7), 1487–1499.

Hirota, N., Edskes, H., & Hall, D. (2019). Unified theoretical description of the kinetics of protein aggregation. *Biophysical Reviews*, *11*(2), 191–208.

Hortschansky, P., Schroeckh, V., Christopeit, T., Zandomeneghi, G., & Fändrich, M. (2005). The aggregation kinetics of Alzheimer's beta-amyloid peptide is controlled by stochastic nucleation. *Protein Science: A Publication of the Protein Society*. *14*(7), 1753–1759.

Ishida, T., & Kinoshita, K. (2007). PrDOS: Prediction of disordered protein regions from amino acid sequence. *Nucleic Acids Research*, *35*(Web Server issue), W460–W464.

Janssen, K., Duran-Romaña, R., Bottu, G., Guharoy, M., Botzki, A., Rousseau, F., & Schymkowitz, J. (2023). SNPeffect 5.0: Large-scale structural phenotyping of protein coding variants extracted from next-generation sequencing data using AlphaFold models. *BMC Bioinformatics*, *24*(1), 287.

Kaganovich, D., Kopito, R., & Frydman, J. (2008). Misfolded proteins partition between two distinct quality control compartments. *Nature*, *454*(7208), 1088–1095.

Kawashima, S., Pokarowski, P., Pokarowska, M., Kolinski, A., Katayama, T., & Kanehisa, M. (2008). AAindex: Amino acid index database, progress report 2008. *Nucleic Acids Research*, *36*(Database issue), D202–D205.

Kayed, R., Head, E., Sarsoza, F., Saing, T., Cotman, C. W., Necula, M., Margol, L., Wu, J., Breydo, L., Thompson, J. L., Rasool, S., Gurlo, T., Butler, P., & Glabe, C. G. (2007). Fibril specific, conformation dependent antibodies recognize a generic epitope common to amyloid fibrils and fibrillar oligomers that is absent in prefibrillar oligomers. *Molecular Neurodegeneration*, *2*, 18.

Kayed, R., Head, E., Thompson, J. L., McIntire, T. M., Milton, S. C., Cotman, C. W., & Glabe, C. G. (2003). Common structure of soluble amyloid oligomers implies common mechanism of pathogenesis. *Science*, *300*(5618), 486–489.

Kayed, R., Pensalfini, A., Margol, L., Sokolov, Y., Sarsoza, F., Head, E., Hall, J., & Glabe, C. (2009). Annular protofibrils are a structurally and functionally distinct type of amyloid oligomer. *The Journal of Biological Chemistry*, *284*(7), 4230–4237.

Kayed, R., Sokolov, Y., Edmonds, B., McIntire, T. M., Milton, S. C., Hall, J. E., & Glabe, C. G. (2004). Permeabilization of lipid bilayers is a common conformation-dependent activity of soluble amyloid oligomers in protein misfolding diseases. *The Journal of Biological Chemistry*, *279*(45), 46363–46366.

Kerner, M. J., Naylor, D. J., Ishihama, Y., Maier, T., Chang, H.-C., Stines, A. P., Georgopoulos, C., Frishman, D., Hayer-Hartl, M., Mann, M., & Ulrich Hartl, F. (2005). Proteome-wide analysis of chaperonin-dependent protein folding in *Escherichia coli*. *Cell*, *122*(2), 209–220.

Kim, C., Choi, J., Lee, S. J., Welsh, W. J., & Yoon, S. (2009). NetCSSP: Web application for predicting chameleon sequences and amyloid fibril formation. *Nucleic Acids Research*, *37*(Web Server issue), W469–W473.

Klaips, C. L., Jayaraj, G. G., & Hartl, F. U. (2018). Pathways of cellular proteostasis in aging and disease. *The Journal of Cell Biology*, *217*(1), 51–63.

Kuriata, A., Iglesias, V., Pujols, J., Kurcinski, M., Kmiecik, S., & Ventura, S. (2019). Aggrescan3D (A3D) 2.0: Prediction and engineering of protein solubility. *Nucleic Acids Research*, *47*(W1), W300–W307.

Kurtishi, A., Rosen, B., Patil, K. S., Alves, G. W., & Møller, S. G. (2019). Cellular proteostasis in neurodegeneration. *Molecular Neurobiology*, *56*(5), 3676–3689.

Ladiwala, A. R. A., Dordick, J. S., & Tessier, P. M. (2011). Aromatic small molecules remodel toxic soluble oligomers of amyloid β through three independent pathways. *The Journal of Biological Chemistry*, *286*(5), 3209–3218.

Lauer, T. M., Agrawal, N. J., Chennamsetty, N., Egodage, K., Helk, B., & Trout, B. L. (2012). Developability index: A rapid in silico tool for the screening of antibody aggregation propensity. *Journal of Pharmaceutical Sciences*, *101*(1), 102–115.

Li, D., & Liu, C. (2020). Structural diversity of amyloid fibrils and advances in their structure determination. *Biochemistry*, *59*(5), 639–646.

Liaw, C., Tung, C.-W., & Ho, S.-Y. (2013). Prediction and analysis of antibody amyloidogenesis from sequences. *PLoS One*, *8*(1), e53235.

López de la Paz, M., de Mori, G. M. S., Serrano, L., & Colombo, G. (2005). Sequence dependence of amyloid fibril formation: insights from molecular dynamics simulations. *Journal of Molecular Biology*, *349*(3), 583–596.

López de la Paz, M., & Serrano, L. (2004). Sequence determinants of amyloid fibril formation. *Proceedings of the National Academy of Sciences of the United States of America*, *101*(1), 87–92.

Lorber, B., Fischer, F., Bailly, M., Roy, H., & Kern, D. (2012). Protein analysis by dynamic light scattering: Methods and techniques for students. *Biochemistry and Molecular Biology Education: A Bimonthly Publication of the International Union of Biochemistry and Molecular Biology*, *40*(6), 372–382.

Louros, N., Konstantoulea, K., De Vleeschouwer, M., Ramakers, M., Schymkowitz, J., & Rousseau, F. (2020). WALTZ-DB 2.0: An updated database containing structural information of experimentally determined amyloid-forming peptides. *Nucleic Acids Research*, *48*(D1), D389–D393.

Louros, N., Schymkowitz, J., & Rousseau, F. (2023). Mechanisms and pathology of protein misfolding and aggregation. *Nature Reviews Molecular Cell Biology*, *24*, 912–933. https://doi.org/10.1038/s41580-023-00647-2

Mahler, H.-C., Friess, W., Grauschopf, U., & Kiese, S. (2009). Protein aggregation: Pathways, induction factors and analysis. *Journal of Pharmaceutical Sciences*, *98*(9), 2909–2934.

Makarov, V. A., Kim Andrews, B., Smith, P. E., & Montgomery Pettitt, B. (2000). Residence times of water molecules in the hydration sites of myoglobin. *Biophysical Journal*, *79*(6), 2966–2974.

Malmos, K., Blancas-Mejia, L. M., Weber, B., Buchner, J., Ramirez-Alvarado, M., Naiki, H., & Otzen, D. (2017). ThT 101: A primer on the use of thioflavin T to investigate amyloid formation. *Amyloid: The International Journal of Experimental and Clinical Investigation: The Official Journal of the International Society of Amyloidosis*, *24*(1), 1–16.

Marchut, A. J., & Hall, C. K. (2007). Effects of chain length on the aggregation of model polyglutamine peptides: molecular dynamics simulations. *Proteins*, *66*(1), 96–109.

Maurer-Stroh, S., Debulpaep, M., Kuemmerer, N., de la Paz, M. L., Martins, I. C., Reumers, J., Morris, K. L., Copland, A., Serpell, L., Serrano, L., Schymkowitz, J. W. H., & Rousseau, F. (2010). Exploring the sequence determinants of amyloid structure using position-specific scoring matrices. *Nature Methods*, *7*(3), 237–242.

Meric, G., Robinson, A. S., & Roberts, C. J. (2017). Driving forces for nonnative protein aggregation and approaches to predict aggregation-prone regions. *Annual Review of Chemical and Biomolecular Engineering*, *8*, 139–159.

Micsonai, A., Wien, F., Kernya, L., Lee, Y.-H., Goto, Y., Réfrégiers, M., & Kardos, J. (2015). Accurate secondary structure prediction and fold recognition for circular dichroism spectroscopy. *Proceedings of the National Academy of Sciences of the United States of America*, *112*(24), E3095–E3103.

Miller, L. M., Bourassa, M. W., & Smith, R. J. (2013). FTIR spectroscopic imaging of protein aggregation in living cells. *Biochimica Et Biophysica Acta*, *1828*(10), 2339–2346.

Miller, S. B. M., Mogk, A., & Bukau, B. (2015). Spatially organized aggregation of misfolded proteins as cellular stress defense strategy. *Journal of Molecular Biology*, *427*(7), 1564–1574.

Morel, B., Varela, L., Azuaga, A. I., & Conejero-Lara, F. (2010). Environmental conditions affect the kinetics of nucleation of amyloid fibrils and determine their morphology. *Biophysical Journal*, *99*(11), 3801–3810.

Morris, A. M., Watzky, M. A., & Finke, R. G. (2009). Protein aggregation kinetics, mechanism, and curve-fitting: A review of the literature. *Biochimica Et Biophysica Acta*, *1794*(3), 375–397.

Morriss-Andrews, A., & Shea, J.-E. (2015). Computational studies of protein aggregation: Methods and applications. *Annual Review of Physical Chemistry*, *66*, 643–666.

Nguyen, P. H., Ramamoorthy, A., Sahoo, B. R., Zheng, J., Faller, P., Straub, J. E., Dominguez, L., Shea, J.-E., Dokholyan, N. V., De Simone, A., Ma, B., Nussinov, R., Najafi, S., Ngo, S. T., Loquet, A., Chiricotto, M., Ganguly, P., McCarty, J., Li, M. S., & Derreumaux, P. (2021). Amyloid oligomers: A joint experimental/computational perspective on Alzheimer's disease, Parkinson's disease, type II diabetes, and amyotrophic lateral sclerosis. *Chemical Reviews, 121*(4), 2545–2647.

O'Donnell, C. W., Waldispühl, J., Lis, M., Halfmann, R., Devadas, S., Lindquist, S., & Berger, B. (2011). A method for probing the mutational landscape of amyloid structure. *Bioinformatics, 27*(13), i34–i42.

Orlando, G., Silva, A., Macedo-Ribeiro, S., Raimondi, D., & Vranken, W. (2020). Accurate prediction of protein beta-aggregation with generalized statistical potentials. *Bioinformatics, 36*(7), 2076–2081.

Ow, S.-Y., & Dunstan, D. E. (2013). The effect of concentration, temperature and stirring on hen egg white lysozyme amyloid formation. *Soft Matter, 9*(40), 9692–9701.

Panda, C., Kumar, S., Gupta, S., & Pandey, L. M. (2023). Structural, kinetic, and thermodynamic aspects of insulin aggregation. *Physical Chemistry Chemical Physics: PCCP.* https://doi.org/10.1039/d3cp03103a

Peng, S., Ding, F., Urbanc, B., Buldyrev, S. V., Cruz, L., Stanley, H. E., & Dokholyan, N. V. (2004). Discrete molecular dynamics simulations of peptide aggregation. *Physical Review. E, Statistical, Nonlinear, and Soft Matter Physics, 69*(4 Pt 1), 041908.

Prabakaran, R., Rawat, P., Kumar, S., & Gromiha, M. M. (2021a). ANuPP: A versatile tool to predict aggregation nucleating regions in peptides and proteins. *Journal of Molecular Biology, 433*(11), 166707.

Prabakaran, R., Rawat, P., Kumar, S., & Gromiha, M. M. (2021b). Evaluation of in silico tools for the prediction of protein and peptide aggregation on diverse datasets. *Briefings in Bioinformatics, 22*(6), bbab240. https://doi.org/10.1093/bib/bbab240.

Prabakaran, R., Rawat, P., Thangakani, A. M., Kumar, S., & Gromiha, M. M. (2021). Protein aggregation: In silico algorithms and applications. *Biophysical Reviews, 13*(1), 71–89.

Prabakaran, R., Rawat, P., Yasuo, N., Sekijima, M., Kumar, S., & Gromiha, M. M. (2022). Effect of charged mutation on aggregation of a pentapeptide: Insights from molecular dynamics simulations. *Proteins, 90*(2), 405–417.

Rawat, P., Kumar, S., & Michael Gromiha, M. (2018). An in-silico method for identifying aggregation rate enhancer and mitigator mutations in proteins. *International Journal of Biological Macromolecules, 118*(Pt A), 1157–1167.

Rawat, P., Prabakaran, R., Kumar, S., & Gromiha, M. M. (2019). AggreRATE-Pred: A mathematical model for the prediction of change in aggregation rate upon point mutation. *Bioinformatics, 36*(5), 1439–1444. https://doi.org/10.1093/bioinformatics/btz764.

Rawat, P., Prabakaran, R., Kumar, S., & Gromiha, M. M. (2021). AbsoluRATE: An in-silico method to predict the aggregation kinetics of native proteins. *Biochimica et Biophysica Acta: Proteins and Proteomics, 1869*(9), 140682.

Rawat, P., Prabakaran, R., Sakthivel, R., Thangakani, A. M., Kumar, S., & Gromiha, M. M. (2020). CPAD 2.0: A repository of curated experimental data on aggregating proteins and peptides. *Amyloid: The International Journal of Experimental and Clinical Investigation: The Official Journal of the International Society of Amyloidosis, 27*(2), 128–133.

Redler, R. L., Shirvanyants, D., Dagliyan, O., Ding, F., Kim, D. N., Kota, P., Proctor, E. A., Ramachandran, S., Tandon, A., & Dokholyan, N. V. (2014). Computational approaches to understanding protein aggregation in neurodegeneration. *Journal of Molecular Cell Biology, 6*(2), 104–115.

Sahoo, B. R., Cox, S. J., & Ramamoorthy, A. (2020). High-resolution probing of early events in amyloid-β aggregation related to Alzheimer's disease. *Chemical Communications, 56*(34), 4627–4639.

Sankar, K., Krystek Jr, S. R., Carl, S. M., Day, T., & Maier, J. K. X. (2018). AggScore: Prediction of aggregation-prone regions in proteins based on the distribution of surface patches. *Proteins, 86*(11), 1147–1156.

Siepen, J. A., & Westhead, D. R. (2002). The fibril_one on-line database: Mutations, experimental conditions, and trends associated with amyloid fibril formation. *Protein Science: A Publication of the Protein Society, 11*(7), 1862–1866.

Sunde, M., & Blake, C. (1997). The structure of amyloid fibrils by electron microscopy and X-ray diffraction. *Advances in Protein Chemistry, 50*, 123–159.

Takács, K., Varga, B., & Grolmusz, V. (2019). PDB_Amyloid: An extended live amyloid structure list from the PDB. *FEBS Open Bio, 9*(1), 185–190.

Tan, K. P., Nguyen, T. B., Patel, S., Varadarajan, R., & Madhusudhan, M. S. (2013). Depth: A web server to compute depth, cavity sizes, detect potential small-molecule ligand-binding cavities and predict the pKa of ionizable residues in proteins. *Nucleic Acids Research, 41*(Web Server issue), W314–W321.

Tartaglia, G. G., Cavalli, A., Pellarin, R., & Caflisch, A. (2005). Prediction of aggregation rate and aggregation-prone segments in polypeptide sequences. *Protein Science, 14*(10), 2723–2734. https://doi.org/10.1110/ps.051471205.

Tartaglia, G. G., & Vendruscolo, M. (2008). The Zyggregator method for predicting protein aggregation propensities. *Chemical Society Reviews, 37*(7), 1395–1401.

Thangakani, A. M., Kumar, S., Nagarajan, R., Velmurugan, D., & Gromiha, M. M. (2014). GAP: Towards almost 100 percent prediction for β-strand-mediated aggregating peptides with distinct morphologies. *Bioinformatics, 30*(14), 1983–1990.

Thangakani, A. M., Prabakaran, R., Sakthivel, R., Thangakani, A. M., Kumar, S., & Gromiha, M. M. (2016). CPAD, Curated protein aggregation database: A repository of manually curated experimental data on protein and peptide aggregation. *PloS One, 11*(4), e0152949.

Thompson, M. J., Sievers, S. A., Karanicolas, J., & Eisenberg, D. (2006). The 3D profile method for identifying fibril-forming segments of proteins. *Proceedings of the National Academy of Sciences of the United States of America, 103*(11), 4074–4078.

Tian, J., Wu, N., Guo, J., & Fan, Y. (2009). Prediction of amyloid fibril-forming segments based on a support vector machine. *BMC Bioinformatics, 10*(Suppl 1), S45.

Tsolis, A. C., Papandreou, N. C., Iconomidou, V. A., & Hamodrakas, S. J. (2013). A consensus method for the prediction of 'aggregation-prone' peptides in globular proteins. *PloS One, 8*(1), e54175.

Tyedmers, J., Mogk, A., & Bukau, B. (2010). Cellular strategies for controlling protein aggregation. *Nature Reviews. Molecular Cell Biology, 11*(11), 777–788.

UniProt Consortium. (2023). UniProt: The universal protein knowledgebase in 2023. *Nucleic Acids Research, 51*(D1), D523–D531.

Varadi, M., Baets, G. D., Vranken, W. F., Tompa, P., & Pancsa, R. (2018). AmyPro: A database of proteins with validated amyloidogenic regions. *Nucleic Acids Research, 46*(D1), D387–D392.

Vu, K. H. P., Blankenburg, G. H., Lesser-Rojas, L., & Chou, C. F. (2022). Applications of single-molecule vibrational spectroscopic techniques for the structural investigation of amyloid oligomers. *Molecules, 27*(19), 6448. https://doi.org/10.3390/molecules27196448.

Walsh, I., Seno, F., Tosatto, Silvio, C. E., & Trovato, A. (2014). PASTA 2.0: An improved server for protein aggregation prediction. *Nucleic Acids Research, 42*(Web Server issue), W301–W307.

Wojciechowski, J. W., & Kotulska, M. (2020). PATH—Prediction of amyloidogenicity by threading and machine learning. *Scientific Reports, 10*(1), 7721.

Wozniak, P. P., & Kotulska, M. (2015). AmyLoad: Website dedicated to amyloidogenic protein fragments. *Bioinformatics, 31*(20), 3395–3397.

Yang, W., Tan, P., Fu, X., & Hong, L. (2019). Prediction of amyloid aggregation rates by machine learning and feature selection. *The Journal of Chemical Physics, 151*(8), 084106.

Yerbury, J. J., Ooi, L., Dillin, A., Saunders, D. N., Hatters, D. M., Beart, P. M., Cashman, N. R., Wilson, M. R., & Ecroyd, H. (2016). Walking the tightrope: Proteostasis and neurodegenerative disease. *Journal of Neurochemistry, 137*(4), 489–505.

Zambrano, R., Jamroz, M., Szczasiuk, A., Pujols, J., Kmiecik, S., & Ventura, S. (2015). AGGRESCAN3D (A3D): Server for prediction of aggregation properties of protein structures. *Nucleic Acids Research, 43*(W1), W306–W313.

Zibaee, S., Makin, O. S., Goedert, M., & Serpell, L. C. (2007). A simple algorithm locates beta-strands in the amyloid fibril core of alpha-synuclein, Abeta, and tau using the amino acid sequence alone. *Protein Science: A Publication of the Protein Society*. Wiley, *16*(5), 906–918.

https://doi.org/10.1142/9789811293269_0002

Chapter 2

Computational resources for understanding and predicting the stability of proteins upon mutations

P. Ramakrishna Reddy[1], A. Kulandaisamy[1], and M. Michael Gromiha[1,*]

[1]*Department of Biotechnology, Bhupat and Jyoti Mehta School of Biosciences, Indian Institute of Technology Madras, Chennai, Tamil Nadu 600036, India*

Abstract

Protein stability refers to the free energy difference between the folded and unfolded states of a protein. Amino acid mutations in a protein may alter the stability and are often associated with various diseases. Experimentally, the effect of mutations on protein stability is determined by thermal and chemical denaturation methods. These experimental data on the stability of proteins and their mutants are accumulated in several databases, and ProThermDB is the primary resource that offers more than 30,000 thermodynamic data (ΔG^{H2O}, ΔG, T_m, $\Delta\Delta G^{H2O}$, $\Delta\Delta G$, and

[*]Corresponding author
Tel: +91-44-2257-4138
Fax: +91-44-2257 4102
MMG: gromiha@iitm.ac.in

ΔT_m) from 1,200 proteins. Utilizing this database, protein stability change upon mutation is related to various protein sequence and structure-based features such as physicochemical properties, inter-residue interactions, solvent accessibility, and energetic parameters. These informations are effectively used for developing computational methods for predicting protein stability upon mutations. In summary, this chapter provides comprehensive details on experimental approaches for determining protein stability and the development of databases and computational tools, as well as insights to understand the complex relationship among protein sequence, structure, function, and stability. In addition, we discussed the effect of protein stability on disease-causing mutations that help for designing the stable mutants and mutation-specific drug design strategies.

2.1 Introduction

Protein stability (ΔG^0) refers to the energy difference between the folded (ΔG_{fold}) and unfolded (ΔG_{unfold}) states. The folded state stability is determined by various covalent and non-covalent interactions, such as hydrophobic, electrostatic, hydrogen bonds, disulfide bonds, and van der Waals interactions, whereas the unfolded state is mainly influenced by conformational entropy (Pace, 1990; Gromiha & Ponnuswamy, 1993). The stability of a protein is relatively small, typically around 5 to 10 kcal/mol (Pace, 1990).

Protein stability is influenced by various factors, including the amino acid sequence, interactions between amino acid residues, interactions with other molecules, and environmental conditions such as pH and temperature (Tokuriki & Tawfik, 2009; Talley & Alexov, 2010). Amino acid mutations in proteins may alter the stability and lead to misfolding, aggregation, loss of function, and diseases (Stefl *et al.*, 2013; Kulandaisamy *et al.*, 2018; Ganesan *et al.*, 2019). For instance, mutation of E22N and D23G in amyloid-β proteins alter protein stability and are associated with Alzheimer's disease (Grant *et al.*, 2007; Stehr *et al.*, 2011). Shi and Moult (2011) reported that amino acid substitutions Y53H, E124Q, E135R, and E142D in Pim-1 kinase destabilize the protein, which lead to cancer initiation and progression. In contrast, a few disease-causing missense mutations stabilize the protein. A typical example is the H101Q mutation in chloride intracellular channel protein 2, which is associated with a mental disorder and increased stability by 1.05 kcal/mol (Witham *et al.*, 2011; Takano *et al.*, 2012).

In this chapter, we provide comprehensive details on experimental approaches for determining protein stability, development of databases, computational tools, and large-scale studies for understanding the relationship among protein sequence, structure, function, and stability. In addition, we discuss the effect of protein stability changes on disease-causing mutations.

2.2 Experimental techniques to study protein stability

Experimentally, the stability of proteins is determined by equilibrium unfolding studies, such as thermal and chemical denaturation methods.

2.2.1 *Thermal denaturation studies for determining protein stability*

Thermal denaturation techniques often involve a protein to monitor the temperature at which it unfolds or loses its native structure, and this temperature is known as the melting temperature (T_m). Various biophysical methods, including circular dichroism (CD), differential scanning calorimetry (DSC), absorbance (Abs), and fluorescence (Fl) are widely utilized to determine thermodynamic parameters such as the T_m, enthalpy change (ΔH), entropy change (ΔS), heat capacity change (ΔC_p), and Gibbs free energy (ΔG). The ΔG value for the equilibrium reaction is referred to as the conformational stability of a protein and is defined as

$$\Delta G = \Delta H - T\Delta S \tag{2.1}$$

Where, T is the temperature.

2.2.2 *Chemical denaturation studies for determining protein stability*

The chemical denaturation methods use chemical/chaotropic agents (urea or GdnHcl) to induce protein unfolding or denaturation by disrupting the protein non-covalent interactions, such as hydrogen bonds and hydrophobic interactions. The stability of the protein is quantified by the concentration of denaturants required to unfold the protein. Further, the thermodynamic parameters, free energy of unfolding in the absence of denaturant (ΔG^{H2O}), slope of the denaturation curve (m), and midpoint of denaturation (C_m) are also derived from the chemical denaturation studies.

2.3 Characterization of protein stability upon mutations

Protein stability is affected by different types of mutations (insertion, deletion, and missense), and among them missense mutations are predominant (Alber, 1989; Studer *et al.*, 2013). These missense mutations can either reduce or enhance protein thermodynamic stability and are associated with genetic disorders, cancers, and neurodegenerative diseases. Experimentally, the effect of mutations on protein stability is determined using a site-directed mutagenesis experiment, followed by denaturation methods. The effect of mutations on protein stability ($\Delta\Delta G$) is quantified as the difference in free energy between wild-type ($\Delta G_{Wild\text{-}type}$) and mutant ($\Delta G_{Mutant}$) proteins. It is computed by

$$\Delta\Delta G = \Delta G_{Mutant} - \Delta G_{Wild\text{-}type} \tag{2.2}$$

2.4 Databases for understanding protein stability

Several computational databases have been developed for accessing protein stability data related to both wild-type and mutant proteins (Kumar *et al.*, 2006; Kulandaisamy *et al.*, 2021; Nikam *et al.*, 2021; Xavier *et al.*, 2021). Among them, ProThermDB (Nikam *et al.*, 2021) is the primary resource that offers a large collection of thermodynamic measures, and it is efficiently used to develop other secondary databases in the literature. These resources assist in conducting a large-scale analysis on exploring the relationship among the sequence, structure, and stability of a protein upon mutations, understanding the effect of disease-causing mutations on protein stability, and developing computational tools for predicting protein stability upon mutations.

2.4.1 *ProThermDB: Primary and comprehensive repository of protein thermodynamic data*

ProThermDB (Kulandaisamy *et al.*, 2021; Nikam *et al.*, 2021) is a comprehensive database that provides thermodynamic data for both wild-type and mutant proteins and it can be accessed at https://web.iitm.ac.in/bioinfo2/prothermdb/. This database has more than 30,000 entries from more than 1,200 proteins, and each entry is accessed by a unique accession number. Notably, ProThermDB is a more remarkable resource for single-point mutations than other types of mutations. For entries with available structural data, it includes Protein Data Bank (PDB) codes for both wild-type and mutant structures. Additionally, it contains information about the experimental method, measurements, temperature, pH, buffer name, and any additional remarks relevant to the experimental setup. Different experimental thermodynamic parameters, including ΔG^{H2O}, ΔG, T_m, m, and C_m are used to represent the stability of wild-type and mutant proteins. For exploring the impact of mutations on protein stability. ProThermDB includes mutational data such as the change in free energy of unfolding in the absence of denaturant ($\Delta\Delta G^{H2O}$), the change in free energy upon thermal denaturation ($\Delta\Delta G$), and the change in melting temperature (ΔT_m). Stabilizing and destabilizing mutations are indicated by positive and negative values for $\Delta\Delta G^{H2O}$, $\Delta\Delta G$, and ΔT_m, respectively. Additionally, ProThermDB allows users to visualize the mutations at the protein 3D structure level using the JSmol interface.

ProThermDB provides the stability of protein based on solvent accessibility and the secondary structure of the mutated residue. Based on the solvent accessibility of the protein structure, 43% of single mutations are in the buried regions, while 29% are partially buried, and 28% are exposed to the surroundings. In the case of secondary structure, 38%

of these single mutations are in α-helices, 27% in coils, and 21% in β-strands.

This comprehensive database not only provides protein stability data that is derived from low-throughput or conventional biophysical technologies but also includes data for various human cell lines (MCF7 breast cancer cells, HEK293T, HepG2 cells, Jurkat lymphoma T cells, K562 cultures) and organism-based information (*Arabidopsis thaliana, Homo sapiens, Thermus thermophilus, Escherichia coli, Saccharomyces cerevisiae, Toxoplasma gondii*) that is determined from high-throughput techniques such as thermal proteome profiling (TPP), limited proteolysis (LiP), and so on. A comprehensive guide for retrieving the contents of ProThermDB is illustrated in Kulandaisamy *et al.* (2021). These developments ensure that ProThermDB remains a valuable resource for studying protein thermodynamics and stability, benefiting the scientific community.

2.4.2 *Other secondary databases for assessing protein stability*

ThermoMutDB (http://biosig.unimelb.edu.au/thermomutdb/) includes experimentally derived thermodynamic parameters for wild-type and mutant proteins (Xavier *et al.*, 2021). This database contains ΔΔG values for about 16,000 mutations. ThermoMutDB obtained the data available in ProTherm by reviewing scientific literature.

Kulandaisamy *et al.* (2021) constructed a specific database for the stability of membrane proteins (MPTherm). It contains more than 7,000 experimental thermodynamic data from 320 membrane proteins, along with protein sequence and structure, mutation details, membrane topology information, experimental conditions and methods, and literature details. This database is accessible at https://www.iitm.ac.in/bioinfo/mptherm/ and information was mainly collated from the literature and ProThermDB.

2.5 Construction of benchmarking datasets on protein stability derived from existing databases

Utilizing the above-mentioned computational databases, a significant number of benchmarking datasets were constructed for conducting the large-scale analysis on relating protein sequence and structure-based features to protein stability. Further, these datasets were used for developing tools for predicting protein stability upon mutations. **Table 2.1** contains information about 35 different datasets (Guerois *et al.*, 2002; Capriotti *et al.*, 2004, 2005, Kortemme *et al.*, 2004; Saraboji *et al.*, 2006; Huang *et al.*, 2007; Capriotti *et al.*, 2008; Masso & Vaisman, 2008; Dehouck *et al.*, 2009; Potapov *et al.*, 2009; Khan & Vihinen, 2010; Zhang *et al.*, 2012; Chen *et al.*, 2013; Yang *et al.*, 2013; Folkman *et al.*, 2016; Getov *et al.*, 2016; Pucci *et al.*, 2016; Quan *et al.*, 2016; Broom *et al.*, 2017; Yang *et al.*, 2018; Chen *et al.*, 2020; Pires *et al.*, 2020; Kulandaisamy *et al.*, 2021) and details such as number of total mutations, number of stabilizing and destabilizing mutations, number of alanine and non-alanine mutations, etc.

Table 2.1 Datasets used to predict protein stability change upon mutation

	Name of the dataset	Total mutations*	Destabilizing mutations*	Stabilizing mutations*	Neutral mutations*	References
Datasets from ProThermDB	S3568	3,568 (150)	2,052 (120)	477 (81)	1,039 (107)	Chen *et al.* (2020)
	S630	630 (39)	364 (26)	74 (18)	192 (28)	Chen *et al.* (2020)
	S983	983 (28)	623 (23)	80 (18)	280 (21)	Getov *et al.* (2016)
	S605	605 (58)	358 (49)	71 (29)	176 (40)	Broom *et al.* (2017)
	S543	543 (55)	342 (43)	52 (28)	149 (39)	Folkman *et al.* (2016)
	S499	491 (34)	258 (28)	72 (18)	161 (29)	Capriotti *et al.* (2008)
	S388	388 (17)	287 (15)	23 (7)	78 (9)	Capriotti *et al.* (2004)
	S2971	2,971 (119)	371 (78)	1,725 (98)	875 (89)	Kortemme *et al.* (2004)
	S2760	2,760 (75)	1,384 (60)	422 (45)	954 (62)	Yang *et al.* (2013)

(*Continued*)

Table 2.1 (*Continued*)

Name of the dataset	Total mutations*	Destabilizing mutations*	Stabilizing mutations*	Neutral mutations*	References
S2648	2,648 (131)	1,598 (107)	295 (74)	755 (97)	Dehouck *et al.* (2009)
S2204	2,204 (89)	1,382 (74)	276 (57)	546 (57)	Saraboji *et al.* (2006)
S2087	2,087 (64)	1,082 (51)	316 (37)	650 (47)	Capriotti *et al.* (2005)
S1962	1,962 (77)	1,282 (65)	200 (50)	480 (50)	Masso and Vaisman, (2008)
S1948	1,948 (58)	1,042 (46)	327 (36)	579 (42)	Capriotti *et al.* (2005)
S1925	1,925 (54)	1029 (44)	283 (33)	578 (40)	Masso and Vaisman, (2008)
S1859	1,859 (64)	978 (50)	292 (36)	589 (47)	Huang *et al.* (2007)
S1810	1,810 (71)	1,388 (61)	422 (45)	0 (0)	Yang *et al.* (2013)
S1681	1,623 (58)	868 (42)	251 (31)	504 (37)	Capriotti *et al.* (2008)
S1676	1,676 (95)	925 (59)	220 (41)	531 (51)	Folkman *et al.* (2016)
S1634	1,634 (55)	850 (40)	244 (30)	482 (37)	Capriotti *et al.* (2004)
S1626	1,626 (101)	178 (41)	640 (57)	328 (44)	Pucci *et al.* (2016)
S1615	1,615 (41)	911 (33)	211 (27)	493 (33)	Capriotti *et al.* (2004)
S1396	1,396 (45)	658 (37)	268 (30)	470 (31)	Saraboji *et al.* (2006)
S1262	1,262 (49)	793 (38)	133 (33)	336 (33)	Getov *et al.* (2016)
S1109	1,109 (60)	689 (51)	105 (32)	315 (38)	Zhang *et al.* (2012)

Table 2.1 (*Continued*)

	Name of the dataset	Total mutations*	Destabilizing mutations*	Stabilizing mutations*	Neutral mutations*	References
	Q3421	3,421 (148)	1,996 (116)	412 (99)	1,013 (106)	Quan *et al.* (2016)
	S1564	1,564 (99)	603 (68)	201 (54)	349 (67)	Chen *et al.* (2013)
	S2156	2,156 (84)	221 (48)	1225 (71)	710 (70)	Potapov *et al.* (2009)
	PON-Tstab_dataset	1,564 (99)	865 (73)	233 (58)	466 (73)	Yang *et al.* (2018)
	M3131	2,881 (136)	1,703 (110)	340 (77)	838 (100)	Chen *et al.* (2013)
	S1784	1,784 (80)	905 (59)	204 (47)	675 (65)	Khan and Vihinen, (2010)
	S625	625 (28)	23 (11)	409 (26)	193 (24)	Guerois *et al.* (2002)
	S339	339 (9)	9 (6)	249 (9)	81 (9)	Guerois *et al.* (2002)
Datasets from MPTherm	mCSM-Membrane	404 (9)	184	70	150	Pires *et al.* (2020)
	MPTherm-pred	929 (144)	539	390	—	Kulandaisamy *et al.* (2021)

*Number of proteins are given in parentheses.

2.5.1 *Distribution of stabilizing and destabilizing mutations in proteins*

The distribution of experimental ΔΔG values (computed using equation 2.2) falls in the range of –17 to 13 kcal/mol with a higher frequency of destabilizing mutations (**Figure 2.1**), and over 50% are associated with a narrow range of –3 to 1 kcal/mol. It is noteworthy that stabilizing mutations are less prominent in the given data. Analysis of the occurrence of amino acid residues in wild-type and mutant proteins showed that alanine mutations have diverse effects on protein stability, resulting in both stabilizing and destabilizing, whereas glycine mutations predominantly led to

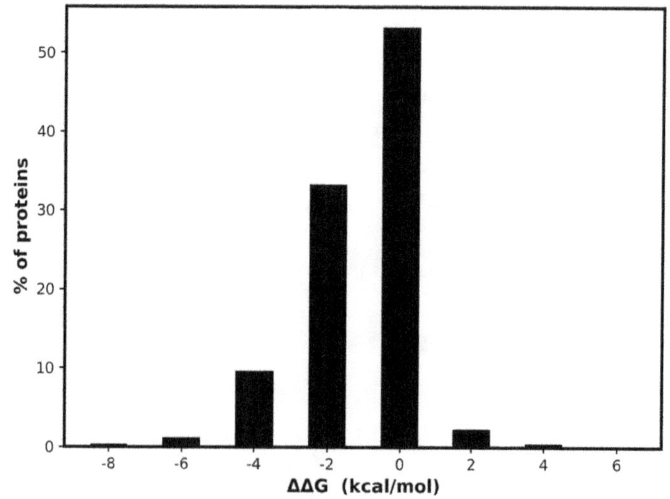

Figure 2.1 Distribution of experimental stability changes (ΔΔG) upon missense mutations in benchmarking datasets

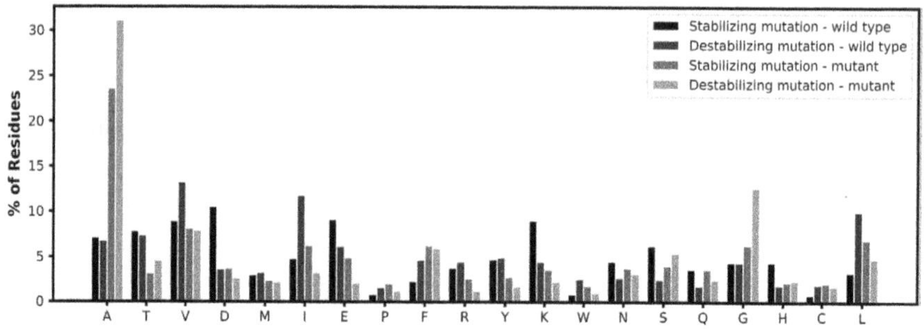

Figure 2.2 The occurrence of the 20 amino acid residues at wild-type and mutant positions of stabilizing and destabilizing mutations

destabilize a protein. Notably, mutations involving charged residues (wild-type), such as Asp and Lys increase the stability of a protein (**Figure 2.2**).

2.6 Important features for predicting protein stability change upon mutation

Identifying relevant features for predicting protein stability change upon mutation helps to understand the factors that are influencing protein stability.

Thus, certain key protein sequence and structure-derived features are widely used to relate to protein stability and are discussed below.

2.6.1 *Protein sequence-based features*

1. **Physicochemical properties:** The properties of wild-type and mutated amino acids, such as size, charge, hydrophobicity, and polar/non-polar nature can significantly correlate with protein stability (Gromiha, 2003). These properties are available in AAindex database and literature (Gromiha, 2005; Kawashima *et al.*, 2007; Kulandaisamy *et al.*, 2020). Large-scale studies on a set of protein sequences and/or structures have been used to derive these properties.
2. **Conservation score:** The tendency of an amino acid residue, which is conserved across related proteins is important for maintaining the structure and function of a protein, while less conserved residues may be more tolerant of variation (Valdar, 2002; Manning *et al.*, 2008). AACons is a Java-based tool for calculating the conservation score of residues in a multiple sequence alignment (MSA) (MacGowan *et al.*, 2020).
3. **Position-specific scoring matrices (PSSMs)/profiles:** A PSSM or profile is a matrix that represents the likelihood of each amino acid occurring at each position within a sequence alignment. The generation of a PSSM involves conducting a BLAST search with an E-value cutoff (≤ 0.001) to find homologous sequences, followed by MSA of the retrieved homologous proteins.

2.6.2 *Protein structure-based features*

1. **Solvent accessibility:** Solvent accessibility refers to the extent to which individual atoms or residues in a protein are exposed to the surrounding solvent. It is measured in terms of the accessible surface area (ASA) of a residue or atom using a spherical probe that rolls over the protein surface. Most often, water is used as the probe with a radius of 1.4 Å for globular proteins, while $-CH_2$ with a radius of 2.0 Å is used for membrane proteins. The residues or regions in the protein that are high and less accessible for solvents are termed as exposed and buried regions, respectively.

2. **Hydrogen bonds:** Hydrogen bonds play a crucial role in maintaining the structural integrity of proteins. A hydrogen bond is an electrostatic attraction between a hydrogen (H) atom covalently bonded to a more electronegative atom (donor) and another electronegative atom that possesses a lone pair of electrons (acceptor).
3. **Ionic interactions:** Ionic interactions involve the attraction or repulsion between charged amino acid residues. Specifically, positively charged residues (R, K, H) can interact with negatively charged ones (D, E).
4. **Hydrophobic interactions:** Amino acid residues, such as A, V, L, I, M, F, W, P, and Y, are hydrophobic and tend to show hydrophobic interactions when they fall within the 5Å range (Tina *et al.*, 2007).
5. **Contact order:** This parameter reflects the relative importance of local and nonlocal contacts to the native structure of a protein (Plaxco *et al.*, 1998). It is defined as

$$CO = \Sigma \Delta Sij/L*N$$

where N represents total number of contacts, ΔS_{ij} sequence separation between residues i and j within a distance of 6Å, and L is the number of residues in the protein.
6. **Long-range order:** It is defined as contacts between two residues that are close in space and far in the sequence (Gromiha & Selvaraj, 2001).

$$LRO = \Sigma nij/N$$
$$n = 1 \text{ if } |i - j| > 12; \text{ else } n = 0;$$

where i and j are the residues within 8 Å and N is the total number of residues in the protein.
7. **Energetic terms:** A set of 23 energy terms can be calculated using the stability module of FoldX software (Schymkowitz *et al.*, 2005). These energy terms refer to the contribution of hydrophobic interactions, electrostatic interactions, different types of hydrogen bonds, van der Waals interactions, and the entropic cost.

2.7 Relationship between amino acid properties and protein stability

Understanding the factors influencing protein stability upon mutations is a challenge and crucial for understanding the mechanisms of protein folding and designing stable mutants. Protein stability is dependent on various protein sequence and structure-based features (Dill, 1990; Rose & Wolfenden, 1993; Ponnuswamy & Gromiha, 1994; Pace, 1995). Utilizing a comprehensive set of 49 amino acid properties, Gromiha *et al.* (1999a, b) computed the changes in property values (ΔP) to understand the relationship between ΔP and protein stability (ΔT_m).

$$\mathbf{\Delta P(i) = P_{mut}(i) - P_{wild}(i)} \tag{2.3}$$

$P_{mut}(i)$ and $P_{wild}(i)$ represent the property values of mutant and wild-type residues, respectively, with i ranging from 1 to N, the total number of mutants.

Features such as chromatographic index (Rf) and surrounding hydrophobicity (Hp) emerged as strong determinants of stability in buried mutations, with correlations (r) of 0.59 and 0.50, respectively. Gromiha (2003) expanded the analysis to include sequence and structural information for predicting thermal stability changes in buried and partially buried mutations. This study suggested that the stability of buried mutants is influenced not only by hydrophobicity and secondary structure but also by shape, flexibility, entropy, and inter-residue contacts. For partially buried mutations, hydrogen bonds, long-range interactions, and hydrophobic interactions contribute significantly to stability (Gromiha *et al.*, 1999b).

Gills and Rooman (1997) used torsion and distance potentials to predict the effect of mutation on the stability of proteins and indicated that distance potentials, primarily governed by hydrophobic interactions play a central role in stabilizing the protein core. In contrast, torsion potentials associated with local interactions can significantly influence the stability of residues at the protein surface. Further, hydrogen bonds and electrostatic interactions are reported to contribute to the stability of partially buried and surface mutations (Akke & Forsén, 1990; Serrano *et al.*, 1990; Nicholson *et al.*, 1991; Sun *et al.*, 1991).

2.8 Computational methods for predicting the stability of proteins upon mutations

The computational methods for predicting the stability of proteins upon mutations are broadly categorized into three groups: (i) mechanistic approaches - rely on established scientific theories and empirical data to make predictions about the stability; (ii) machine learning methods-use machine learning techniques to predict the stability without relying on a deep understanding of the underlying mechanisms; and (iii) integrated approaches - which are the combinations of both mechanistic and machine learning approaches. Each method is constructed by dividing the dataset into training and test set and further evaluated with different statistical indices (sensitivity, specificity, accuracy, Area under the curve (AUC), Matthews correlation coefficient (MCC), F1-score, correlation, mean absolute error, etc.) and validation methods (5/10/20-fold cross validation, leave-out, etc.). A list of methods (Schymkowitz *et al.*, 2005; Parthiban *et al.*, 2006; Dehouck *et al.*, 2009; Teng *et al.*, 2010; Kellogg *et al.*, 2011; Chen *et al.*, 2013; Giollo *et al.*, 2014; Pires *et al.*, 2014a, b; Fariselli *et al.*, 2015; Laimer *et al.*, 2015; Folkman *et al.*, 2016; Laimer *et al.*, 2016; Quan *et al.*, 2016; Pandurangan *et al.*, 2017; Rodrigues *et al.*, 2018; Yang *et al.*, 2018; Cao *et al.*, 2019; Montanucci *et al.*, 2019; Chen *et al.*, 2020; Li *et al.*, 2021) for predicting the protein stability change upon mutation (ΔΔG) is presented in **Table 2.2**.

2.8.1 *Mechanistic models*

This category includes models that are based on mechanistic principles and utilize statistical potentials or energy functions to predict ΔΔG. DDGUN and DDGUN3D are untrained models that use evolutionary information and structure-based terms, such as difference in the interaction energy and relative solvent accessibility of the residue, for predicting ΔΔG. FoldX is an empirical force field that calculates the free energy

Table 2.2 Methods for predicting the protein stability change ($\Delta\Delta G$) upon mutation

Name	Base algorithm	Features	URL	References
Mechanistic				
Sequence based				
DDGUN	Untrained and anti-symmetric	Evolutionary score, sequence-based statistical potential score, hydrophobicity score	https://folding.biofold.org/ddgun/	Montanucci *et al.* (2019)
Structure based				
DDGUN3D	Untrained and anti-symmetric	Evolutionary score, sequence- and structure-based statistical potential score, hydrophobicity score	https://folding.biofold.org/ddgun/	Montanucci *et al.* (2019)
FoldX	Linear combination of energy terms	Empirical free energy terms, including entropy, Van der Walls forces, hydrogen bonds, and electrostatic interactions	https://foldxsuite.crg.eu/	Schymkowitz *et al.* (2005)
CUPSAT	Combination of statistical potentials	Structural environment-specific atom potentials and torsion angle potentials, solvent accessibility	https://cupsat.brenda-enzymes.org/	Parthiban *et al.* (2006)
SDM2/SDM	Statistical potential energy function	Environment-specific amino acid substitution tables, residue depth and packing density, and other interaction parameters	http://marid.bioc.cam.ac.uk/sdm2	Pandurangan *et al.* (2017)
POPMuSiC	Combination of statistical potentials	13 statistical potentials, two volume-dependent terms	https://soft.dezyme.com/	Dehouck *et al.* (2009)
Rosetta	Rosetta energy function	Difference in energy between the wild-type and mutant structure	https://www.rosettacommons.org/software	Kellogg *et al.* (2011)

(Continued)

Table 2.2 (*Continued*)

Name	Base algorithm	Features	URL	References
Machine Learning				
Sequence based				
SAAFEC-SEQ	XGBoost	Sequence features, physicochemical properties, and evolutionary information	http://compbio.clemson.edu/SAAFEC-SEQ/	Li *et al.* (2021)
EASE-MM	SVM	Evolutionary conservation score, amino acid parameters, structural properties (secondary structures and accessible surface area)	https://sparks-lab.org/server/ease-mm/	Folkman *et al.* (2016)
INPS	Support Vector Regression	Substitution score from the BLOSUM62, hydrophobicity, evolutionary information, and relative solvent accessibility	https://inps.biocomp.unibo.it/inpsSuite/default/index	Fariselli *et al.* (2015)
MuStab	SVM	Substitution score from the BLOSUM62, hydrophobicity, evolutionary information, and relative solvent accessibility	—	Teng *et al.* (2010)
Structure based				
PON-tstab	Random Forest	Conservation features, variation type, neighborhood features, and other protein feature	http://structure.bmc.lu.se/PON-Tstab/	Yang *et al.* (2018)
MAESTROweb	SVM, ANN, multiple linear regression	Statistical energy functions, sequence, and structure information	https://biwww.che.sbg.ac.at/maestro/web	Laimer *et al.* (2015), Laimer *et al.* (2016)

STRUM	Gradient Boosting Regressor	Sequence and structure profile scores, and different energy functions	http://zhanglab.ccmb.med.umich.edu/STRUM/	Quan *et al.* (2016)
mCSM	Random Forest, Gaussian Process	Use graph-based signature and pharmacophore changes.	http://biosig.unimelb.edu.au/mcsm/	Pires *et al.* (2014)
NeEMO	Neural Network	Uses the residue–residue interaction network (RIN) to predict the effects of mutations	http://protein.bio.unipd.it/neemo/	Giollo *et al.* (2014)
Integrated Structure based				
iStable2.0	XGBoost	Combines various models such as DUET, SDM, SDM2, mCSM, CUPSAT, PoPMuSiC, iPTREE-STAB, MUpro, I-Mutant2.0, AUTO-MUTE2.0, and MAESTRO	http://ncblab.nchu.edu.tw/iStable2	Chen *et al.* (2020)
DeepDDG	Neural Network	mCSM, SDM, and DUET	http://protein.org.cn/ddg.html	Cao *et al.* (2019)
DynaMut	Random Forest	DUET, mCSM, SDM, and ENCoM	http://biosig.unimelb.edu.au/dynamut/	Rodrigues *et al.* (2018)
DUET	SVM	(mCSM and SDM) and obtained by combining the results using SVM	http://biosig.unimelb.edu.au/duet/stability	Pires *et al.* (2014)
iStable	SVM	AUTO-MUTE, MUPRO, PoPMuSiC2.0, CUPSAT, and I-Mutant2.0.	http://predictor.nchu.edu.tw/iStable/	Chen *et al.* (2013)

change based on different energy contributions. CUPSAT employs statistical potentials such as torsion and distance potentials for prediction. POPMuSiC is a structure-based method and uses a linear combination of statistical potentials to predict ΔΔG.

2.8.2 *Machine learning models*

In this category, methods use machine learning algorithms to predict ΔΔG. SAAFEC-SEQ is based on the XGBoost algorithm and uses physico-chemical properties and evolutionary information for ΔΔG prediction. MAESTROweb employs SVM, ANN, and multiple linear regression and uses mutation site, mutation sensitivity profiles, and disulfide bond score for ΔΔG predictions. STRUM is a structure-based method for predicting stability changes caused by single-point mutations using profile scores and energy functions. INPS employs support vector regression using mutation type and evolutionary information to predict ΔΔG. mCSM uses the concept of graph-based structural signatures and random forest to predict ΔΔG. NeEMO uses a residue interaction network (RIN) to extract features related to the structural environment of mutant residue and employs a neural network algorithm to predict the change in protein stability.

2.8.3 *Integrated methods*

The integrated models combine multiple methods or tools for improved prediction accuracy. iStable2.0 combines 11 sequence- and structure-based prediction tools and integrates them through a machine learning framework. DeepDDG is a structure-based method that combines mCSM, SDM, and DUET. DynaMut combines DUET, mCSM, SDM, and ENCoM. DUET integrates two methods, namely SDM and mCSM.

2.8.4 *Membrane protein-specific stability prediction methods*

Kroncke *et al.* (2016) evaluated the ability of existing methods to predict the impact of amino acid mutations on the stability of membrane proteins.

All methods performed significantly worse, yielding a correlation of less than 0.4 for membrane proteins, and highlighted the necessity of membrane protein-specific stability prediction methods. Thus, a few machine-learning methods were specifically developed to characterize the stability of membrane proteins.

MPTherm-pred (Kulandaisamy *et al.*, 2021) is a method for predicting the effect of mutations on membrane protein thermal stability. Sequence and structure-based features are related to change in thermal stability (ΔT_m) using multiple regression techniques. mCSM-membrane (Pires *et al.*, 2020) analyzes the impacts of mutations on membrane protein stability using graph-based signatures to model protein geometry and physicochemical properties. Recently developed TMH Stab-pred (Reddy *et al.*, 2023) predicts the stability of α-helical membrane proteins using sequence and structure-based features.

2.9 Effect of disease-causing mutations on the thermodynamic stability of proteins

Mutations in proteins may alter protein stability and are often associated with various diseases. Experimental studies have demonstrated that most of the disease-causing amino acid mutations tend to make proteins less stable. This destabilization effect has been observed in various studies (Tokuriki & Tawfik, 2009; Zhang *et al.*, 2011; Chaturvedi & Mahalakshmi, 2013; Grothe *et al.*, 2013). The effect of destabilization is pronounced when mutations introduce significant changes in physicochemical properties at the mutation site. In many cases, destabilization is accompanied by structural changes in the protein (Zhang *et al.*, 2011; Chaturvedi & Mahalakshmi, 2013). For instance, Grothe *et al.* (2013) examined the mutants R124C and R555W in growth factor beta-induced protein (TGFBIp) and reported that these mutations lead to corneal dystrophies and are less stable than the wild-type. Chaturvedi and Mahalakshmi (2013) investigated the influence of methionine replacement with leucine in the outer membrane protein X (OmpX) protein and found that Met→Leu (M18L; M21L; and M118L) substitutions led to Alzheimer's disease and a decrease in unfolding free energy by (~8.5 kJ/mol). Shi and

Moult (2011) investigated the impact of amino acid substitutions Y53H, E124Q, E135R, and E142D on the structural stability and activity of Pim-1 kinase, a protein linked to cancer initiation and progression, and revealed that the wild-type Pim-1 was more stable than the mutants; specifically, ΔG of the E135R variant is 3.5-fold less than that of the wild type. The decreased stability in Pim-1 variants was primarily attributed to a decrease in the change in the solvent-exposed surface area upon unfolding.

However, some disease-causing amino acid mutations can lead to protein stabilization (Witham *et al.*, 2011; Zhang *et al.*, 2011; Takano *et al.*, 2012). For example, Witham *et al.* (2011) identified a missense mutation, H101Q, in the CLIC2 gene associated with X-linked intellectual disability (XLID), which increased the stability of CLIC2 compared to the wild type due to the formation of hydrogen bonds by Gln.

Martelli *et al.* (2016) found that mutations associated with diseases tend to have a more significant impact on protein stability compared to harmless genetic variations. Around 44% of disease-related mutations and 20% of harmless genetic variations were predicted to cause substantial changes in protein stability, with $|\Delta\Delta G| > 1$ kcal/mol. Interestingly, a portion of disease-related mutations (approximately 47%) that had minor effects on stability ($|\Delta\Delta G| \leq 1$ kcal/mol) were in exposed regions of the protein. This suggests that both disease-causing and harmless mutations can have low $|\Delta\Delta G|$ values when they occur in solvent-accessible parts of the protein. Recently, Pandey *et al.* (2022) revealed that 89.2% of cancer-causing mutations are destabilizing.

2.10 Conclusions

In conclusion, the study of protein stability includes experimental and computational approaches, contributing valuable insights about the protein sequence, structure, function, and stability. The experimental data on protein stability is accumulated in databases, and ProThermDB is a unique, well-maintained, and reliable one. The computational methods mainly utilize machine learning techniques and mechanistic models, which are mainly based on physicochemical properties, inter-residue interactions, contact potentials, and energetic parameters. Further, by elucidating the relationship between protein stability and disease-causing mutations, researchers can identify potential therapeutic targets and design interventions to modulate protein stability. The availability of high-quality datasets, addressing biases, and relating stability to diseases are key steps for further advancement.

Acknowledgments

We thank the members of the Protein Bioinformatics Lab for their valuable suggestions and inputs to improve the work. We acknowledge the Indian Institute of Technology Madras for its computational facilities. PRR is supported by PMRF. This work is partially supported by the Department of Biotechnology, Government of India to MMG (BT/PR40309/BTIS/137/57/2023).

References

Akke, M., & Forsén, S. (1990). Protein stability and electrostatic interactions between solvent exposed charged side chains. *Proteins: Structure, Function, and Bioinformatics, 8*(1), 23–29.

Alber, T. (1989). Mutational effects on protein stability. *Annual Review of Biochemistry, 58*(1), 765–792.

Broom, A., Jacobi, Z., Trainor, K., & Meiering, E. M. (2017). Computational tools help improve protein stability but with a solubility tradeoff. *Journal of Biological Chemistry, 292*(35), 14349–14361.

Cao, H., Wang, J., He, L., Qi, Y., & Zhang, J. Z. (2019). DeepDDG: Predicting the stability change of protein point mutations using neural networks. *Journal of Chemical Information and Modeling, 59*(4), 1508–1514.

Capriotti, E., Fariselli, P., & Casadio, R. (2004). A neural-network-based method for predicting protein stability changes upon single point mutations. *Bioinformatics, 20*(Suppl. 1), i63–i68.

Capriotti, E., Fariselli, P., & Casadio, R. (2005). I-Mutant2. 0: Predicting stability changes upon mutation from the protein sequence or structure. *Nucleic Acids Research, 33*(Suppl. 2), W306–W310.

Capriotti, E., Fariselli, P., Rossi, I., & Casadio, R. (2008). A three-state prediction of single point mutations on protein stability changes. *BMC Bioinformatics, 9*(2), 1–9.

Chaturvedi, D., & Mahalakshmi, R. (2013). Methionine mutations of outer membrane protein X influence structural stability and beta-barrel unfolding. *PLoS One, 8*(11), e79351.

Chen, C. W., Lin, J., & Chu, Y. W. (2013, January). iStable: Off-the-shelf predictor integration for predicting protein stability changes. In: *BMC bioinformatics* (Vol. 14, No. 2, pp. 1–14). BioMed Central.

Chen, C. W., Lin, M. H., Liao, C. C., Chang, H. P., & Chu, Y. W. (2020). iStable 2.0: Predicting protein thermal stability changes by integrating various characteristic modules. *Computational and Structural Biotechnology Journal, 18*, 622–630.

Dehouck, Y., Grosfils, A., Folch, B., Gilis, D., Bogaerts, P., & Rooman, M. (2009). Fast and accurate predictions of protein stability changes upon mutations using statistical potentials and neural networks: PoPMuSiC-2.0. *Bioinformatics, 25*(19), 2537–2543.

Dill, K. A. (1990). Dominant forces in protein folding. *Biochemistry, 29*(31), 7133–7155.

Fariselli, P., Martelli, P. L., Savojardo, C., & Casadio, R. (2015). INPS: Predicting the impact of non-synonymous variations on protein stability from sequence. *Bioinformatics, 31*(17), 2816–2821.

Folkman, L., Stantic, B., Sattar, A., & Zhou, Y. (2016). EASE-MM: Sequence-based prediction of mutation-induced stability changes with feature-based multiple models. *Journal of Molecular Biology, 428*(6), 1394–1405.

Ganesan, K., Kulandaisamy, A., Binny Priya, S., & Gromiha, M. M. (2019). HuVarBase: A human variant database with comprehensive information at gene and protein levels. *PLoS One, 14*(1), e0210475.

Getov, I., Petukh, M., & Alexov, E. (2016). SAAFEC: Predicting the effect of single point mutations on protein folding free energy using a knowledge-modified MM/PBSA approach. *International Journal of Molecular Sciences, 17*(4), 512.

Gilis, D., & Rooman, M. (1997). Predicting protein stability changes upon mutation using database-derived potentials: Solvent accessibility determines the importance of local versus non-local interactions along the sequence. *Journal of Molecular Biology, 272*(2), 276–290.

Giollo, M., Martin, A. J., Walsh, I., Ferrari, C., & Tosatto, S. C. (2014). NeEMO: A method using residue interaction networks to improve prediction of protein stability upon mutation. *BMC Genomics, 15*(4), 1–11.

Grant, M. A., Lazo, N. D., Lomakin, A., Condron, M. M., Arai, H., Yamin, G., Rigby, A. C., & Teplow, D. B. (2007). Familial Alzheimer's disease mutations alter the stability of the amyloid β-protein monomer folding nucleus. *Proceedings of the National Academy of Sciences, 104*(42), 16522–16527.

Gromiha, M. M. (2003). Factors influencing the thermal stability of buried protein mutants. *Polymer, 44*(14), 4061–4066.

Gromiha, M. M. (2005). A statistical model for predicting protein folding rates from amino acid sequence with structural class information. *Journal of Chemical Information and Modeling, 45*(2), 494–501.

Gromiha, M. M., Oobatake, M., Kono, H., Uedaira, H., & Sarai, A. (1999a). Relationship between amino acid properties and protein stability: Buried mutations. *Journal of Protein Chemistry, 18*, 565–578.

Gromiha, M. M., Oobatake, M., Kono, H., Uedaira, H., & Sarai, A. (1999b). Role of structural and sequence information in the prediction of protein stability changes: Comparison between buried and partially buried mutations. *Protein Engineering, 12*(7), 549–555.

Gromiha, M. M., & Ponnuswamy, P. K. (1993). Relationship between amino acid properties and protein compressibility. *Journal of Theoretical Biology, 165*(1), 87–100.

Gromiha, M. M., & Selvaraj, S. (2001). Comparison between long-range interactions and contact order in determining the folding rate of two-state proteins: Application of long-range order to folding rate prediction. *Journal of Molecular Biology, 310*(1), 27–32.

Grothe, H. L., Little, M. R., Sjogren, P. P., Chang, A. A., Nelson, E. F., & Yuan, C. (2013). Altered protein conformation and lower stability of the dystrophic transforming growth factor beta-induced protein mutants. *Molecular Vision, 19*, 593.

Guerois, R., Nielsen, J. E., & Serrano, L. (2002). Predicting changes in the stability of proteins and protein complexes: A study of more than 1000 mutations. *Journal of Molecular Biology, 320*(2), 369–387.

Huang, L. T., Gromiha, M. M., & Ho, S. Y. (2007). iPTREE-STAB: Interpretable decision tree based method for predicting protein stability changes upon mutations. *Bioinformatics, 23*(10), 1292–1293.

Kawashima, S., Pokarowski, P., Pokarowska, M., Kolinski, A., Katayama, T., & Kanehisa, M. (2007). AAindex: Amino acid index database, progress report 2008. *Nucleic Acids Research, 36*(Suppl. 1), D202–D205.

Kellogg, E. H., Leaver-Fay, A., & Baker, D. (2011). Role of conformational sampling in computing mutation-induced changes in protein structure and stability. *Proteins: Structure, Function, and Bioinformatics, 79*(3), 830–838.

Khan, S., & Vihinen, M. (2010). Performance of protein stability predictors. *Human Mutation, 31*(6), 675–684.

Kortemme, T., Kim, D. E., & Baker, D. (2004). Computational alanine scanning of protein–protein interfaces. *Science's STKE, 2004*(219), pl2–pl2.

Kroncke, B. M., Duran, A. M., Mendenhall, J. L., Meiler, J., Blume, J. D., & Sanders, C. R. (2016). Documentation of an imperative to improve methods for predicting membrane protein stability. *Biochemistry, 55*(36), 5002–5009.

Kulandaisamy, A., Binny Priya, S., Sakthivel, R., Tarnovskaya, S., Bizin, I., Hönigschmid, P., Frishman, D., & Gromiha, M. M. (2018). MutHTP: Mutations in human transmembrane proteins. *Bioinformatics, 34*(13), 2325–2326.

Kulandaisamy, A., Nikam, R., Harini, K., Sharma, D., & Gromiha, M. M. (2021). Illustrative tutorials for ProThermDB: Thermodynamic database for proteins and mutants. *Current Protocols, 1*(11), e306.

Kulandaisamy, A., Sakthivel, R., & Gromiha, M. M. (2021). MPTherm: Database for membrane protein thermodynamics for understanding folding and stability. *Briefings in Bioinformatics, 22*(2), 2119–2125.

Kulandaisamy, A., Zaucha, J., Frishman, D., & Gromiha, M. M. (2021). MPTherm-pred: Analysis and prediction of thermal stability changes upon mutations in transmembrane proteins. *Journal of Molecular Biology, 433*(11), 166646.

Kulandaisamy, A., Zaucha, J., Sakthivel, R., Frishman, D., & Michael Gromiha, M. (2020). Pred-MutHTP: Prediction of disease-causing and neutral mutations in human transmembrane proteins. *Human Mutation, 41*(3), 581–590.

Kumar, M. S., Bava, K. A., Gromiha, M. M., Prabakaran, P., Kitajima, K., Uedaira, H., & Sarai, A. (2006). ProTherm and ProNIT: Thermodynamic databases for proteins and protein–nucleic acid interactions. *Nucleic Acids Research, 34*(Suppl. 1), D204–D206.

Laimer, J., Hiebl-Flach, J., Lengauer, D., & Lackner, P. (2016). MAESTROweb: A web server for structure-based protein stability prediction. *Bioinformatics, 32*(9), 1414–1416.

Laimer, J., Hofer, H., Fritz, M., Wegenkittl, S., & Lackner, P. (2015). MAESTRO-multi agent stability prediction upon point mutations. *BMC Bioinformatics, 16*(1), 1–13.

Li, G., Panday, S. K., & Alexov, E. (2021). SAAFEC-SEQ: A sequence-based method for predicting the effect of single point mutations on protein thermodynamic stability. *International Journal of Molecular Sciences, 22*(2), 606.

MacGowan, S. A., Madeira, F., Britto-Borges, T., Warowny, M., Drozdetskiy, A., Procter, J. B., & Barton, G. J. (2020). The Dundee resource for sequence analysis and structure prediction. *Protein Science, 29*(1), 277–297.

Manning, J. R., Jefferson, E. R., & Barton, G. J. (2008). The contrasting properties of conservation and correlated phylogeny in protein functional residue prediction. *BMC Bioinformatics, 9*(1), 1–16.

Martelli, P. L., Fariselli, P., Savojardo, C., Babbi, G., Aggazio, F., & Casadio, R. (2016). Large scale analysis of protein stability in OMIM disease related human protein variants. *BMC Genomics, 17*(2), 239–247.

Masso, M., & Vaisman, I. I. (2008). Accurate prediction of stability changes in protein mutants by combining machine learning with structure based computational mutagenesis. *Bioinformatics, 24*(18), 2002–2009.

Montanucci, L., Capriotti, E., Frank, Y., Ben-Tal, N., & Fariselli, P. (2019). DDGun: An untrained method for the prediction of protein stability changes upon single and multiple point variations. *BMC Bioinformatics, 20*, 1–10.

Nicholson, H., Anderson, D. E., Dao Pin, S., & Matthews, B. W. (1991). Analysis of the interaction between charged side chains and the alpha-helix dipole using designed thermostable mutants of phage T4 lysozyme. *Biochemistry, 30*(41), 9816–9828.

Nikam, R., Kulandaisamy, A., Harini, K., Sharma, D., & Gromiha, M. M. (2021). ProThermDB: Thermodynamic database for proteins and mutants revisited after 15 years. *Nucleic Acids Research, 49*(D1), D420–D424.

Pace, C. N. (1990). Conformational stability of globular proteins. *Trends in Biochemical Sciences, 15*(1), 14–17.

Pace, C. N. (1995). [24] Evaluating contribution of hydrogen bonding and hydrophobic bonding to protein folding. In: *Methods in enzymology* (Vol. 259, pp. 538–554). Academic Press.

Pandey, M., Anoosha, P., Yesudhas, D., & Gromiha, M. M. (2022). Identification of potential driver mutations in glioblastoma using machine learning. *Briefings in Bioinformatics, 23*(6), bbac451.

Pandurangan, A. P., Ochoa-Montano, B., Ascher, D. B., & Blundell, T. L. (2017). SDM: A server for predicting effects of mutations on protein stability. *Nucleic Acids Research, 45*(W1), W229–W235.

Parthiban, V., Gromiha, M. M., & Schomburg, D. (2006). CUPSAT: Prediction of protein stability upon point mutations. *Nucleic Acids Research, 34*(Suppl. 2), W239–W242.

Pires, D. E., Ascher, D. B., & Blundell, T. L. (2014a). DUET: A server for predicting effects of mutations on protein stability using an integrated computational approach. *Nucleic Acids Research, 42*(W1), W314–W319.

Pires, D. E., Ascher, D. B., & Blundell, T. L. (2014b). mCSM: Predicting the effects of mutations in proteins using graph-based signatures. *Bioinformatics, 30*(3), 335–342.

Pires, D. E., Rodrigues, C. H., & Ascher, D. B. (2020). mCSM-membrane: Predicting the effects of mutations on transmembrane proteins. *Nucleic Acids Research, 48*(W1), W147–W153.

Plaxco, K. W., Simons, K. T., & Baker, D. (1998). Contact order, transition state placement and the refolding rates of single domain proteins. *Journal of Molecular Biology, 277*(4), 985–994.

Ponnuswamy, P. K., & Gromiha, M. M. (1994). On the conformational stability of folded proteins. *Journal of Theoretical Biology*, *166*(1), 63–74.

Potapov, V., Cohen, M., & Schreiber, G. (2009). Assessing computational methods for predicting protein stability upon mutation: Good on average but not in the details. *Protein Engineering, Design & Selection, 22*(9), 553–560.

Pucci, F., Bourgeas, R., & Rooman, M. (2016). Predicting protein thermal stability changes upon point mutations using statistical potentials: Introducing HoTMuSiC. *Scientific Reports*, *6*(1), 23257.

Quan, L., Lv, Q., & Zhang, Y. (2016). STRUM: Structure-based prediction of protein stability changes upon single-point mutation. *Bioinformatics, 32*(19), 2936–2946.

Reddy, P. R., Kulandaisamy, A., & Gromiha, M. M. (2023). TMH Stab-pred: Predicting the stability of α-helical membrane proteins using sequence and structural features. *Methods, 218*, 118–124.

Rodrigues, C. H., Pires, D. E., & Ascher, D. B. (2018). DynaMut: Predicting the impact of mutations on protein conformation, flexibility and stability. *Nucleic Acids Research, 46*(W1), W350–W355.

Rose, G. D., & Wolfenden, R. (1993). Hydrogen bonding, hydrophobicity, packing, and protein folding. *Annual Review of Biophysics and Biomolecular Structure, 22*(1), 381–415.

Saraboji, K., Gromiha, M. M., & Ponnuswamy, M. N. (2006). Average assignment method for predicting the stability of protein mutants. *Biopolymers: Original Research on Biomolecules, 82*(1), 80–92.

Schymkowitz, J., Borg, J., Stricher, F., Nys, R., Rousseau, F., & Serrano, L. (2005). The FoldX web server: An online force field. *Nucleic Acids Research, 33*(Suppl. 2), W382–W388.

Serrano, L., Horovitz, A., Avron, B., Bycroft, M., & Fersht, A. R. (1990). Estimating the contribution of engineered surface electrostatic interactions to protein stability by using double-mutant cycles. *Biochemistry, 29*(40), 9343–9352.

Shi, Z., & Moult, J. (2011). Structural and functional impact of cancer-related missense somatic mutations. *Journal of Molecular Biology, 413*(2), 495–512.

Stefl, S., Nishi, H., Petukh, M., Panchenko, A. R., & Alexov, E. (2013). Molecular mechanisms of disease-causing missense mutations. *Journal of Molecular Biology, 425*(21), 3919–3936.

Stehr, H., Jang, S. H. J., Duarte, J. M., Wierling, C., Lehrach, H., Lappe, M., & Lange, B. M. (2011). The structural impact of cancer-associated missense mutations in oncogenes and tumor suppressors. *Molecular Cancer, 10*(1), 1–10.

Studer, R. A., Dessailly, B. H., & Orengo, C. A. (2013). Residue mutations and their impact on protein structure and function: Detecting beneficial and pathogenic changes. *Biochemical Journal, 449*(3), 581–594.

Sun, D. P., Sauer, U., Nicholson, H., & Matthews, B. W. (1991). Contributions of engineered surface salt bridges to the stability of T4 lysozyme determined by directed mutagenesis. *Biochemistry, 30*(29), 7142–7153.

Takano, K., Liu, D., Tarpey, P., Gallant, E., Lam, A., Witham, S., Alexov, E., Chaubey, A., Stevenson, R. E., Schwartz, C. E., Board, P. G., & Dulhunty, A. F. (2012). An X-linked channelopathy with cardiomegaly due to a CLIC2 mutation enhancing ryanodine receptor channel activity. *Human Molecular Genetics, 21*(20), 4497–4507.

Talley, K., & Alexov, E. (2010). On the pH-optimum of activity and stability of proteins. *Proteins: Structure, Function, and Bioinformatics, 78*(12), 2699–2706.

Teng, S., Srivastava, A. K., & Wang, L. (2010). Sequence feature-based prediction of protein stability changes upon amino acid substitutions. *BMC Genomics, 11*(2), 1–8.

Tina, K. G., Bhadra, R., & Srinivasan, N. (2007). PIC: Protein interactions calculator. *Nucleic Acids Research, 35*(Suppl. 2), W473–W476.

Tokuriki, N., & Tawfik, D. S. (2009). Stability effects of mutations and protein evolvability. *Current Opinion in Structural Biology, 19*(5), 596–604.

Valdar, W. S. (2002). Scoring residue conservation. *Proteins: Structure, Function, and Bioinformatics, 48*(2), 227–241.

Witham, S., Takano, K., Schwartz, C., & Alexov, E. (2011). A missense mutation in CLIC2 associated with intellectual disability is predicted by in silico modeling to affect protein stability and dynamics. *Proteins: Structure, Function, and Bioinformatics, 79*(8), 2444–2454.

Xavier, J. S., Nguyen, T. B., Karmarkar, M., Portelli, S., Rezende, P. M., Velloso, J. P., Ascher, D. B., & Pires, D. E. (2021). ThermoMutDB: A thermodynamic database for missense mutations. *Nucleic Acids Research, 49*(D1), D475–D479.

Yang, Y., Chen, B., Tan, G., Vihinen, M., & Shen, B. (2013). Structure-based prediction of the effects of a missense variant on protein stability. *Amino Acids, 44*, 847–855.

Yang, Y., Urolagin, S., Niroula, A., Ding, X., Shen, B., & Vihinen, M. (2018). PON-tstab: Protein variant stability predictor. Importance of training data quality. *International Journal of Molecular Sciences, 19*(4), 1009.

Zhang, Z., Norris, J., Schwartz, C., & Alexov, E. (2011). In silico and in vitro investigations of the mutability of disease-causing missense mutation sites in spermine synthase. *PloS one, 6*(5), e20373.

Zhang, Z., Wang, L., Gao, Y., Zhang, J., Zhenirovskyy, M., & Alexov, E. (2012). Predicting folding free energy changes upon single point mutations. *Bioinformatics, 28*(5), 664–671.

https://doi.org/10.1142/9789811293269_0003

Chapter 3

Exploring important factors influencing the folding rates of proteins and their mutants

Liang-Tsung Huang[1,*]

[1]*Department of Medical Informatics, Tzu Chi University, Hualien 970, Taiwan*

Abstract

Understanding the importance of sequence and structure-based parameters of proteins for determining the folding rates of proteins and accurately predicting protein folding rates, as well as changes in folding rates upon mutations, provides deep insights into protein folding kinetics and design. In this chapter, we briefly outline the databases available for protein folding rates, followed by the development of algorithms for predicting protein folding rates using sequence and structure information and changes in folding rates upon mutation.

3.1 Introduction

Correct protein folding ensures that proteins can perform their designated biological functions. Understanding the protein folding mechanism

*Corresponding author
LTH: LTH@mail.tcu.edu.tw

contributes not only to fundamental scientific knowledge but also holds practical applications in both applied science and medical research. Therefore, the study of protein folding rate, which dictates the slow or fast folding of proteins, is valuable for biological functions, disease, misfolding, drug design, biotechnology, protein engineering, and evolutionary biology.

For an accurate prediction of protein folding rate, information from protein structures and corresponding amino acid sequences has been derived as important parameters. However, it is necessary to develop more effective factors for enhancing prediction performance. By characterizing the topological characteristics of proteins, it is shown that the logarithm of the experimental protein folding rate depends on both the local geometry and the topology of the protein's native state (Wang & Panagiotou, 2022). Besides, an effective cumulative backbone torsion angle (CBTAeff) was proposed to describe the size of the conformational space and used to predict protein folding rate with a high correlation (Li *et al.*, 2020).

The protein folding rate is highly sensitive to temperature. Starting from the assumption that protein folding is an event of quantum transition between molecular conformations, the folding rate for all two-state proteins was calculated to analyze their temperature dependencies (Lv & Luo, 2014). Then, the impact of hydrodynamic interactions (HI) on protein folding was investigated by using a coarse-grained protein model, which is a structure-based model and allows to study the ideal energy landscape of a protein. Additionally, the effect of HI on protein folding appears to have a crossover behavior about the folding temperature (Zegarra *et al.*, 2018).

On the other hand, mRNA sequence is regarded as genetic language, which may provide critical information for protein folding. Several studies have focused on the relationship between mRNA sequences and corresponding protein folding rates. The influences of palindromes in mRNA have been analyzed, and an extremely significant negative linear correlation was reported in the relationship between palindrome densities and protein folding rates (Li *et al.*, 2020). Furthermore, the vocabulary and phraseology of mRNA sequences and the corresponding protein folding rates were analyzed, and the results indicate that some information for regulating protein folding rates must be derived from the mRNA sequences (Li *et al.*, 2021).

These studies give remarkable evidence for determining folding rates and point to potential parameters for a more accurate prediction of protein folding rates. In this chapter, we will discuss the databases related to protein folding rates, potential influencing factors, and methods for predicting folding rates.

3.2 Database for protein folding rates

Databases for experimental data on folding kinetics are important and necessary for developing methods to predict folding/unfolding rates and their variation on mutation. The following describes several previous databases in chronological order. **Table 3.1** lists the available databases for protein folding rates.

Protein folding database (PFD 2.0) (Fulton *et al.*, 2007) collects thermodynamic and kinetic data into a searchable and structured repository that conforms to the standards of the International Foldeomics Consortium. It allows registered users to deposit their folding data, and the form is divided into several logical sections: protein, construct, publication,

Table 3.1 Databases for protein folding rates*

Name	URL	Year	References
PFD 2.0[#]	http://www.foldeomics.org/pfd/	2007	Fulton *et al.* (2007)
KineticDB[#]	http://kineticdb.protres.ru/db/index.pl	2009	Bogatyreva *et al.* (2009)
ACPro	https://www.ats.amherst.edu/protein	2014	Wagaman *et al.* (2014)
PFDB	http://lee.kias.re.kr/~bala/PFDB	2019	Manavalan *et al.* (2019)
K-Pro	https://folding.biofold.org/k-pro	2023	Turina *et al.* (2023)

*Last accessed on December 15, 2023; [#]Not available.

mutations, equilibrium method, kinetic method, equilibrium data, kinetic data, and other data and comments. KineticDB (Bogatyreva *et al.*, 2009) accumulates folding kinetics data for about 90 unique proteins, including single-domain proteins, separate protein domains, and short peptides without disulfide bonds in their native structure. Each entry includes a single protein folding kinetics measurement extracted from the original paper, protein details with its best available tertiary structure, experimental conditions, reference to the original paper, and experimental results.

The Amherst College Protein Folding Kinetics (ACPro) Database (Wagaman *et al.*, 2014) combines kinetics with the sequence and structural information of 126 verified proteins, consisting of 83 two-state and 43 multistate folding classes. The entry contains the verified experimental information along with additional associated structural information retrieved from the Protein Data Bank (PDB) (Berman *et al.*, 2000). PFDB (Manavalan *et al.*, 2019) is a standardized PFD with temperature correction in which the logarithmic rate constants of proteins are calculated at the standard temperature (25°C). The database consists of 141 single-domain globular proteins comprising 89 two-state and 52 non-two-state folding classes.

K-Pro (Turina *et al.*, 2023) is a database designed for collecting and storing experimental kinetic data on monomeric proteins with a two-state folding mechanism. The database consists of 1529 records from 62 proteins corresponding to 65 structures including different kinetic parameters such as the logarithm of the folding and unfolding rates in water ($\ln(k_f^{H2O})$, $\ln(k_u^{H2O})$) and/or in the presence of denaturant ($\ln(k_f^{DEN})$, $\ln(k_u^{DEN})$), the equilibrium free energy change at zero denaturant (ΔG^{H2O}), the folding and unfolding slopes of the Chevron plot (m_f, m_u), the concentration of denaturant at 50% of unfolded protein (C_m), Stanford's β, the folding ϕ values, and partly thermodynamic parameters associated with the kinetic data.

3.3 Prediction of protein folding rates from sequence

Several web-based tools have been developed for predicting protein folding rates using sequence information and structural parameters. We review these methods from different perspectives, comprising prediction performance, usability, and utility, as well as development and validation methodologies (Chang *et al.*, 2015). **Table 3.2** provides details on available methods for predicting protein folding rates.

Protein sequences offer several advantages over structural parameters in the prediction of protein folding rates. Protein sequence data is more

Table 3.2 Computational methods for predicting protein folding rates and changes in folding rates upon mutations*

Method	URL	Year	References
FOLD-RATE Q	http://bioinformatics.tcu.edu.tw/FOLDRATE20r/foldrate20.htm	2007	Huang and Gromiha (2008)
SFoldRate	http://gila.bioe.uic.edu/lab/tools/foldingrate/fr0.html	2008	Ouyang & Liang (2008)
Pred-PFR	http://www.csbio.sjtu.edu.cn/bioinf/FoldingRate/	2009	Shen *et al.* (2009)
FoldRate	http://www.csbio.sjtu.edu.cn/bioinf/FoldRate/	2009	Chou & Shen (2009)
SeqRate[#]	http://casp.rnet.missouri.edu/fold_rate/index.html	2010	Lin *et al.* (2010)
Swfoldrate[#]	http://www.jci-bioinfo.cn/swfrate/input.jsp	2013	Cheng *et al.* (2013)
Folding RaCe	http://www.iitm.ac.in/bioinfo/proteinfolding/foldingrace.html	2015	Chaudhary *et al.* (2015)
FRTpred[#]	http://thegleelab.org/FRTpred	2022	Manavalan and Lee (2022)
FOLD-RATE	https://www.iitm.ac.in/bioinfo/fold-rate/	2006	Gromiha *et al.* (2006)
K-Fold	https://folding.biofold.org/k-fold/	2007	Capriotti and Casadio (2007)
PRORATE[#]	http://sunflower.kuicr.kyoto-u.ac.jp/~sjn/folding/	2010	Song *et al.* (2010)
FREEDOM	http://bioinformatics.tcu.edu.tw/FREEDOMr/freedom.htm	2010	Huang and Gromiha (2010)
FORA	http://bioinformatics.tcu.edu.tw/FORAr/fora.htm	2012	Huang and Gromiha (2012)

*Last accessed on December 15, 2023; [#]Not available.

abundant and easily accessible compared to experimental three-dimensional protein structures. Furthermore, analyzing protein sequences is computationally less demanding compared to predicting three-dimensional protein structures. Moreover, genome-wide association studies (GWAS) and other large-scale genomic analyses often rely on protein sequence data to identify potential links between genetic variations and diseases.

Many studies have reported that protein folding rate is influenced by the characteristics of amino acid sequences. FOLD-RATE Q (Huang & Gromiha, 2008) systematically analyzed the relationship between protein folding rates and the physical–chemical, energetic, and conformational properties of amino acid residues. Based on quadratic response surface models (QRSM), the correlation for predicting folding rate is 0.99, 0.98, and 0.96 for all-alpha, all-beta, and mixed-class proteins, respectively. Based only on the amino acid sequence and geometric contact number defined by a Voronoi criterion, SFoldRate (Ouyang & Liang, 2008) was developed for predicting the folding rate of diverse classes of proteins with a correlation coefficient of 0.82. Pred-PFR (Shen *et al.*, 2009) focuses on multiple sequential features that can be derived from the amino acid sequential information, including amino acid properties, protein size effects, and information derived from secondary structure (SS). The correlation coefficient for predicting protein folding is up to 0.88.

Owing to the fact that smaller proteins usually fold faster than larger ones and that α-helix and β-sheet are the two most major structural elements, FoldRate (Chou & Shen, 2009) is designed for predicting folding rate based on three corresponding parameters, that is, protein size or length effect, α-helix effect, and its β-sheet effect. The overall Pearson correlation coefficient (PCC) value yielded by the ensemble predictor is 0.88. SeqRate (Lin *et al.*, 2010) is a non-linear machine learning method (support vector machine classification and regression) for predicting both protein folding kinetic type (two-state versus multistate) and real-value folding rates. The features are extracted only from sequence information, including protein sequence length, amino acid composition, predicted long-range contact order (LRCO), predicted long-range contact number (LRCN), predicted α-helical content and β-sheet content, and amino acid composition. The PCC between predicted and experimental folding rates is 0.81 and 0.80 for two-state and three-state protein folders, respectively.

Introducing both a back-propagation neural network (BPNN) and a genetic algorithm, Guo and Rao (2011) proposed a combined method for predicting protein folding rate from amino acid sequence. The correlation coefficient is 0.80, and the standard error (SE) is 2.65 for 93 proteins. The results also suggest that the sequence order information contributes to the determination of protein folding rate. Swfoldrate (Cheng *et al.*, 2013) uses the long-range and short-range contacts in a protein to derive an extended version of the pseudo amino acid composition (PseAAC) based on a sliding window. By using the nonlinear support vector machine (SVM) regression model, the method obtained a correlation coefficient of 0.9313 and a SE of 2.2692.

Folding RaCe (Chaudhary *et al.*, 2015) is a knowledge-based methodology to predict the changes in folding rates upon mutations. In this method, mutants were classified according to SS, accessible surface area, and position along the primary sequence. Further, three prime amino acid features were shortlisted for each class, along with an optimized window length. The Jack-knife test resulted in a mean absolute error (MAE) of 0.42 s^{-1} and a PCC of 0.73. Based on deep learning neural networks, a model based on fuzzy cognitive map (FCM) along with the Levenberg–Marquardt (LM) algorithm is proposed to predict the protein folding rate (Liu *et al.*, 2017). The nine properties of protein sequence are used as follows: Alpha helix of C terminus (α_c); beta sheets (P_β); compression capability of protein sequence (K^0); react ability of proteins in solvents (R_α); contact surface of the unfolded chain and solvent (ΔASA); Gibbs free energy of hydration in denatured proteins (ΔG_{hD}); the average range of amino acid contact (N_m); the polarity value of amino acids (P); and the number of torsion angles of the side chain (n). The correlation coefficient reaches 0.94 and 0.9 in two independent tests.

Integrating 10 representative feature extraction methods and three machine-learning algorithms, FRTpred (Manavalan & Lee, 2022) predicts the logarithmic protein folding rate constant, $\ln(k_f)$, and folding type from amino acid sequence by combining the predicted values of the 30 baseline models. The PCC is up to 0.826 for predicting $\ln(k_f)$ without folding class information, that is, non-two-state (N2S) or two-state (2S), and the balanced accuracy (BACC) is 0.843 for predicting the folding type. The roles of amino acids in proteins can vary significantly in different kinetic mechanisms. Huang *et al.* (2014) analyzed relevant issues of protein folding in two-state or multistate kinetic mechanisms. Many residues that are easy to form into regular SS (α helices, β sheets, and turns) can promote two-state folding, and most of hydrophilic residues can speed up multistate folding. These findings can provide more effective features of amino acid sequence to develop accurate models for predicting folding rate.

3.4 Relationship between protein folding rates and structure-based parameters

Discovering novel knowledge from existing data is crucial for advancing our understanding of complex biological processes such as protein folding. Predicting protein folding rate using protein structures offers several advantages over relying solely on sequential parameters. Protein structures provide a three-dimensional context to allow the identification of specific folding motifs, SS, and tertiary interactions. The relationship between protein function and structure is often better understood through the analysis of 3D structures. Furthermore, protein structures offer insights into the dynamic aspects of folding, including intermediate states and transitional structures. These dynamic parameters can be crucial for a more accurate prediction of the folding process. Consequently, several structure-based parameters, including contact order (Plaxco *et al.*, 1998), long-range order (Gromiha & Selvaraj, 2001), and multiple contact index (Gromiha, 2009), have been proposed to relate protein folding rates. The evaluation of these parameters along with other methods has been discussed recently (Nithiyanandam *et al.*, 2023).

FOLD-RATE (Manavalan & Lee, 2022) is developed for predicting the folding rates of proteins from their amino acid sequences. The results also indicate that the classification of proteins into different structural classes shows an excellent correlation between amino acid properties and folding rates. Therefore, the correlation coefficients can be improved up to 0.99, 0.97, and 0.90 for all-alpha, all-beta, and mixed-class proteins, respectively. K-Fold (Capriotti & Casadio, 2007) is a tool for predicting the protein folding kinetic order and the value of the logarithm of folding rate from the protein structure through its PDB code. Based on SVM, the correlation coefficient between the predicted and experimental data is 0.74 with a SE of 1.2.

PRORATE (Song *et al.*, 2010) is a method for predicting protein folding rate for two-state and multistate protein folding kinetics that combines structural topology and complex network properties that are calculated from protein three-dimensional structures. They calculated eight topology measures and built two kinds of network models based on two length scales: the protein contact network (PCN) and the long-range interaction network (LIN). For leave-one-out cross-validation (LOOCV) tests, the PCC was reported to be 0.88, 0.90, and 0.90 for two-state, multistate, and combined protein folding kinetics, respectively.

For assessing the impact of more structural parameters, 10 different ML algorithms for predicting folding rate are evaluated with eight different structural parameters and five different network centrality measures based on PFDB database (Nithiyanandam *et al.*, 2023). Further, the kinetics of folding are likely to be influenced to a greater extent by a small number of crucial structural features. Huang *et al.* (2015) indicated that SS can determine folding rates of only large, multistate folding proteins and fail to predict those for small, two-state proteins. Additionally, the importance of SS for protein folding is ordered. Besides, nucleation-based prediction of the protein folding rate is developed based on a dynamic programming method (Galzitskaya & Glyakina, 2012).

While protein structures provide valuable information, it is essential to note that combining both sequential and structural parameters in predictive models can lead to more comprehensive and accurate predictions of protein folding.

3.5 Prediction of protein folding rates upon mutations

Amino acid substitutions in proteins alter their folding rates, which affect protein functions, and some of them may lead to diseases. Hence, it is necessary to predict the change in protein folding rates upon mutations.

FREEDOM (Huang & Gromiha, 2010) is the first study for predicting protein folding rate change upon point mutation. The proposed method for discriminating the effect of folding rates is based on quasi-regression models with diverse features, such as amino acid properties, second structure, solvent accessibility (SA), conservation score of residues and long-range contacts (LRC). LRC refer to interactions between amino acid residues that are distant in terms of sequence but come into close spatial proximity in the three-dimensional structure of a protein.

FORA (Huang & Gromiha, 2012) is a method based on quadratic regression models (QRMs) for predicting the real value of protein folding rate change upon point mutation. The method combined both sequence and structure properties. The former includes wild-type, mutant, and three neighboring residues, and the latter involves the SS and LRC of the wild type. By 10-fold cross validation, correlation is 0.72 between the experimental and predicted changes in protein folding rates.

PON-Fold (Yang *et al.*, 2023) is a machine-learning-based method for predicting the folding rate effects of amino acid substitutions in two-state folding proteins using the LightGBM framework. It facilitates the implementation of a gradient-boosting decision tree algorithm using a total of 1,161 biological features of six types, such as 688 amino acid features, three conservation features, 436 variation-type features, 25 neighborhood features, one protein-type feature, and eight structural features.

Huang *et al.* (2014) proposed several rules for discriminating protein folding rate change upon single-point mutation and developed a decision tree algorithm and a decision table algorithm for discrimination.

3.6 Conclusions

Understanding protein folding rates is fundamental for elucidating biological functions, disease mechanisms, drug design, and various applications in biotechnology and protein engineering. Prediction of protein folding rates using sequence information offers advantages, as it is more abundant and computationally less demanding than analyzing three-dimensional structures. On the other hand, predicting folding rates from protein structures provides a three-dimensional context, allowing the identification of specific folding motifs and dynamic aspects of folding. Additionally, the integration of both structural and sequential parameters enhances the overall comprehensiveness and accuracy of protein folding rate predictions. This chapter covers these aspects, such as the development of databases for protein folding rates, the prediction of protein folding rates using sequence and structural information as well change in folding rates of proteins upon mutations.

Acknowledgments

I would like to express my sincere gratitude to Tzu Chi University, Taiwan, for their support of my research in the field of biomedical informatics. The university's generous funding has allowed me to develop the bioinformatics tools online (http://bioinformatics.tcu.edu.tw) website, which provides services for protein analysis and applications. I am also deeply indebted to Professor M. Michael Gromiha, whose expert advice and guidance have been instrumental in my research journey. His long-standing dedication to protein informatics research inspires me immensely, while his unwavering professionalism and infectious passion for the field of bioinformatics are truly admirable.

References

Berman, H. M., Westbrook, J., Feng, Z., Gilliland, G., Bhat, T. N., Weissig, H., Shindyalov, I. N., & Bourne, P. E. (2000). The protein data bank. *Nucleic Acids Research, 28*(1), 235–242.

Bogatyreva, N. S., Osypov, A. A., & Ivankov, D. N. (2009). KineticDB: A database of protein folding kinetics. *Nucleic Acids Research, 37*(Suppl. 1), D342–D346.

Capriotti, E., & Casadio, R. (2007). K-Fold: A tool for the prediction of the protein folding kinetic order and rate. *Bioinformatics, 23*(3), 385–386.

Chang, C. C., Tey, B. T., Song, J., & Ramanan, R. N. (2015). Towards more accurate prediction of protein folding rates: A review of the existing Web-based bioinformatics approaches. *Briefings in Bioinformatics, 16*(2), 314–324.

Chaudhary, P., Naganathan, A. N., & Gromiha, M. M. (2015). Folding RaCe: A robust method for predicting changes in protein folding rates upon point mutations. *Bioinformatics, 31*(13), 2091–2097.

Cheng, X., Xiao, X., Wu, Z.-C., Wang, P., & Lin, W.-Z. (2013). Swfoldrate: Predicting protein folding rates from amino acid sequence with sliding window method. *Proteins, 81*(1), 140–148.

Chou, K., & Shen, H. (2009). FoldRate: A web-server for predicting protein folding rates from primary sequence. *The Open Bioinformatics Journal, 3*, 31–50.

Fulton, K. F., Bate, M. A., Faux, N. G., Mahmood, K., Betts, C., & Buckle, A. M. (2007). Protein folding database (PFD 2.0): An online environment for the International Foldeomics Consortium. *Nucleic Acids Research, 35*(Suppl. 1), D304–D307.

Galzitskaya, O. V., & Glyakina, A. V. (2012). Nucleation-based prediction of the protein folding rate and its correlation with the folding nucleus size. *Proteins, 80*(12), 2711–2727.

Gromiha, M. M. (2009). Multiple contact network is a key determinant to protein folding rates. *Journal of Chemical Information and Modeling, 49*(4), 1130–1135.

Gromiha, M. M., & Selvaraj, S. (2001). Comparison between long-range interactions and contact order in determining the folding rate of two-state proteins: Application of long-range order to folding rate prediction. *Journal of Molecular Biology, 310*(1), 27–32.

Gromiha, M. M., Thangakani, A. M., & Selvaraj, S. (2006). FOLD-RATE: Prediction of protein folding rates from amino acid sequence. *Nucleic Acids Research, 34*(Web Server issue), W70–W74.

Guo, J., & Rao, N. (2011). Predicting protein folding rate from amino acid sequence. *Journal of Bioinformatics and Computational Biology, 9*(1), 1–13.

Huang, J. T., Huang, W., Huang, S. R., & Li, X. (2014). How the folding rates of two- and multistate proteins depend on the amino acid properties. *Proteins, 82*(10), 2375–2382.

Huang, J. T., Wang, T., Huang, S. R., & Li, X. (2015). Prediction of protein folding rates from simplified secondary structure alphabet. *Journal of Theoretical Biology, 383*, 1–6.

Huang, L. T. (2014). Finding simple rules for discriminating folding rate change upon single mutation by statistical and learning methods. *Protein & Peptide Letters, 21*(8), 743–751.

Huang, L. T., & Gromiha, M. M. (2008). Analysis and prediction of protein folding rates using quadratic response surface models. *Journal of Computational Chemistry, 29*(10), 1675–1683.

Huang, L. T., & Gromiha, M. M. (2010). First insight into the prediction of protein folding rate change upon point mutation. *Bioinformatics, 26*(17), 2121–2127.

Huang, L. T., & Gromiha, M. M. (2012). Real value prediction of protein folding rate change upon point mutation. *Journal of Computer-Aided Molecular Design, 26*(3), 339–347.

Li, R., Li, H., Feng, X., Zhao, R., & Cheng, Y. (2021). Study on the influence of mRNA, the genetic language, on protein folding rates. *Frontiers in Genetics, 12*, 635250.

Li, R., Li, H., Yang, S., & Feng, X. (2020). The influences of palindromes in mRNA on protein folding rates. *Protein & Peptide Letters, 27*(4), 303–312.

Li, Y., Zhang, Y., & Lv, J. (2020). An effective cumulative torsion angles model for prediction of protein folding rates. *Protein & Peptide Letters, 27*(4), 321–328.

Lin, G. N., Wang, Z., Xu, D., & Cheng, J. (2010). SeqRate: Sequence-based protein folding type classification and rates prediction. *BMC Bioinformatics, 11*(Suppl 3), S1.

Liu, L., Ma, M., & Cui, J. (2017). A novel model-based on FCM-LM algorithm for prediction of protein folding rate. *Journal of Bioinformatics and Computational Biology, 15*(4), 1750012.

Lv, J., & Luo, L. (2014). Statistical analyses of protein folding rates from the view of quantum transition. *Science China Life Sciences, 57*(12), 1197–212.

Manavalan, B., Kuwajima, K., & Lee, J. (2019). PFDB: A standardized protein folding database with temperature correction. *Scientific Reports, 9*(1), 1588.

Manavalan, B., & Lee, J. (2022). FRTpred: A novel approach for accurate prediction of protein folding rate and type. *Computers in Biology and Medicine, 149*, 105911.

Nithiyanandam, S., Sangaraju, V. K., Manavalan, B., & Lee, G. (2023). Computational prediction of protein folding rate using structural parameters and network centrality measures. *Computers in Biology and Medicine, 155*, 106436.

Ouyang, Z., & Liang, J. (2008). Predicting protein folding rates from geometric contact and amino acid sequence. *Protein Science, 17*(7), 1256–1263.

Plaxco, K. W., Simons, K. T., & Baker, D. (1998). Contact order, transition state placement and the refolding rates of single domain proteins. *Journal of Molecular Biology, 277*(4), 985–994.

Shen, H.-B., Song, J.-N., & Chou, K.-C. (2009). Prediction of protein folding rates from primary sequence by fusing multiple sequential features. *Journal of Biomedical Science and Engineering, 2*(03), 136.

Song, J., Takemoto, K., Shen, H., Tan, H., Gromiha, M. M., & Akutsu, T. (2010). Prediction of protein folding rates from structural topology and complex network properties. *IPSJ Transactions on Bioinformatics, 3*, 40–53.

Turina, P., Fariselli, P., & Capriotti, E. (2023). K-Pro: Kinetics data on proteins and mutants. *Journal of Molecular Biology, 435*(20), 168245.

Wagaman, A. S., Coburn, A., Brand-Thomas, I., Dash, B., & Jaswal, S. S. (2014). A comprehensive database of verified experimental data on protein folding kinetics. *Protein Science, 23*(12), 1808–1812.

Wang, J., & Panagiotou, E. (2022). The protein folding rate and the geometry and topology of the native state. *Scientific Reports, 12*(1), 6384.

Yang, Y., Chong, Z., & Vihinen, M. (2023). PON-Fold: Prediction of substitutions affecting protein folding rate. *International Journal of Molecular Sciences, 24*(16), 13023.

Zegarra, F. C., Homouz, D., Eliaz, Y., Gasic, A. G., & Cheung, M. S. (2018). Impact of hydrodynamic interactions on protein folding rates depends on temperature. *Physical Review E, 97*(3), 032402.

https://doi.org/10.1142/9789811293269_0004

Chapter 4

Computational approaches for understanding protein disorder upon mutation: databases, algorithms, and applications

Dhanusha Yesudhas[1,*], Ambuj Srivastava[2], S. Lekshmi[3], and M. Michael Gromiha[3,*]

[1]*Genitourinary Malignancies Branch, National Cancer Institute, National Institute of Health, Bethesda, MD, USA*
[2]*Department of Molecular Biology, University of California, San Diego, CA, USA*
[3]*Department of Biotechnology, Bhupat and Jyoti Mehta School of Biosciences, Indian Institute of Technology Madras, Chennai, Tamil Nadu 600036, India*

Abstract

Intrinsically disordered regions/proteins (IDRs/IDPs) are highly flexible and lack well-defined three-dimensional structures in solution. The flexibility of IDRs is crucial for their functions, such as binding multiple partners, forming membrane-less compartments, and functional

*Corresponding authors
DY: dhanusha.yesudhas@nih.gov
MMG: gromiha@iitm.ac.in

moonlighting. Disordered regions can be experimentally determined by nuclear magnetic resonance (NMR), X-ray crystallography, cryogenic electron microscopy (cryo-EM), and fluorescence resonance energy transfer (FRET) methods. Mutations in IDRs could lead to diseases including neurodegenerative disorders and cancer. Although disease-causing missense mutations more commonly occur in evolutionarily conserved structured regions, more than 20% of human disease mutations are found in IDRs. In this chapter, we will survey the list of databases available for IDRs in proteins, explore computational tools for identifying IDRs, and the influence of mutations on IDRs. Further, the relationship between IDRs and disease-causing mutations will be discussed.

4.1 Introduction

In protein sequences, intrinsically disordered regions (IDRs) lack stable structures under normal physiological conditions and often carry out their functions without fully folding into a specific conformation. Regions characterized by disorder contain a higher concentration of charged and polar amino acids while being deficient in large hydrophobic residues (Uversky *et al.*, 2000). This results in a diminished hydrophobic effect, which typically serves as the primary driving force for the folding of polypeptides into their tightly packed tertiary structure, known as the native folded state (Ahmed *et al.*, 2022).

Numerous intrinsically disordered proteins/regions (IDPs/IDPRs) have the capability to undergo structural folding upon binding to their partners and induced folding varies across different systems, resulting in the formation of complexes with diverse structural and functional characteristics. Intrinsic disorder forms the foundation for dynamic "on-off" switch-like interactions commonly encountered in signaling networks, where IDPs/IDPRs bind to their partners with low affinity but high specificity (Uversky *et al.*, 2019). In addition, these IDRs exhibit the ability to morph and change their shape in response to different binding partners, leading to distinct folding patterns. IDPs/IDPRs can establish exceedingly tight complexes with affinities close to picomolar levels (Borgia *et al.*, 2018), retaining their inherent disordered state and preserving long-range flexibility as well as dynamic nature. This phenomenon is believed to hinge on the long-range electrostatic attraction between the charge-rich segments of the polypeptide chains rather than relying on well-defined structural binding sites (Borgia *et al.*, 2018; Uversky *et al.*, 2019).

Peng *et al.* (2014) reported that IDRs are prevalent in nature, especially in eukaryotes, where they are found in over 25% of proteins. Additionally, proteins containing IDRs have been linked to the development of various human diseases, including cancer and neurodegenerative disorders (Uversky *et al.*, 2014). This relationship has spurred interest in exploring these proteins as potential drug targets.

The experimentally characterized IDRs are only limited, and hence several computational methods have been developed for predicting these regions from protein sequences and investigating their functions. These methods are based on physicochemical properties such as polarizability and isoelectric point, predicted secondary structure (Tang *et al.*, 2022), evolutionary information (Erdős *et al.*, 2021), as well as substitution matrices along with information on neighboring residues along the sequence (Anoosha *et al.*, 2015). Tang *et al.* (2020) leveraged a natural language processing (NLP) model to forecast disorder residues in a protein.

In this chapter, we outline experimental methods for detecting IDRs in proteins, reported databases in the literature, as well as computational tools for predicting IDRs from amino acid sequences and the effect of mutations. Further, we explore the role of IDRs in disease-causing mutations.

4.2 Experimental studies for identifying disordered proteins/regions

X-ray crystallography is the primary method for deciphering the three-dimensional structures of macromolecules, which provide information on disordered regions (Shi, 2014). Cryogenic electron microscopy (cryo-EM) is a powerful imaging technique in structural biology that allows researchers to visualize biological macromolecules and complexes at near-atomic resolution (Nwanochie *et al.*, 2019; Ma *et al.*, 2022). In addition, ion mobility spectrometry coupled with mass spectrometry has been employed in the investigation of IDPs, offering potential advantages due to its ability to elucidate structure and stoichiometry. Stuchfield *et al.* (2018) used ion mobility spectrometry–mass spectrometry (IM–MS) coupled with electrospray ionization to study the conformational heterogeneity in disordered proteins. α-Synuclein has been the subject of thorough investigation utilizing mass spectrometry and various other methods, owing to its pivotal role in the advancement of Parkinson's disease and its role as a prime example of an IDP in biophysical research (Mensch *et al.*, 2017). Phillips *et al.* (2015) employed hybrid mass spectrometry methods, which included ion mobility spectrometry coupled with mass spectrometry (IM–MS), to unveil the extensive conformational adaptability of α-synuclein.

4.2.1 *Computational analysis of disordered proteins/regions*

IDRs and IDPs are enriched in charged amino acids, preventing hydrophobic collapse. Residues in IDRs interact with various biomolecules such as DNA, RNA, lipids, metals, and carbohydrates to undergo disorder-to-order transition (DOT) and perform different functions (Han *et al.*, 2023). Shimizu *et al.* (2009) reported that interactions between disordered proteins form a significant part of protein–protein interaction networks in humans. Deryusheva *et al.* (2019) observed that 15% of all disordered proteins are known to interact with lipids due to the interaction between polar head groups and charged amino acid residues in the disordered regions. These interactions formed by disordered proteins with various biomolecules provide profound insights to target these regions for therapeutic purposes owing to the link between disordered regions in proteins and diseases. Hence, several tools have been reported in the last two decades for the identification of disorders in proteins.

4.3 Identification of disordered regions in proteins

Disordered proteins are identified from protein structures determined by X-ray crystallography, nuclear magnetic resonance (NMR), and cryo-EM using the following criteria. In X-ray crystallography, non-resolved amino acids are listed in missing residues in Protein Data Bank (PDB) files, and a continuous segment of at least three missing residues is considered disordered. In NMR experiments, secondary structure changes are detected in proteins, and the dynamic nature of disordered regions is easily distinguishable from ordered regions (Dyson & Wright, 2021). The disordered regions gaining a stable three-dimensional structure upon binding to their partners are called DOT regions, which are obtained by comparing disordered regions in free proteins and protein complexes. Srivastava *et al.* (2021) have identified DOT residues in protein–RNA complexes and showed their role in binding and activity.

4.4 Databases for disordered proteins

Several databases have been developed for IDPs, which are based on manually curated experimental data and predicted IDRs (Piovesan *et al.*, 2022). **Table 4.1** lists the available databases for IDRs in proteins and IDPs.

Mutual folding induced by binding (MFIB) compiles data on protein–protein complexes formed by disordered proteins, which undergo DOT on binding with each other (Fichó *et al.*, 2017). They annotated disordered proteins based on direct evidence and the existence of a close homologue

Table 4.1 Databases for intrinsically disordered proteins and regions

Name	Data	URL	References
MFIB	Protein–protein interactions facilitated by residues undergoing disorder-to-order transition	http://mfib.enzim.ttk.mta.hu/	Fichó *et al.* (2017)
DIBS	Complex with at least one disordered protein along with thermodynamic information	http://dibs.enzim.ttk.mta.hu/	Schad *et al.* (2018)
ELM	Database and tools for simple linear motifs (SLiMs) of disordered regions	http://elm.eu.org	Kumar *et al.* (2020)

(Continued)

Table 4.1 (*Continued*)

Name	Data	URL	References
DisProt	Experimental and predicted disordered proteins and disorder-to-order transition information	https://disprot.org	Hatos *et al.* (2020)
Mobi-DB	Complexes containing at least one disordered protein and their thermodynamic and functional information	https://mobidb.org	Piovesan *et al.* (2021)
IDEAL	Interaction network of disordered proteins	https://www.ideal-db.org	Fukuchi *et al.* (2014)
FuzDB	Binding affinity, activity, and disordered residue information	https://fuzdb.org	Hatos *et al.* (2022)

of disordered regions/proteins from Pfam. MFIB is available at: http://mfib.enzim.ttk.mta.hu/. The Disordered Binding Site (DIBS) database (Schad *et al.*, 2018) enlists complexes formed between disordered proteins and ordered proteins. Apart from structural and functional annotation, DIBS offers information on dissociation constants of interactions, post-translational modifications, and linear motifs. The dissociation constants of complexes between IDPs and ordered proteins fall in the range of 10^{-3} M and 10^{-11} M, which reveals the presence of transient complexes along with permanent ones between IDPs and ordered proteins. The complexes are classified as “confirmed” if direct evidence of complex formation is available, “inferred from homology” if a close homologue that lacks an intrinsic structure is available, and “inferred from motif” if the interface is known to have a short motif. The DIBS database is available at http://dibs.enzim.ttk.mta.hu/.

Eukaryotic linear motifs (ELM) is a combination of a database and web server that provides information on short linear motifs (SLiMs), molecular recognition features, and miniMotifs of disordered regions that are present in protein interaction interfaces. Further, the motif classes are grouped based on their functions as ligands (LIG), targeting (TRG), docking (DOC), degradation (DEG), modification (MOD), or cleavage (CLV) motifs (Kumar *et al.*, 2020). ELM is available at: http://elm.eu.org).

Protein Disorder database (DisProt) encompasses information on sequence construct features including mutations and post-translational modifications, disorder ontology, and improvements in the annotation format (Hatos *et al.*, 2020). Annotations of amino acid repeats displaying IDP properties have been added to the updated version of DisProt based on indirect evidence, and 14% of the annotations have been reviewed and validated by the biocurator. A new functional class named "biological condensation" includes proteins undergoing liquid–liquid phase separation (LLPS) and cellular protein condensates irrespective of their physiological and pathological states. Disprot is available at https://disprot.org.

The Mobi-DB database annotates the binding modes of disordered proteins based on whether it undergoes DOT or not (Piovesan *et al.*, 2021). It has information on disordered regions, which have post-translational modifications and undergo LLPS. The webserver is available at https://mobidb.org/. Intrinsically Disordered proteins with Extensive Annotation and Literature (IDEAL) provides illustrations of protein–protein interactions with IDPs as nodes and the interactions with other proteins as edges (Fukuchi *et al.*, 2014). It is available at https://www.ideal-db.org/.

IDPs may not form an ordered structure even in complex with binding partners like protein, DNA, RNA, or small molecule, and the "fuzziness" that is retained in the complex might have significance in complex formation, function, or regulation. The FuzDB database has a collection of structural evidence of fuzziness and biochemical evidence like binding affinity, specificity, and transcriptional activity (Hatos *et al.*, 2022). The fuzzy complexes are classified into two topological classes in the database, which are "polymorphic" and "dynamic." The polymorphic class includes IDRs, which fold with the same binding partner to form alternate structures, whereas the dynamic topological class has fuzzy proteins, which exhibit interconversion between various conformations in the complex. The database is available at https://fuzdb.org/.

4.5 Computational methods for identifying disordered proteins and regions

Although experimental techniques are highly precise and accurate in identifying disordered regions, they are time and resource intensive.

Hence, several computational methods have been developed for identifying IDRs in proteins, which are shown to be effective and essential for achieving high-throughput predictions. These methods are mainly based on scoring-functions and machine learning-based methods. The available computational tools are listed in **Table 4.2**.

Table 4.2 Tools for predicting disordered regions in proteins

Predictor	Features	Link	Reference
IUPred	Biophysical properties and energy estimation	https://iupred.elte.hu/	Dosztányi *et al.* (2018)
IUPRed2A	Integration of IUPred and ANCHOR2	http://iupred2a.elte.hu	Mészáros *et al.* (2018)
IUPred3	Evolutionary information and energy estimation	https://iupred3.elte.hu	Erdős *et al.* (2021)
ANCHOR2	Pairwise energy estimation	https://iupred2a.elte.hu/	Dosztányi *et al.* (2009)
POODLE-I	Evolutionary information and amino acid composition	http://cblab.my-pharm.ac.jp/poodle/poodle.html	Hirose *et al.* (2010)
SPOT-Disorder2	Evolutionary and structural information	https://sparkslab.org/server/spot-disorder2	Hanson *et al.* (2019)
IDP-Seq2Seq	Evolutionary information, predicted secondary structure, physicochemical, and spatial structure properties	http://bliulab.net/IDP-Seq2Seq	Tang *et al.* (2020)
RFPR-IDP	Physicochemical features and evolutionary information	http://bliulab.net/RFPR-IDP/server	Liu *et al.* (2021)
DeepIDP-2L	One-hot coding, evolutionary information, predicted secondary structure, and physicochemical properties	http://bliulab.net/DeepIDP-2L	Tang *et al.* (2022)

Table 4.2 (*Continued*)

Predictor	Features	Link	Reference
Trans-DFL	Secondary structure and solvent accessibility, in addition to RFPR-IDP features	http://bliulab.net/TransDFL	Pang *et al.* (2023)
IDPFusion	Residue-profile features include physicochemical properties, structural features like secondary structure and solvent accessibility, and evolutionary features	http://bliulab.net/IDP-Fusion/	Tang *et al.* (2023)
flDPnn	Sequence-profile and protein-level features	http://biomine.cs.vcu.edu/servers/flDPnn	Hu *et al.* (2021)
CLIP	Multiple sequence alignment and physicochemical properties	http://biomine.cs.vcu.edu/servers/CLIP	Peng *et al.* (2023)
DisoMine	Secondary structure, protein backbone and side-chain dynamics, early folding propensity	https://www.bio2byte.be/b2btools/disomine	Orlando *et al.* (2022)
DEPICTER	Consensus: IUPredshort, IUPredlong, and SPOT-Disorder2	http://biomine.cs.vcu.edu/servers/DEPICTER	Barik *et al.* (2020)
DEPICTER2	Consensus scores	http://biomine.cs.vcu.edu/servers/DEPICTER2	Basu *et al.* (2023)

IUPred (Dosztányi *et al.*, 2018) calculates the favorable energetic contributions at the residue level based on statistical potential obtained from globular proteins to classify disordered and ordered residues. ANCHOR2, an updated version of ANCHOR (Dosztányi *et al.*, 2009), predicts disordered residues based on the ability to form interactions with ordered segments of other proteins. IUPred2A (Mészáros *et al.*, 2018) is an integration of these two methods, IUPred and ANCHOR2. IUPred2A is available at: http://iupred2a.elte.hu/. Erdős *et al.* (2021) further modified the IUPred predictor to develop IUPred3, which utilizes an energy-estimation approach along with filtering ordered residues and adding information about experimentally validated disordered regions and evolutionary information. It is available at: https://iupred3.elte.hu/.

POODLE-I is a length-dependent predictor built on a meta-approach. This setup works seamlessly as a coordinated process, using multiple POODLE programs to predict disordered regions on a single server. This design allows for a cohesive integration of detailed algorithms from each sub-module program, enhancing its predictive capabilities (Hirose *et al.*, 2010).

Hanson *et al.* (2019) developed SPOT-Disorder2, which utilizes a combination of deep squeeze-and-excitation residual inception, long short-term memory (LSTM), and input from evolutionary information to identify disordered regions in proteins. It is available at https://sparks-lab.org/server/spot-disorder2/). Tang *et al.* (2020) developed IDP-Seq2Seq, which encodes FASTA sequence information in a NLP model for the prediction of disorder residues. The sequence space is converted into semantic space using NLP, and this semantic space contains information about structural context, such as possible residue–residue contacts, which is used to predict IDRs in proteins. The web server is available at http://bliulab.net/IDP-Seq2Seq/.

Liu *et al.* (2021) developed RFPR-IDP, which considers not only disordered proteins but also fully ordered proteins, for training this sequence-based predictor model to reduce false positives. The predictor is based on a convolutional neural network and bidirectional long-short-term memory for predicting both fully ordered and disordered proteins. RFPR-IDP uses two sequential features, including seven physicochemical features and PSSM features obtained using the PSI-BLAST. It is available at http://bliulab.net/RFPR-IDP/server). This method has been refined, and different methods were developed such as DeepIDP-2L (Tang *et al.*, 2022), Trans-DFL (Pang *et al.*, 2023), and IDP-Fusion (Tang *et al.*, 2023). For the development of Trans-DFL, secondary structure features generated by the SPIDER tool and solvent accessibility features from the SABLE tool have been added, which are useful for the prediction of disordered flexible linkers (DFLs). For DeepIDP-2L, a hierarchical attention network captures features of long disorder regions (LDRs) by setting higher attention weights to termini where LDRs are predominant. DeepIDP-2L uses a convolutional attention network to capture the local features of adjacent residues. One-hot coding, evolutionary information obtained from position-specific scoring matrix (PSSM), predicted secondary structure, and physicochemical properties like steric parameter, polarizability, volume, hydrophobicity, isoelectric point, helix probability, and sheet probability are the features incorporated in DeepIDP-2L.

"flDPnn" is a tool that utilizes deep neural networks (Hu *et al.*, 2021), and it is based on sequence-profile and protein-level feature encoding. "flDPnn" being one of the fastest disorder predictor tools built on deep neural networks, predicts disorder and disorder functions from input protein sequences (webserver is available at http://biomine.cs.vcu.edu/servers/flDPnn/). Peng *et al.* (2023) developed a predictor of linear interacting peptide that is based on co-evolutionary information obtained from multiple sequence alignments, physicochemical properties, and disorder predictions. It is available at: http://biomine.cs.vcu.edu/servers/CLIP/. DisoMine (Orlando *et al.*, 2022) uses a recurrent neural network model that does not consider the evolutionary information of proteins to avoid bias toward well-studied proteins. It performs well in the prediction of long-disorder of at least 10 residues in length and not for short-disordered regions in loops. It is available at: https://www.bio2byte.be/b2btools/disomine/.

DEPICTER (Barik *et al.*, 2020) is an example of a consensus-based predictor of protein disorder using the outputs from $IUPred_{short}$, $IUPred_{long}$, and SPOT-Disorder2 http://biomine.cs.vcu.edu/servers/DEPICTER/). Later, this approach has been expanded for predicting the binding of disordered regions to DNA, RNA, protein, and lipid (Basu *et al.*, 2023). The web server is available at: http://biomine.cs.vcu.edu/servers/DEPICTER2/.

Critical Assessment of Intrinsic protein Disorder (CAID) is a prediction portal (Del Conte *et al.*, 2023), which offers a comparison of the performance of a list of different tools in predicting disordered proteins or IDRs in proteins. This tool assesses the methods both based on performance and the techniques used to predict when input is given as a single sequence based on a user-defined threshold. CAID enables the identification of high-confidence residues that are disordered in a protein by combining different intrinsic disorder and binding prediction methods. It is available at: https://caid.idpcentral.org/.

4.6 Prediction of change in disorder upon mutation

Amino acid mutations in disordered regions affect the structure and function of a protein by binding with other molecules, and some of them lead to diseases. Forcelloni and Giansanti (2020) reported that long disordered regions are more tolerant to mutations, and it might be due to less evolutionary pressure (Afanasyeva *et al.*, 2018). Meyer (2018) employed a peptide-based proteomic screening approach to examine the effects of mutations in IDRs on protein–protein interactions and showed that mutations occurring in the disordered cytosolic regions of three transmembrane proteins (GLUT1, ITPR1, and CACNA1H) result in an enhanced binding affinity with clathrin. Further, mutations in disordered regions promote or block the DOT (Dembinski *et al.*, 2014). It also influences change in the structural stability of proteins and leads to aggregation (Babu, 2016).

Deleterious mutations often occur in the post-translational or regulation sites, leading to disruption of protein–protein interaction networks. Wong *et al.* (2020) showed that disease-causing mutations are frequently found at the core of protein–protein interfaces and are enriched by disordered regions. Mészáros *et al.* (2021) showed that mutations in the post-transcription sites often drive the disease phenotypes in cancer, and many of these mutations occur in IDRs. Further, mutations in IDRs could lead to altered pathways and imbalance downstream signaling (Vacica & Iakoucheva, 2012).

Anoosha *et al.* (2015) developed an innovative approach, DIM-Pred, for identifying protein missense mutations that cause the order-to-disorder transition of the mutated position. This method integrates amino acid properties, substitution matrices, and considers the influence of neighboring residues along the sequence. DIM-Pred is available at http://www.iitm.ac.in/bioinfo/DIM_Pred/.

Ali *et al.* (2014) examined the reliability and suitability of disorder prediction programs in determining alterations in protein disorder due to amino acid substitutions. They have developed a method, PON-Diso to identify extended disordered regions within protein sequences, and it is available at http://structure.bmc.lu.se/PON-Diso.

4.7 Importance of disorder on the structure and function of proteins

IDPs show high flexibility, which promotes attaining different shapes and more than one function in proteins. Moreover, IDPs bind to multiple partners and modulate the affinity of binding with the same partners.

Srivastava *et al.* (2022) reported that DOTs occur in the disordered regions of proteins on binding to RNA, and these regions contribute significantly toward binding affinity of the protein for the RNA molecule (**Figure 4.1**). Specifically, the study on archaeal methyl tRNA transferase 5 (Srivastava *et al.*, 2021) revealed that the DOT region in this protein

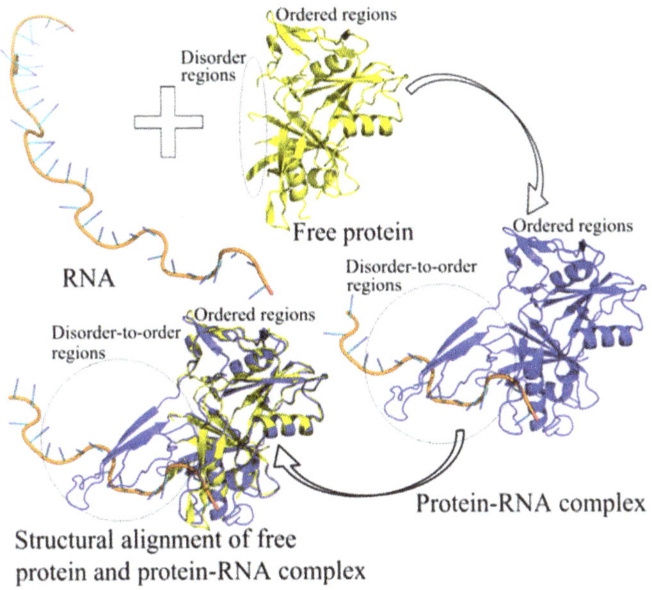

Figure 4.1 Disorder-to-order transition in the CRISPR-Cas RNA Silencing CMR Complex. The free protein, tRNA, and protein–RNA complex are shown in golden, orange, and in blue, respectively

forms an extended structure on binding to tRNA, and it is essential to stabilize the protein–tRNA complex.

Yesudhas *et al.* (2021a) studied the dynamics of the Sars-Cov2 spike protein and found that the IDR in the receptor binding domain of the coronavirus spike protein undergo DOT and stabilize the binding of angiotensin converting enzyme (ACE-2) via interactions mediated by hydrophilic and aromatic residues in the DOT region. Moreover, in SARS-CoV-2, the presence of disordered regions in spike proteins could be responsible for better shape complementarity of the receptor binding domain (RBD) with ACE-2 (Yesudhas *et al.*, 2021b).

4.8 Diseases related to disordered proteins

The impairment of functions (dysregulation, misfolding, or misexpression) due to mutations in IDRs or IDPs leads to several diseases, including cancer, neurodegenerative disorders, diabetes, and cardiovascular diseases (Pajkos *et al.*, 2011; Uversky, 2019; Tsoi *et al.*, 2023). Mészáros *et al.* (2021) showed that IDPs are part of regulation pathways and metabolic networks that can be a driver of cancer. Mészáros *et al.* (2021) identified 47 disordered regions in 145 proteins that are associated with cancer in the human proteome. These cancer "driver" mutations are present in IDRs such as SLiMs and linker motifs and in intrinsically disordered domains (IDDs) as larger functional modules. For example, mutations in the IDDs in calreticulin (CALR) alter Ca^{2+} binding and protein localizations, which lead to cancer (Mészáros *et al.*, 2021). Further, mutations in simple linear motifs (SLiMs) and post-translation sites result in diseases such as cancer (Narayan *et al.*, 2016; Meyer *et al.*, 2018).

Uversky (2010) analyzed amino acid residues in proteins related to neurodegenerative diseases and observed that the majority of them are intrinsic disorder-promoting residues. The mechanism of fibril formation, which leads to neurodegeneration is different in disordered and ordered proteins. Partial folding leads to fibril formation in disordered proteins, whereas partial unfolding is observed in ordered proteins prior to fibril formation. Breydo *et al.* (2012) reviewed the process of aggregation of α-synuclein, an IDP that is abundant in Lewy bodies resulting in neurodegeneration in Parkinson's disease. Another example is amyloid-beta, or Aβ peptide, which is intrinsically disordered and adopts different conformations, and the aggregation of amyloid-beta peptide is followed by tau hyperphosphorylation in Alzheimer's disease.

Disordered driver mutations also affect receptor activity, regulation of transcription, and protein degradation, which lead to diseases. Meyer *et al.* (2018) explored the mutations in disordered regions that resulted in the mislocation of the glucose receptors GLUT1. Gsponer *et al.* (2008) performed proteome-wide studies and explored that disordered proteins are tightly regulated and their dysregulation results in diseases (Uversky, 2014). Yesudhas *et al.* (2021) reported the presence of IDRs in coronaviruses including SARS-CoV, coronavirus HKU1, and SARS-CoV-2.

The prion protein PrP^C is characterized by a large intrinsically disordered N-terminal domain, and these proteins are implicated in Creutzfeldt-Jakob disease (Prusiner, 2001). The disordered domain of this prion protein determines its downstream signaling processes and can modulate the activities of other proteins such as amyloid-beta and α-synuclein. The ability of IDPs to adopt different conformations imparts the potential to interact with other proteins, play a significant role in physiological processes, and form aberrant complexes that result in diseases. Proteins with IDRs also interact with polyQ (polyglutamine) aggregates, which are present in inclusion bodies associated with neurodegenerative diseases (Wear *et al.*, 2015). Another example is the rare variant of TREM2 (triggering receptor expressed on myeloid cells 2), in which the specific mutation G145W is associated with familial dementia (Karsak *et al.*, 2020).

Cheng *et al.* (2006) found that proteins related to cardiovascular diseases are enriched in disorder-promoting residues and depleted in order-promoting residues in a dataset of 457 proteins obtained from SWISS-PROT. The aggregation of islet amyloid polypeptide (IAPP) leading to its deposition (Westermark *et al.*, 2008) in the extracellular space of the islets of Langerhans causes degeneration of β cells, which causes type 2 diabetes (Mukherjee *et al.*, 2017).

4.9 Conclusions

IDRs possess immense significance for biological functions, although they lack defined structures. These regions showcase remarkable flexibility, enabling proteins to interact with various partners and perform multifaceted roles within cells. IDPs are abundant and are crucial components of various cellular pathways. Hence, several databases and tools have been developed to explore the role of IDRs and IDPs as well as their mutants. Further, investigations into disordered regions are vital for understanding the molecular mechanisms of diseases, including cancer and neurodegenerative disorders, and provide insights into potential therapeutic interventions.

Acknowledgments

The authors thank the Department of Biotechnology, IIT Madras for its computational facilities.

References

Afanasyeva, A., Bockwoldt, M., Cooney, C. R., Heiland, I., & Gossmann, T. I., (2018). Human long intrinsically disordered protein regions are frequent targets of positive selection. *Genome Research*, *28*(7), 975–982.

Ahmed, S. S., Rifat, Z. T., Lohia, R., Campbell, A. J., Dunker, A. K., Rahman, M. S., & Iqbal, S. (2022). Characterization of intrinsically disordered regions in proteins informed by human genetic diversity. *PLOS Computational Biology*, *18*(3), e1009911.

Ali, H., Urolagin, S., Gurarslan, Ö., & Vihinen, M. (2014). Performance of protein disorder prediction programs on amino acid substitutions. *Human mutation*, *35*(7), 794–804.

Anoosha, P., Sakthivel, R., & Gromiha, M. M. (2015). Prediction of protein disorder on amino acid substitutions. *Analytical biochemistry*, *491*, 18–22.

Barik, A., Katuwawala, A., Hanson, J., Paliwal, K., Zhou, Y., & Kurgan, L. (2020). DEPICTER: intrinsic disorder and disorder function prediction server. *Journal of Molecular Biology*, *432*(11), 3379–3387.

Babu, M. M. (2016). The contribution of intrinsically disordered regions to protein function, cellular complexity, and human disease. *Biochemical Society Transactions*, *44*(5), 1185–1200.

Basu, S., Gsponer, J., & Kurgan, L. (2023). DEPICTER2: A comprehensive webserver for intrinsic disorder and disorder function prediction. *Nucleic Acids Research*, gkad330.

Borgia, A., Borgia, M. B., Bugge, K., Kissling, V. M., Heidarsson, P. O., Fernandes, C. B., Sottini, A., Soranno, A., Buholzer, K. J., Nettels, D., Kragelund, B. B., Best, R. B., & Schuler, B. (2018). Extreme disorder in an ultrahigh-affinity protein complex. *Nature*, *555*, 61–66.

Breydo, L., Wu, J. W., & Uversky, V. N. (2012). α-Synuclein misfolding and Parkinson's disease. *Biochimica et Biophysica Acta (BBA) — Molecular Basis of Disease*, *1822*(2), 261–285.

Cheng, Y., LeGall, T., Oldfield, C. J., Dunker, A. K., & Uversky, V. N. (2006). Abundance of intrinsic disorder in protein associated with cardiovascular disease. *Biochemistry*, *45*(35), 10448–10460.

Del Conte, A., Bouhraoua, A., Mehdiabadi, M., Clementel, D., Monzon, A. M., Tosatto, S. C., & Piovesan, D. (2023). CAID prediction portal: A comprehensive service for predicting intrinsic disorder and binding regions in proteins. *Nucleic Acids Research*, gkad430.

Dembinski, H., Wismer, K., Balasubramaniam, D., Gonzalez, H. A., Alverdi, V., Iakoucheva, L. M., & Komives, E. A. (2014). Predicted disorder-to-order transition mutations in IκBα disrupt function. *Physical Chemistry Chemical Physics*, *16*(14), 6480–6485.

Deryusheva, E., Nemashkalova, E., Galloux, M., Richard, C. A., Eléouët, J. F., Kovacs, D., Van Belle, K., Tompa, P., Uversky, V., & Permyakov, S. (2019). Does intrinsic disorder in proteins favor their interaction with lipids? *Proteomics*, *19*(6), 1800098.

Dosztányi, Z. (2018). Prediction of protein disorder based on IUPred. *Protein science*, *27*(1), 331–340.

Dosztányi, Z., Mészáros, B., & Simon, I. (2009). ANCHOR: Web server for predicting protein binding regions in disordered proteins. *Bioinformatics*, *25*(20), 2745–2746.

Dyson, H. J., & Wright, P. E. (2021). NMR illuminates intrinsic disorder. *Current Opinion in Structural Biology*, *70*, 44–52.

Erdős, G., Pajkos, M., & Dosztányi, Z. (2021). IUPred3: Prediction of protein disorder enhanced with unambiguous experimental annotation and visualization of evolutionary conservation. *Nucleic Acids Research*, *49*(W1), W297–W303.

Fichó, E., Reményi, I., Simon, I., & Mészáros, B. (2017). MFIB: A repository of protein complexes with mutual folding induced by binding. *Bioinformatics*, *33*(22), 3682–3684.

Forcelloni, S., & Giansanti, A. (2020). Evolutionary forces and codon bias in different flavors of intrinsic disorder in the human proteome. *Journal of Molecular Evolution*, *88*(2), 164–178.

Fukuchi, S., Amemiya, T., Sakamoto, S., Nobe, Y., Hosoda, K., Kado, Y., Murakami, S. D., Koike, R., Hiroaki, H., & Ota, M. (2014). IDEAL in 2014 illustrates interaction networks composed of intrinsically disordered proteins and their binding partners. *Nucleic Acids Research*, *42*(D1), D320–D325.

Gsponer, J., Futschik, M. E., Teichmann, S. A., & Babu, M. M. (2008). Tight regulation of unstructured proteins: From transcript synthesis to protein degradation. *Science*, *322*(5906), 1365–1368.

Han, B., Ren, C., Wang, W., Li, J., & Gong, X. (2023). Computational prediction of protein intrinsically disordered region related interactions and functions. *Genes*, *14*(2), 432.

Hanson, J., Paliwal, K. K., Litfin, T., & Zhou, Y. (2019). SPOT-Disorder2: Improved protein intrinsic disorder prediction by ensembled deep learning. *Genomics, Proteomics & Bioinformatics*, *17*(6), 645–656.

Hatos, A., Hajdu-Soltész, B., Monzon, A. M., Palopoli, N., Álvarez, L., Aykac-Fas, B., Bassot, C., Benítez, G. I., Bevilacqua, M., Chasapi, A., Chemes, L., Davey, N. E., Davidović, R., Dunker, A. K., Elofsson, A., Gobeill, J., Foutel, N. S. G., Sudha, G., Guharoy, M., & Piovesan, D. (2020). DisProt: Intrinsic protein disorder annotation in 2020. *Nucleic Acids Research*, *48*(D1), D269–D276.

Hatos, A., Monzon, A. M., Tosatto, S. C., Piovesan, D., & Fuxreiter, M. (2022). FuzDB: A new phase in understanding fuzzy interactions. *Nucleic Acids Research*, *50*(D1), D509–D517.

Hirose, S., Shimizu, K., & Noguchi, T. (2010). POODLE-I: Disordered region prediction by integrating POODLE series and structural information predictors based on a workflow approach. *In Silico Biology*, *10*(3–4), 185–191.

Hu, G., Katuwawala, A., Wang, K., Wu, Z., Ghadermarzi, S., Gao, J., & Kurgan, L. (2021). flDPnn: Accurate intrinsic disorder prediction with putative propensities of disorder functions. *Nature Communications*, *12*(1), 4438.

Karsak, M., Glebov, K., Scheffold, M., Bajaj, T., Kawalia, A., Karaca, I., Rading, S., Kornhuber, J., Peters, O., Diez-Fairen, M., Frölich, L., Hüll, M.,

Wiltfang, J., Scherer, M., Riedel-Heller, S., Schneider, A., Heneka, M. T., Fliessbach, K., Sharaf, A., . . . Ramirez, A. (2020). A rare heterozygous TREM2 coding variant identified in familial clustering of dementia affects an intrinsically disordered protein region and function of TREM2. *Human Mutation*, *41*(1), 169–181.

Kumar, M., Gouw, M., Michael, S., Sámano-Sánchez, H., Pancsa, R., Glavina, J., Diakogianni, A., Valverde, J. A., Bukirova, D., Čalyševa, J., Palopoli, N., Davey, N. E., Chemes, L. B., & Gibson, T. J. (2020). ELM — the eukaryotic linear motif resource in 2020. *Nucleic Acids Research*, *48*(D1), D296–D306.

Liu, Y., Wang, X., & Liu, B. (2021). RFPR-IDP: Reduce the false positive rates for intrinsically disordered protein and region prediction by incorporating both fully ordered proteins and disordered proteins. *Briefings in Bioinformatics*, *22*(2), 2000–2011.

Ma, H., Jia, X., Zhang, K., & Su, Z. (2022). Cryo-EM advances in RNA structure determination. *Signal Transduction and Targeted Therapy*, *7*(1), 58.

Mensch, C., Konijnenberg, A., Van Elzen, R., Lambeir, A. M., Sobott, F., & Johannessen, C. (2017). Raman optical activity of human α-synuclein in intrinsically disordered, micelle-bound α-helical, molten globule, and oligomeric β-sheet state. *Journal of Raman Spectroscopy*, *48*(7), 910–918.

Mészáros, B., Erdős, G., & Dosztányi, Z. (2018). IUPred2A: Context-dependent prediction of protein disorder as a function of redox state and protein binding. *Nucleic Acids Research*, *46*(W1), W329–W337.

Mészáros, B., Hajdu-Soltész, B., Zeke, A., & Dosztányi, Z. (2021). Mutations of intrinsically disordered protein regions can drive cancer but lack therapeutic strategies. *Biomolecules*, *11*(3), 381.

Meyer, K., Kirchner, M., Uyar, B., Cheng, J. Y., Russo, G., Hernandez-Miranda, L. R., Szymborska, A., Zauber, H., Rudolph, I. M., Willnow, T. E., & Akalin, A. (2018). Mutations in disordered regions can cause disease by creating dileucine motifs. *Cell*, *175*(1), 239–253.

Mukherjee, A., Morales-Scheihing, D., Salvadores, N., Moreno-Gonzalez, I., Gonzalez, C., Taylor-Presse, K., Mendez, N., Shahnawaz, M., Gaber, A. O., Sabek, O. M., Fraga, D. W., & Soto, C. (2017). Induction of IAPP amyloid deposition and associated diabetic abnormalities by a prion-like mechanism. *Journal of Experimental Medicine*, *214*(9), 2591–2610.

Narayan, S., Bader, G. D., & Reimand, J. (2016). Frequent mutations in acetylation and ubiquitination sites suggest novel driver mechanisms of cancer. *Genome Medicine*, *8*(1), 1–13.

Nwanochie, E., & Uversky, V. N. (2019). Structure determination by single-particle cryo-electron microscopy: Only the sky (and intrinsic disorder) is the limit. *International Journal of Molecular Sciences*, *20*(17), 4186.

Orlando, G., Raimondi, D., Codice, F., Tabaro, F., & Vranken, W. (2022). Prediction of disordered regions in proteins with recurrent neural networks and protein dynamics. *Journal of Molecular Biology*, *434*(12), 167579.

Pajkos, M., Mészáros, B., Simon, I., & Dosztányi, Z. (2011). Is there a biological cost of protein disorder? Analysis of cancer-associated mutations. *Molecular BioSystems*, *8*, 296–307.

Pang, Y., & Liu, B. (2023). TransDFL: Identification of disordered flexible linkers in proteins by transfer learning. *Genomics, Proteomics & Bioinformatics*, *21*(2), 359–369.

Peng, Z., Li, Z., Meng, Q., Zhao, B., & Kurgan, L. (2023). CLIP: Accurate prediction of disordered linear interacting peptides from protein sequences using co-evolutionary information. *Briefings in Bioinformatics*, *24*(1), bbac502.

Peng, Z., Mizianty, M. J., & Kurgan, L. (2014). Genome-scale prediction of proteins with long intrinsically disordered regions. *Proteins: Structure, Function, and Bioinformatics*, *82*(1), 145–158.

Phillips, A. S., Gomes, A. F., Kalapothakis, J. M., Gillam, J. E., Gasparavicius, J., Gozzo, F. C., Kunath, T., MacPhee, C., & Barran, P. E. (2015). Conformational dynamics of α-synuclein: Insights from mass spectrometry. *Analyst*, *140*(9), 3070–3081.

Piovesan, D., Monzon, A. M., Quaglia, F., & Tosatto, S. C. (2022). Databases for intrinsically disordered proteins. *Acta Crystallographica Section D: Structural Biology*, *78*(2), 144–151.

Piovesan, D., Necci, M., Escobedo, N., Monzon, A. M., Hatos, A., Mičetić, I., Quaglia, F., Paladin, L., Ramasamy, P., Dosztányi, Z., Vranken, W. F., Davey, N. E., Parisi, G., Fuxreiter, M., & Tosatto, S. C. (2021). MobiDB: Intrinsically disordered proteins in 2021. *Nucleic Acids Research*, *49*(D1), D361–D367.

Prusiner, S. B. (2001). Neurodegenerative diseases and prions. *New England Journal of Medicine*, *344*(20), 1516–1526.

Schad, E., Ficho, E., Pancsa, R., Simon, I., Dosztányi, Z., & Meszaros, B. (2018). DIBS: A repository of disordered binding sites mediating interactions with ordered proteins. *Bioinformatics*, *34*(3), 535–537.

Shi, Y. (2014). A glimpse of structural biology through X-ray crystallography. *Cell*, *159*(5), 995–1014.

Shimizu, K., & Toh, H. (2009). Interaction between intrinsically disordered proteins frequently occurs in a human protein–protein interaction network. *Journal of Molecular Biology*, *392*(5), 1253–1265.

Srivastava, A., Yesudhas, D., Ahmad, S., & Gromiha, M. M. (2021). Deciphering the role of residues involved in disorder-to-order transition regions in archaeal tRNA methyltransferase 5. *Genes*, *12*(3), 399.

Srivastava, A., Yesudhas, D., Ahmad, S., & Gromiha, M. M. (2022). Understanding disorder-to-order transitions in protein–RNA complexes using molecular dynamics simulations. *Journal of Biomolecular Structure and Dynamics*, *40*(17), 7915–7925.

Stuchfield, D., & Barran, P. (2018). Unique insights to intrinsically disordered proteins provided by ion mobility mass spectrometry. *Current Opinion in Chemical Biology*, *42*, 177–185.

Tang, Y. J., Pang, Y. H., & Liu, B. (2020). IDP-Seq2Seq: Identification of intrinsically disordered regions based on sequence to sequence learning. *Bioinformatics*, *36*(21), 5177–5186.

Tang, Y. J., Pang, Y. H., & Liu, B. (2022). DeepIDP-2L: Protein intrinsically disordered region prediction by combining convolutional attention network and hierarchical attention network. *Bioinformatics*, *38*(5), 1252–1260.

Tang, Y. J., Yan, K., Zhang, X., Tian, Y., & Liu, B. (2023). Protein intrinsically disordered region prediction by combining neural architecture search and multi-objective genetic algorithm. *BMC Biology*, *21*(1), 188.

Tsoi, P. S., Quan, M. D., Ferreon, J. C., & Ferreon, A. C. M. (2023). Aggregation of disordered proteins associated with neurodegeneration. *International Journal of Molecular Sciences*, *24*(4), 3380.

Uversky, V. N. (2019). Intrinsically disordered proteins and their "mysterious" (meta) physics. *Frontiers in Physics*, *7*, 10.

Uversky, V. N. (2010). Targeting intrinsically disordered proteins in neurodegenerative and protein dysfunction diseases: another illustration of the D2 concept. *Expert Review of Proteomics*, *7*(4), 543–564.

Uversky, V. N., Davé, V., Iakoucheva, L. M., Malaney, P., Metallo, S. J., Pathak, R. R., & Joerger, A. C. (2014). Pathological unfoldomics of uncontrolled chaos: Intrinsically disordered proteins and human diseases. *Chemical Reviews*, *114*, 6844–6879.

Uversky, V. N., Gillespie, J. R., & Fink, A. L. (2000). Why are "natively unfolded" proteins unstructured under physiologic conditions? *Proteins*, *41*(3), 415–427.

Vacic, V., & Iakoucheva, L. M. (2012). Disease mutations in disordered regions — exception to the rule? *Molecular BioSystems*, *8*(1), 27–32.

Wear, M. P., Kryndushkin, D., O'Meally, R., Sonnenberg, J. L., Cole, R. N., & Shewmaker, F. P. (2015). Proteins with intrinsically disordered domains are preferentially recruited to polyglutamine aggregates. *PLoS One*, *10*(8), e0136362.

Westermark, G. T., Westermark, P., Berne, C., & Korsgren, O. (2008). Widespread amyloid deposition in transplanted human pancreatic islets. *New England journal of Medicine*, *359*(9), 977–979.

Wong, E. T., So, V., Guron, M., Kuechler, E. R., Malhis, N., Bui, J. M., & Gsponer, J. (2020). Protein–protein interactions mediated by intrinsically disordered protein regions are enriched in missense mutations. *Biomolecules*, *10*(8), 1097.

Yesudhas, D., Srivastava, A., & Gromiha, M. M. (2021a). COVID-19 outbreak: History, mechanism, transmission, structural studies and therapeutics. *Infection*, *49*, 199–213.

Yesudhas, D., Srivastava, A., Sekijima, M., & Gromiha, M. M. (2021b). Tackling Covid-19 using disordered-to-order transition of residues in the spike protein upon angiotensin-converting enzyme 2 binding. *Proteins: Structure, Function, and Bioinformatics*, *89*(9), 1158–1166.

Part II

Protein function and binding affinity

https://doi.org/10.1142/9789811293269_0005

Chapter 5

Binding affinity changes upon mutation in protein–protein complexes

Rahul Nikam[1], Fathima Ridha[1], Sherlyn Jemimah[1,2], Kumar Yugandhar[1,3], and M. Michael Gromiha[1,*]

[1]*Department of Biotechnology, Bhupat and Jyoti Mehta School of Biosciences, Indian Institute of Technology Madras, Chennai, Tamil Nadu 600036, India*

[2]*Department of Biomedical Engineering, Khalifa University, Abu Dhabi, United Arab Emirates*

[3]*Department of Computational Biology, Cornell University, New York, NY, USA*

Abstract

Protein–protein interactions (PPIs) are central to the functioning of living organisms, orchestrating various biological processes. Intermolecular forces and binding interfaces between proteins govern the specificity and strength of these interactions. Genetic mutations may disrupt or alter the strength of binding between proteins. Understanding the effect of mutations on binding affinity in protein–protein complexes is essential

*Corresponding author
MMG: gromiha@iitm.ac.in

for elucidating the structure–function relationship of proteins, predicting disease-associated phenotypes, and designing therapeutic interventions. This chapter provides a comprehensive overview of the impact of mutations on PPIs, encompassing experimental approaches, available databases, factors influencing binding affinity change, and specific computational tools for predicting the change in affinity upon mutation. Finally, this chapter highlights the necessity for continuous innovation in experimental and computational techniques to fully unravel the complex mechanisms underlying binding affinity changes and harness this knowledge for the development of novel therapeutic strategies.

5.1 Introduction

Protein–protein interactions (PPIs) are crucial for various biological processes such as signal transduction, cellular regulation, and enzymatic activities (Keskin *et al.*, 2008). These interactions rely on a complex interplay of intermolecular forces and complementary binding sites between the involved proteins, determining their specificity and strength. Genetic mutations that disrupt or modify these interactions can have a profound impact on protein function and cellular physiology (Reva *et al.*, 2011).

Understanding the effect of mutations on the binding affinity of protein–protein complexes is essential for several reasons: (i) providing valuable insights into the structure–function relationship of proteins, such as elucidating the key amino acid residues responsible for binding and stabilizing the complex (Kulandaisamy *et al.*, 2017); (ii) predicting the consequences of mutations on cellular signaling pathways and on potential disease-associated phenotypes (Vidal *et al.*, 2011; Ozturk *et al.*, 2022); and (iii) offering valuable information for rational drug design and therapeutic interventions. Earlier studies showed that most of the mutations associated with diseases decrease binding affinity (Kucukkal *et al.*, 2015; Jubb *et al.*, 2017; Jemimah & Gromiha, 2020), enabling the targeting of specific protein interactions to modulate cellular processes.

This chapter focuses on four key aspects: (i) experimental approaches for identifying changes in binding affinity upon mutation, (ii) available databases containing experimental binding affinity data, (iii) factors

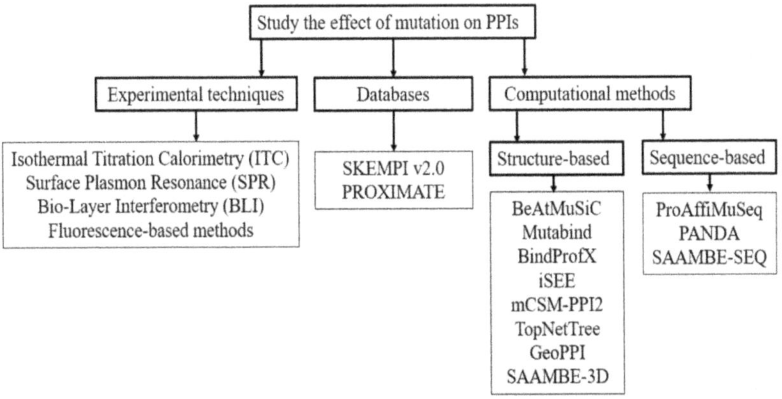

Figure 5.1 Overall workflow illustrating the contents of the chapter

influencing the binding affinity change, and (iv) computational tools for predicting the change in affinity upon mutation. **Figure 5.1** illustrates the major themes covered in this chapter. The chapter also discusses case studies from various biological systems, highlighting the outcomes of mutations on binding affinities and their implications for cellular functions. Integrating experimental and computational approaches will provide a holistic perspective on the mechanism of binding affinity changes induced by mutations.

5.2 Protein–protein binding affinity and effect of mutations on binding affinity

Binding affinity is the measure of the strength of the interaction between two or more proteins that bind to form a complex (Schreiber *et al.*, 2009). As the biological function of a protein is defined by its cellular interactions, and the misregulation of these interactions can lead to various diseases, an in-depth understanding of binding affinity and its underlying contributing factors is necessary. It is expressed as the dissociation constant (K_d), which is the concentration of free protein that occupies half of the total sites of the second protein at equilibrium (Kastritis *et al.*, 2012).

A simple reversible reaction between two interacting proteins, A and B, can be represented as

$$A + B \rightleftharpoons AB$$

or

$$[A]+[B] \underset{k_{\text{off}}}{\overset{k_{\text{on}}}{\rightleftharpoons}} [AB]$$

where $[A]$ and $[B]$ signify the concentrations of free proteins, while $[AB]$ indicates the concentration of the complex in its bound state. The parameter k_{on} stands for the association rate constant (expressed in $M^{-1}\ s^{-1}$), while k_{off} represents the dissociation rate constant (measured in s^{-1}).

The dissociation constant K_d usually denotes the binding affinity of the protein–protein complex and is given by

$$K_d = \frac{k_{\text{off}}}{k_{\text{on}}} = \frac{[A][B]}{[AB]}$$

Smaller values of K_d show high binding affinity, and vice versa. Stable complexes have high and weak transient complexes have low binding affinity. The corresponding equilibrium dissociation constant K_d, can be empirically translated into the Gibbs free energy of binding ΔG, which is commonly used to describe the affinity of an interaction.

$$\Delta G = -RT\, ln\left(\frac{1}{K_d}\right)$$

where R is the gas constant (0.0019 kcal $K^{-1}mol^{-1}$) and T is the absolute temperature (K).

Mutations in protein–protein complexes can alter the binding affinity by perturbing the interactions and potentially leading to functional disruptions or diseases (Yates & Sternberg, 2013; Li *et al.*, 2014). Quantitatively, the influence of mutations is measured by the change in binding free energy ($\Delta\Delta G$) denoted as

$$\Delta\Delta G = \Delta G_{\text{mut}} - \Delta G_{\text{wt}}$$

where ΔG_{mut} and ΔG_{wt} are the binding free energy of mutant and wild-type proteins, respectively (**Figure 5.2**). Positive and negative values of $\Delta\Delta G$

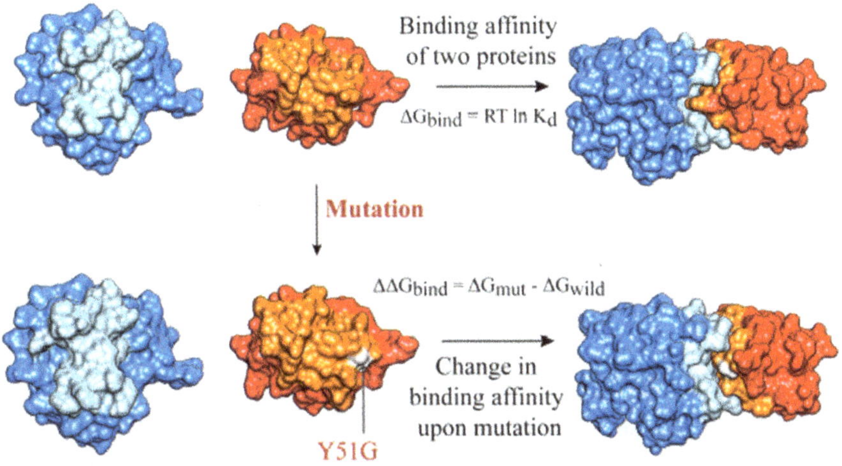

Figure 5.2 Binding affinity of a protein–protein complex and change in affinity upon mutation (Adapted from Gromiha *et al.*, 2020)

correspond to mutations decreasing and increasing binding affinity, respectively.

5.3 Experimental methods for change in binding affinity of protein–protein complexes

Several experimental methods have been employed to investigate the effect of mutations on the binding affinity of protein–protein complexes. Site-directed mutagenesis is a commonly used technique for introducing specific mutations, while site-saturation mutagenesis allows the introduction of a large range of point mutations at specific sites. In both cases, mutagenic primers, which contain mismatches with the parental sequence in the positions to be mutated, are used for polymerase chain reactions (PCR) with the parental sequence as the template and the amplification products are the mutated variants (Sellés Vidal *et al.*, 2023). Experimental techniques such as isothermal titration calorimetry (ITC), surface plasmon resonance (SPR), fluorescence resonance energy transfer (FRET), and microscale thermophoresis (MST) allow for direct measurement of the binding constants (K_d) and changes in the thermodynamic parameters upon mutation.

In addition, structure determination techniques, such as X-ray crystallography and nuclear magnetic resonance (NMR) spectroscopy, provide valuable information on the conformational changes induced by mutations in the complex.

Complementing experimental approaches, computational methods such as molecular dynamics simulations and free energy calculations offer atomistic insights into the binding process and help to predict the effects of mutations on the stability and affinity of a protein–protein complex.

5.3.1 *Isothermal titration calorimetry*

ITC is the common calorimetric technique utilized to examine PPIs. In this method, the sample cell contains the protein of interest, while the injection syringe contains binding partner. As the syringe is introduced into the cell, ITC monitors heat changes during molecular interactions in real time. Similar experiment with a mutant protein, and the difference reveals the impact of mutation in binding affinity. One of the advantages of ITC is that it is label-free and requires no modification of the binding partners. In addition, it has the ability to determine multiple thermodynamic binding parameters, such as enthalpy, entropy, and heat capacity, in addition to assessing binding affinity. However, it cannot be used to study very low- or very high-affinity PPIs efficiently (Kastritis *et al.*, 2012).

5.3.2 *Surface plasmon resonance*

SPR is one of the most commonly used techniques to study PPIs in which the protein (termed 'receptor') is immobilized on a sensor surface. The interacting protein (termed 'analyte') is then introduced into the solution and flowed over the chip. Performing the experiment using wild-type protein and mutant results in a change in the refractive index that can be optically detected, and the signal is plotted in resonance or response units. The resulting SPR response curves provide insights into binding and dissociation kinetics. By comparing these curves to the reference data from the wild-type interaction, the impact of the mutation on binding affinity can be precisely determined. The main advantage of SPR is its ability to measure the binding affinities and association/dissociation kinetics of complexes in real-time, in a label-free environment, and using relatively small sample quantities. However, the complex and costly experimental setup remains a potential drawback of this technique.

5.3.3 *Bio-layer interferometry*

It is an optical biosensor-based technique to quantify biomolecular interactions. It analyzes the interference pattern of white light reflected from two surfaces: a layer of immobilized protein on the fiber optic sensor tip and an internal reference layer. The binding between the protein of interest immobilized on the biosensor tip surface and its binding partner in solution produces an increase in optical thickness at the biosensor tip, resulting in a wavelength shift, which causes a shift in the interference pattern that can be detected in real-time (Song *et al.*, 2015). By monitoring these changes in real-time, bio-layer interferometry (BLI) generates binding curves that reveal the kinetics of the interaction. Comparing the binding kinetics of the mutant protein with that of the wild type provides crucial insights into how the mutation affects the binding affinity and dynamics of the molecular interaction. The important advantage is that it is label-free and does not require modification of the interaction partners with fluorescent dyes.

5.3.4 *Fluorescence-based methods*

Fluorescence-based methods such as FRET, fluorescence anisotropy (FA), and fluorescence polarization (FP) play a pivotal role in exploring changes in binding affinity resulting from mutations in the protein–protein complex. FRET is used to measure the binding affinity by monitoring the energy transfer between a donor fluorophore attached to the ligand and an acceptor fluorophore attached to both wild-type and mutant receptors. It is possible to determine the binding affinity change by testing FRET efficiency separately with labeled ligands and receptors (Sridharan *et al.*, 2014). However, using labels may change the binding behavior and make it more challenging to calculate binding affinity. Also, contaminants that may promote binding must be properly addressed during measurements (Kastritis *et al.*, 2012).

5.4 Databases for change in binding affinity upon mutation

Compiling experimental data about binding affinities holds significant importance as it facilitates the exploration of factors that impact binding affinity and the subsequent development of computational tools. There exist only a few reliable databases in the literature that contain changes in binding energy, binding kinetics, and thermodynamics upon mutations. These databases encompass PROXiMATE (Jemimah *et al.*, 2017) and SKEMPI2 (Jankauskaite *et al.*, 2019).

PROXiMATE (Jemimah *et al.*, 2017) encompasses 6,480 mutation data across 183 heterodimeric protein complexes along with sequence, structure, and functional information. Additionally, PROXiMATE includes binding affinity data for homodimer complexes. It is available at https://www.iitm.ac.in/bioinfo/PROXiMATE/. The SKEMPI v2.0 database (Jankauskaite *et al.*, 2019) compiles experimentally measured binding affinity values for wild-type and mutant protein–protein complexes. SKEMPI 2.0, the latest version, contains binding affinity data for 7,085 mutations from 345 protein complexes. It can be accessed at https://life.bsc.es/pid/skempi2. Both SKEMPI and PROXiMATE are widely used as benchmark datasets for numerous binding affinity prediction studies. Recently, Ridha *et al.* (2023) developed membrane protein complex binding affinity database (MPAD; https://web.iitm.ac.in/bioinfo2/mpad/), a database specific for membrane proteins, reporting the experimental binding affinities of membrane protein–protein complexes and their mutants.

These databases can be utilized to identify mutations that have the potential to result in enhanced binding energy with their interacting counterparts and for the rapid development of algorithms that predict binding energy changes upon missense mutation.

5.5 Factors influencing the change in binding affinity of protein–protein complexes upon mutation

Several key factors contribute to the binding affinity of protein–protein complexes, particularly when mutations are involved. Amino acid properties, such as charge, size, and hydrophobicity, play a pivotal role

(Schreiber *et al.*, 2009). For instance, substituting a charged residue with a hydrophobic one can disrupt electrostatic interactions and hydrophobic packing, directly influencing binding affinity (Zhou & Pang, 2018). The location of amino acid substitutions is also important for understanding binding affinity. Amino acid substitutions situated at the binding interface or active site cleft could block the entrance to the active site, change recognition, alter specificity, or affect binding affinity. For example, the G2019S substitution in leucine-rich repeat kinase 2 (LRRK2) has been shown to be associated with familial and sporadic Parkinson's disease (Aasly *et al.*, 2005).

Furthermore, the flexibility of the mutated residue and its neighboring amino acids can impact binding affinity, with rigid residues potentially causing more significant structural changes. Size and shape compatibility are critical, as mutations altering residue size or bulkiness may disrupt the binding interface's complementarity. Hydrogen bonds are essential for the affinity and stability of a complex, and mutations can disrupt or create new hydrogen bonding interactions at the interface, thereby affecting binding affinity (Patil *et al.*, 2010; Jemimah & Gromiha, 2020). Mutations may also impact salt bridges, electrostatic interactions, and the distribution of hydrophobic residues at the interface, further influencing binding strength. Conformational changes induced by mutations can lead to improved or hindered binding affinity (Ghosh *et al.*, 2014). **Figure 5.3** illustrates an example of a change in affinity upon mutation (E244A) in

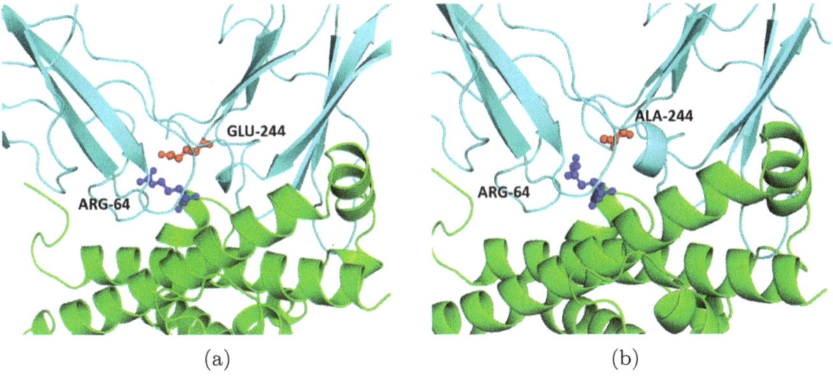

Figure 5.3 Structures of the growth hormone-receptor complex (PDB code: 1A22): (a) wild-type (E244) and (b) mutant (E244A)

the growth hormone-receptor complex. There is a decrease in binding affinity upon mutating Glu to Ala that may be attributed to the loss of an electrostatic interaction between Arg64 and Glu244 and possibly water-mediated hydrogen bonds. Further, the removal of the Glu and the presence of a hydrophobic group (Ala) in the vicinity may create a void, which reduces binding affinity.

Further, mutations can influence the cooperativity between binding partners; positive cooperativity can enhance affinity, while negative cooperativity may reduce it. Energetic contributions to binding, including changes in binding enthalpy and entropy, can be altered by mutations, affecting affinity. Kinetics, represented by the association and dissociation rate, is another essential factor: faster association and slower dissociation generally result in higher affinity (Schreiber *et al.*, 2009). The structural integrity of the proteins involved is also crucial; mutations destabilizing protein structure may lead to structural changes or unfolding, thus reducing binding affinity. When multiple mutations coexist in one or both interacting proteins, the combined effects may not be straightforward to predict, leading to complex changes in affinity (Usmanova *et al.*, 2018). Alterations in solvent and pH conditions can influence binding affinity; mutations may modify the electrostatic properties of the proteins, making their interaction pH-dependent (Zhou & Pang, 2018). Finally, mutations can impact allosteric sites or modulatory regions on proteins, indirectly influencing binding affinity by leading to conformational shifts that affect the binding interface (Weinkam *et al.*, 2013).

5.6 Predicting the change in binding affinity upon mutation

Experimental methods provide accurate measures of binding affinity change; however, they are laborious, expensive, and time-consuming and are not feasible for large-scale analysis. These drawbacks necessitate the development of fast and reliable computational methods for predicting protein–protein $\Delta\Delta G$.

Computational methods that assess the impact of mutations on protein–protein binding have emerged as indispensable tools across a spectrum of biomedical applications (Li *et al.*, 2014). These methods enable us to delve into the intricate molecular mechanisms that underlie disease processes, facilitate drug development, and inform the design of precision therapies. A key focus lies in evaluating binding affinity and interaction energy changes, enabling researchers to gain profound insights into these molecular interactions.

A fundamental approach to understanding protein–protein recognition involves dissecting local geometric effects. This is achieved by quantifying the alterations in interaction energy resulting from mutations within the protein–protein interface. The introduction of a binding affinity index, proposed by Wang and Sarai (1994), enhances our understanding to evaluate binding activity. This index provides a nuanced perspective on the impacts of mutations within these critical interactions, allowing for a detailed and refined assessment.

Several computational tools have been developed for predicting the binding affinity of protein–protein complexes using structure- and sequence-based features. **Table 5.1** lists the currently available methods for predicting the binding affinity change upon mutation.

Table 5.1 Computational methods to predict the change in binding affinity upon mutation

Name	Features	URL	References
Structure-based			
BeAtMuSiC	Statistical potentials derived from sequence- and structure-based features	http://babylone.ulb.ac.be/beatmusic/index.php	Dehouck *et al.* (2013)
BindProf	Interface structural profiles, shape complementarity	http://zhanglab.ccmb.med.umich.edu/BindProf/	Brender and Zhang (2015)
MutaBind	van der Waals energy, solvation energy, free energy changes due to unfolding, SASA, conservation score	http://www.ncbi.nlm.nih.gov/projects/mutabind/	Li *et al.* (2016)
BindProfX	Updated version of BindProf, statistical energy derived from Boltzmann distribution	https://zhanglab.ccmb.med.umich.edu/BindProfX/	Xiong *et al.* (2017)

(*Continued*)

Table 5.1 (*Continued*)

Name	Features	URL	References
iSEE	Interface structure, evolution, and energy-based	https://github.com/haddocking/iSee	Geng *et al.* (2019)
mCSM-PPI2	Graph-based structure features	https://biosig.lab.uq.edu.au/mcsm_ppi2/	Rodrigues *et al.* (2019)
TopNetTree	Persistent homology, CNN	https://codeocean.com/capsule/2202829/tree/v1	Wang *et al.* (2020)
GeoPPI	Graph-based features	https://github.com/Liuxg16/GeoPPI	Liu *et al.* (2021)
SAAMBE-3D	Mutation-based features	http://compbio.clemson.edu/saambe_webserver/	Pahari *et al.* (2020)
Sequence-based			
ProAffiMuSeq	Amino acid properties, PSSM, interface-specific indices, and protein functional classes	https://web.iitm.ac.in/bioinfo2/proaffimuseq/	Jemimah *et al.* (2020)
PANDA	Amino acid composition, conservation score, and physicochemical properties	https://pandaaffinity.pythonanywhere.com/	Abbasi *et al.* (2021)
SAAMBE-SEQ	Amino acid properties, PSSM, mutation-based properties	http://compbio.clemson.edu/saambe_webserver/indexSEQ.php	Li *et al.* (2021)

5.6.1 *Structure-based methods*

BeAtMuSic is a coarse-grained predictor that depends on statistical potentials obtained from protein three-dimensional structures (Dehouck *et al.*, 2013). BindProf adopts a multi-scale approach using a structural profile score, reflecting the likelihood of a given sequence being found in the ensemble of structurally similar protein–protein complexes (Brender & Zhang, 2015). Leveraging the 3D structure of the molecular complex, MutaBind2 computes interaction energies and conservation scores for predicting changes in binding affinity (Zhang *et al.*, 2020). BindProfX

(Brender & Zhang, 2015; Xiong *et al.*, 2017) includes statistical energy functions based on Boltzmann distributions for predicting binding affinity. iSEE is based on 31 features including position-specific scoring matrix (PSSM), structure interface profile, and energy-based features, and utilizes a random forest model to predict $\Delta\Delta G$ caused by a given mutation (Geng *et al.*, 2019).

mCSM-PPI2 method combines the graph-based signature framework of mCSM (Pires *et al.*, 2014) with non-covalent interactions, PSSM scores, and energetic terms (Rodrigues *et al.*, 2019). TopNetTree, which integrates topological features and a deep learning algorithm, is represented by a topology-based network tree (Wang *et al.*, 2020). GeoPPI is a deep learning-based framework that uses deep geometric representations of protein complexes to model the effects of mutations on binding affinity. SAAMBE-3D uses the structural properties of protein complexes to predict the change in binding affinity upon mutation (Pahari *et al.*, 2020).

The utility of the structure-based methods is limited by the availability of relatively fewer experimentally known structures of protein–protein complexes compared to sequence information.

5.6.2 *Sequence-based methods*

ProAffiMuSeq is a sequence-based method specifically designed to forecast alterations in binding affinity caused by mutations (Jemimah *et al.*, 2020). ProAffiMuSeq utilized amino acid properties, the PSSM matrix, and protein functional class to predict the $\Delta\Delta G$. PANDA used amino acid properties similar to ProAffiMuSeq. Additionally, they utilized k-mer composition (Leslie *et al.*, 2002). SAAMBE-SEQ uses PSSM scores, conservation, and utilized mutation-based properties such as net volume, net hydrophobicity, mutation type, net flexibility, chemical property, size, polarity, and hydrogen bonds (Li *et al.*, 2021).

These computational methodologies collectively contribute to enhancing the understanding of the influence of mutations on protein–protein binding and provide invaluable insights for the domains of drug discovery, structural biology, and disease research. Researchers can strategically select the most appropriate method based on the structural or sequence data as well as a consensus method for predicting the change in binding affinity.

The limitation of these computational methods is their ineptitude with accuracy and speed in predicting the effect of binding affinity upon mutation. For example, BeAtMuSiC (Dehouck *et al.*, 2013) predicts the $\Delta\Delta G$ quickly, but it is prone to errors, as reported in the literature (Gromiha *et al.*, 2017). MutaBind takes a longer time for prediction (Li *et al.*, 2016). BindProf needs interacting chain information; therefore, it is biased in predicting the effect of mutation. Another issue with computational methods is that they require time-to-time maintenance. Some of the available tools are not user-friendly to work with, so it limits the usability of the tools. The need for the developer is to tackle bottlenecks such as speed, accuracy, and user-friendly tool.

5.7 Conclusion

In conclusion, this chapter has explored the impact of mutations on PPIs, encompassing experimental approaches, available databases, factors influencing binding affinity change, and computational prediction tools. By understanding the intricate relationship between mutations and PPIs, researchers can gain valuable insights into various biological processes and develop novel therapeutic strategies. Although computational methods offer the potential to predict affinity change upon mutations at scale, there is still scope for improvement as their accuracy is limited by various factors such as the availability of training data, prediction speed, and the underlying algorithms. Future efforts should focus on increasing the size and diversity of training datasets, reducing dataset bias, and developing new prediction methods that incorporate structural information and effective learning methodologies such as deep learning, which could further enhance the accuracy and reliability of mutation predictions. The remarkable progress of AI-based algorithms in protein structure prediction has opened up the possibility of expanding the dataset beyond experimentally determined structures, allowing for the exploration of various protein-based aspects at a larger scale. Thus, by addressing the challenges and incorporating the improvements, it is possible to develop a comprehensive approach to predicting binding affinity changes induced by mutations in

PPIs, with the ultimate goal of accelerating drug discovery and developing novel therapeutic approaches.

Acknowledgments

We thank the members of the Protein Bioinformatics Lab for their valuable discussions and contributions to this work, and the Department of Biotechnology at the Indian Institute of Technology Madras and the High-Performance Computing Environment (HPCE) for providing the necessary computational facilities.

References

Aasly, J. O., Toft, M., Fernandez-Mata, I., Kachergus, J., Hulihan, M., White, L. R., & Farrer, M. (2005). Clinical features of LRRK2-associated Parkinson's disease in central Norway. *Annals of Neurology, 57*(5), 762–765.

Abbasi, W. A., Abbas, S. A., & Andleeb, S. (2021). PANDA: Predicting the change in proteins binding affinity upon mutations by finding a signal in primary structures. *Journal of Bioinformatics and Computational Biology, 19*(4), 2150015.

Brender, J. R., & Zhang, Y. (2015). Predicting the effect of mutations on protein–protein binding interactions through structure-based interface profiles. *PLoS Computational Biology, 11*(10), e1004494.

Dehouck, Y., Kwasigroch, J. M., Rooman, M., & Gilis, D. (2013). BeAtMuSiC: Prediction of changes in protein–protein binding affinity on mutations. *Nucleic Acids Research, 41*(Web Server issue), W333–W339.

Geng, C., Vangone, A., Folkers, G. E., Xue, L. C., & Bonvin, A. M. J. J. (2019). iSEE: Interface structure, evolution, and energy-based machine learning predictor of binding affinity changes upon mutations. *Proteins, 87*(2), 110–119.

Ghosh, D., Dey, S. K., & Saha, C. (2014). Mutation induced conformational changes in genomic DNA from cancerous K562 cells influence drug-DNA binding modes. *PloS One, 9*(1), e84880.

Jankauskaite, J., Jiménez-García, B., Dapkunas, J., Fernández-Recio, J., & Moal, I. H. (2019). SKEMPI 2.0: An updated benchmark of changes in protein–protein binding energy, kinetics and thermodynamics upon mutation. *Bioinformatics (Oxford, England), 35*(3), 462–469.

Jemimah, S., & Gromiha, M. M. (2018). Exploring additivity effects of double mutations on the binding affinity of protein–protein complexes. *Proteins, 86*(5), 536–547.

Jemimah, S., & Gromiha, M. M. (2020). Insights into changes in binding affinity caused by disease mutations in protein–protein complexes. *Computers in Biology and Medicine, 123*, 103829.

Jemimah, S., Sekijima, M., & Gromiha, M. M. (2020). ProAffiMuSeq: Sequence-based method to predict the binding free energy change of protein–protein complexes upon mutation using functional classification. *Bioinformatics (Oxford, England), 36*(6), 1725–1730.

Jemimah, S., Yugandhar, K., & Gromiha, M. M. (2017). PROXiMATE: A database of mutant protein–protein complex thermodynamics and kinetics. *Bioinformatics (Oxford, England), 33*(17), 2787–2788.

Jemimah, S., Yugandhar, K., & Gromiha, M. M. (2020). Binding affinity of protein–protein complexes: Experimental techniques, databases and computational methods. In M. M. Gromiha (Ed.), *Protein Interactions: Computational Methods, Analysis and Applications* (pp 87–108). World Scientific.

Jubb, H. C., Pandurangan, A. P., Turner, M. A., Ochoa-Montaño, B., Blundell, T. L., & Ascher, D. B. (2017). Mutations at protein–protein interfaces: Small changes over big surfaces have large impacts on human health. *Progress in Biophysics and Molecular Biology, 128*, 3–13.

Kastritis, P. L., & Bonvin, A. M. (2012). On the binding affinity of macromolecular interactions: Daring to ask why proteins interact. *Journal of the Royal Society, Interface, 10*(79), 20120835.

Keskin, O., Gursoy, A., Ma, B., & Nussinov, R. (2008). Principles of protein–protein interactions: What are the preferred ways for proteins to interact? *Chemical Reviews, 108*(4), 1225–1244.

Kucukkal, T. G., Petukh, M., Li, L., & Alexov, E. (2015). Structural and physicochemical effects of disease and non-disease nsSNPs on proteins. *Current Opinion in Structural Biology, 32*, 18–24.

Kulandaisamy, A., Lathi, V., ViswaPoorani, K., Yugandhar, K., & Gromiha, M. M. (2017). Important amino acid residues involved in folding and binding of protein–protein complexes. *International Journal of Biological Macromolecules, 94*(Pt A), 438–444.

Leslie, C., Eskin, E., & Noble, W. S. (2002). The spectrum kernel: A string kernel for SVM protein classification. *Pacific Symposium on Biocomputing. Pacific Symposium on Biocomputing*, *7*, 564–575.

Li, G., Pahari, S., Murthy, A. K., Liang, S., Fragoza, R., Yu, H., & Alexov, E. (2021). SAAMBE-SEQ: A sequence-based method for predicting mutation effect on protein–protein binding affinity. *Bioinformatics (Oxford, England), 37*(7), 992–999.

Li, M., Petukh, M., Alexov, E., & Panchenko, A. R. (2014). Predicting the impact of missense mutations on protein–protein binding affinity. *Journal of Chemical Theory and Computation, 10*(4), 1770–1780.

Li, M., Simonetti, F. L., Goncearenco, A., & Panchenko, A. R. (2016). MutaBind estimates and interprets the effects of sequence variants on protein–protein interactions. *Nucleic Acids Research, 44*(W1), W494–W501.

Liu, X., Luo, Y., Li, P., Song, S., & Peng, J. (2021). Deep geometric representations for modeling effects of mutations on protein–protein binding affinity. *PLoS Computational Biology, 17*(8), e1009284.

Ozturk, K., & Carter, H. (2022). Predicting functional consequences of mutations using molecular interaction network features. *Human Genetics, 141*(6), 1195–1210.

Pahari, S., Li, G., Murthy, A. K., Liang, S., Fragoza, R., Yu, H., & Alexov, E. (2020). SAAMBE-3D: Predicting effect of mutations on protein–protein interactions. *International Journal of Molecular Sciences, 21*(7), 2563.

Patil, R., Das, S., Stanley, A., Yadav, L., Sudhakar, A., & Varma, A. K. (2010). Optimized hydrophobic interactions and hydrogen bonding at the target-ligand interface leads the pathways of drug-designing. *PloS One, 5*(8), e12029.

Pires, D. E., Ascher, D. B., & Blundell, T. L. (2014). mCSM: predicting the effects of mutations in proteins using graph-based signatures. *Bioinformatics (Oxford, England), 30*(3), 335–342.

Reva, B., Antipin, Y., & Sander, C. (2011). Predicting the functional impact of protein mutations: application to cancer genomics. *Nucleic Acids Research, 39*(17), e118.

Ridha, F., Kulandaisamy, A., & Michael Gromiha, M. (2023). MPAD: A database for binding affinity of membrane protein–protein complexes and their mutants. *Journal of Molecular Biology, 435*(14), 167870.

Rodrigues, C. H. M., Myung, Y., Pires, D. E. V., & Ascher, D. B. (2019). mCSM-PPI2: Predicting the effects of mutations on protein–protein interactions. *Nucleic Acids Research, 47*(W1), W338–W344.

Schreiber, G., Haran, G., & Zhou, H. X. (2009). Fundamental aspects of protein–protein association kinetics. *Chemical Reviews, 109*(3), 839–860.

Sellés Vidal, L., Isalan, M., Heap, J. T., & Ledesma-Amaro, R. (2023). A primer to directed evolution: Current methodologies and future directions. *RSC Chemical Biology, 4*(4), 271–291. Published 2023, January 27.

Song, C., Zhang, S., & Huang, H. (2015). Choosing a suitable method for the identification of replication origins in microbial genomes. *Frontiers in Microbiology, 6*, 1049.

Sridharan, R., Zuber, J., Connelly, S. M., Mathew, E., & Dumont, M. E. (2014). Fluorescent approaches for understanding interactions of ligands with G protein coupled receptors. *Biochimica et Biophysica Acta, 1838*(1 Pt A), 15–33.

Usmanova, D. R., Bogatyreva, N. S., Ariño Bernad, J., Eremina, A. A., Gorshkova, A. A., Kanevskiy, G. M., Lonishin, L. R., Meister, A. V., Yakupova, A. G., Kondrashov, F. A., & Ivankov, D. N. (2018). Self-consistency test reveals systematic bias in programs for prediction change of stability upon mutation. *Bioinformatics (Oxford, England), 34*(21), 3653–3658.

Vidal, M., Cusick, M. E., & Barabási, A. L. (2011). Interactome networks and human disease. *Cell, 144*(6), 986–998.

Wang, M., Cang, Z., & Wei, G. W. (2020). A topology-based network tree for the prediction of protein–protein binding affinity changes following mutation. *Nature Machine Intelligence, 2*(2), 116–123.

Wang, Y., & Sarai, A. (1994). A simple method for assessing the mutational effect on the protein–DNA interaction: Application to amino acid substitutions. *Protein Engineering, 7*(9), 1083–1087.

Weinkam, P., Chen, Y. C., Pons, J., & Sali, A. (2013). Impact of mutations on the allosteric conformational equilibrium. *Journal of Molecular Biology, 425*(3), 647–661.

Xiong, P., Zhang, C., Zheng, W., & Zhang, Y. (2017). BindProfX: Assessing mutation-induced binding affinity change by protein interface profiles with pseudo-counts. *Journal of Molecular Biology, 429*(3), 426–434.

Yates, C. M., & Sternberg, M. J. (2013). The effects of non-synonymous single nucleotide polymorphisms (nsSNPs) on protein–protein interactions. *Journal of Molecular Biology, 425*(21), 3949–3963.

Zhang, N., Chen, Y., Lu, H., Zhao, F., Alvarez, R. V., Goncearenco, A., Panchenko, A. R., & Li, M. (2020). MutaBind2: Predicting the impacts of single and multiple mutations on protein–protein interactions. *iScience, 23*(3), 100939.

Zhou, H. X., & Pang, X. (2018). Electrostatic interactions in protein structure, folding, binding, and condensation. *Chemical Reviews, 118*(4), 1691–1741.

https://doi.org/10.1142/9789811293269_0006

Chapter 6

Bioinformatics approaches for understanding the consequences of mutations to the binding affinity of protein–DNA complexes

K. Harini[1], Amit Phogat[1], and M. Michael Gromiha[1,2,*]

[1]*Department of Biotechnology, Bhupat and Jyoti Mehta School of Biosciences, Indian Institute of Technology Madras, Chennai 600036, India*

[2]*International Research Frontiers Initiative, School of Computing, Tokyo Institute of Technology, Yokohama 226-8501, Japan*

Abstract

Protein–nucleic acid interactions are inevitable in maintaining the homeostasis of cells. It is important to have a quantitative understanding of these interactions, generally described in terms of the dissociation constant or free energy change of protein–DNA and protein–RNA complexes. These interactions are impaired in the presence of mutations in nucleic acids or their interacting proteins, leading to numerous diseases.

*Corresponding author
Tel: +914422574138
Fax: +914422574102
MG: gromiha@iitm.ac.in

Hence, it is important to understand the binding affinity change upon mutation in protein–nucleic acid complexes. Different experimental techniques are available to study the binding affinities, although they are accurate, it is time and labor-intensive. On the other hand, computational techniques are emerging with numerous databases and computational tools to study protein–DNA complexes. In this chapter, we discuss various databases for the binding affinity of complexes and change upon mutation and the tools available to extract different structural and interaction features from the complexes. Further, we provide details on prediction methods reported for predicting the change in binding affinity upon mutation, along with hotspot residue prediction in protein–DNA complexes.

Keywords: Protein–DNA complexes; binding affinity change upon mutation; prediction methods; computational tools; machine learning; hotspot

6.1 Introduction

DNA is a versatile biomolecule that encodes important information about the cell and performs numerous functions in association with other macromolecules, including proteins. Protein–DNA interactions control/regulate replication, repair, methylation, and transcription and maintain genome stability (Wu *et al.*, 2016). Numerous studies have been performed to unravel complexes and vital interactions, which include the identification of DNA-binding proteins, predicting the binding site residues, binding affinity of protein–DNA complexes, and recognition mechanism of protein–nucleic acid complexes (Gromiha & Nagarajan, 2013; Nagarajan *et al.*, 2013; Gromiha, 2020; Zhang *et al.*, 2022).

The function of a protein–DNA complex is dictated by its binding affinity (Crocker *et al.*, 2016), which is quantitatively measured using the dissociation constant (K_d) and binding free energy (ΔG). Understanding the affinity of protein–DNA interactions is important to study the effects on gene expressions and delineating the recognition mechanism based on interactions between protein and DNA (Ladbury, 1995; Rastogi *et al.*, 2018). It has a broad spectrum of applications, such as designing complexes with the desired affinities, developing prediction methods for the target sites, and quantitative simulation of gene regulation networks.

On the other hand, binding sites in protein–nucleic acid complexes are vital for influencing their specificity and affinity. Alteration in binding sites leads to the loss of binding and/or change in affinity, which may impact the function of the complex. It is reported that a single amino acid mutation in a protein–nucleic acid interface causes deadly diseases, including cancer. For instance, mutants G79C, F83C, and M125I in hepatocyte nuclear factor 4 alpha protein reduced the binding affinity to ApoB promoters, which reduced the transcriptional activity and increased the risk of liver tumorigenesis (Taniguchi *et al.*, 2018). Hence, it is important to understand the change in binding affinity upon mutation (ΔΔG).

Experimentally, the binding affinities of protein–nucleic acid complexes are studied with electrophoretic mobility shift assays, filter binding assays, fluorescence spectroscopy, isothermal titration calorimetry, and surface plasmon resonance. Although determining the binding affinity is accurate, it is time-consuming and labor-intensive to perform a large-scale analysis. Computational databases and prediction methods are growing as an efficient alternative (Yang & Deng, 2020; Harini *et al.*, 2022; Harini *et al.*, 2023). In this chapter, we will discuss different databases and tools available for analyzing protein-nucleic acid complexes, which can be a useful resource for developing computational tools for analyzing the hotspot residues and predicting the change in binding affinity upon mutation.

6.2 Protein–DNA complexes associated with diseases

The role of protein-DNA complexes in cellular mechanisms is pivotal, not only for normal cellular functions but also for understanding their involvement in various diseases. The mutations in DNA-binding proteins affect their binding with DNA, which plays a key role in the initiation and progression of various diseases, such as cancer and neurodegenerative diseases. Multiple studies showed the role of TAR DNA-binding protein 43 (TDP-43) involved in exon splicing and gene expression regulation, as a key protein in the development of amyotrophic lateral sclerosis (ALS) and frontotemporal dementia (FTD) (Arai *et al.*, 2006; Neumann *et al.*, 2006). A significant number of transcription factors are involved in the development of various types of cancer.

The Forkhead box (FOX) protein family is responsible for the expression of genes involved in cell proliferation and cell growth. Mutations in FOX proteins have been implicated in liver, breast, and prostate cancers and chronic myelogenous leukemia (Shiroma *et al.*, 2020; Herman *et al.*, 2021). The c-Myc protein regulates the transcription of Cdc25A, glutamine synthetase, and hTERT genes (Galaktionov *et al.*, 1996; Wu *et al.*, 1999; Bott *et al.*, 2015). Dysregulation of c-Myc activity due to a loss of function mutation in p53 has been reported as the leading cause of breast cancer (Santoro *et al.*, 2019). The high-mobility group (HMG) box containing a protein called SOX2, which plays a crucial role in maintaining the pluripotency of embryonic stem cells, has been implicated in the development and maintenance of cancer cells. SOX2 provides resistance to drug therapies and is constitutively expressed in cancer stem cells (Novak *et al.*, 2020; Shiroma *et al.*, 2020). The human telomeric repeat-binding factor (TRF1) binds to TTAGGG repeats at telomeres and resists the binding of telomerases, preventing telomere elongation. TRF1 binding to telomeric ends is crucial for telomere protection, *Trf1* gene deletion leads to continuous activation of the DNA damage response, obstructing cell division and consequently inducing apoptosis or senescence. *Trf1* overexpression has been linked to breast cancer, renal cell carcinoma, liver, lung, and colon cancers (Bejarano *et al.*, 2017; Shiroma *et al.*, 2020).

The highly conserved DNA binding domain in Y-box binding protein 1 (YB1) regulates the expression of genes participating in cell cycle progression and DNA repair mechanisms. YB1 has been reported as a biomarker for hepatocellular carcinoma, lung, and breast cancer (Yasen *et al.*, 2005; Sangermano *et al.*, 2020). The tumor suppressor TP53, frequently mutated in all cancer types, contains a DNA-binding domain through which it regulates the expression of genes involved in apoptosis, cell cycle, and metabolism. P53 is widely studied for its role in cancer development by interacting with multiple proteins and acting as a transcription factor for multiple genes (Brázda & Fojta, 2019). The SWI/SNF chromatin remodeling complex contains a DNA-binding protein known as AT-rich interaction domain 1A (ARID1A). The binding of this protein to DNA initiates chromatin remodeling and involved in DNA repair mechanisms (Shen *et al.*, 2015; Pulice *et al.*, 2016). The aberrations in the ARID1A gene lead to loss of function, helping in the development of hepatocellular carcinoma, melanomas, and cancers of the stomach, colon, pancreas, lungs, breast, and ovaries (Mullen *et al.*, 2021). Hypoxia-inducible factor 1 alpha (HIF1-α), part of the HIF complex, activated in hypoxic conditions (less oxygen), contains a helix-loop-helix DNA binding motif and target genes for vascular endothelial growth factor (VEGF) and erythropoietin (EPO). In the tumor microenvironment, hypoxia conditions lead to activation of HIF1-α with localization to the nucleus along with HIF1-β, consequently, expression of target genes that leads to angiogenesis in the tumor microenvironment (Semenza, 2003).

In essence, protein-DNA complexes play a significant role in the transcription and expression of various genes involved in crucial pathways of cell division, apoptosis, and metabolism. Aberrations or mutations in DNA-binding proteins cause various cancers, neurodegenerative diseases, and genetic diseases. Hence, it is necessary to study the protein–DNA complexes to understand their role in the development of diseases as well as the relationship among sequence, structure, binding, and function.

6.3 Databases for binding affinity change upon mutation in protein–DNA complexes

Binding affinity is a parameter that can quantitatively provide information on the strength of interactions between proteins and nucleic acids. Mutations in either protein or nucleic acid alter the binding affinity, represented in terms of the change in affinity upon mutation (ΔΔG). Experimentally determined binding affinity details and their changes upon mutation are compiled in different databases.

ProNIT (Prabakaran *et al.*, 2001) is the first thermodynamic database developed for protein–nucleic acid complexes. The database has details of the binding affinity (K_d and/or ΔG) of wild-type and mutant complexes along with their structures, experimental conditions and literature. The mutations in both proteins and nucleic acids are included in the database.

dbAMEPNI (Liu *et al.*, 2018) is a database with binding affinity data specific for alanine mutations in protein-nucleic acid complexes. Each entry is provided with a dissociation constant (K_d), Gibbs free energy change upon mutation ($\Delta\Delta G$), experimental conditions, and PubMed ID. The database is available at http://zhulab.ahu.edu.cn/dbAMEPNI.

ProNAB (Harini *et al.*, 2022) is currently the largest protein–nucleic acid binding affinity database available in the literature, with more than 20,000 data. It has both wild-type and mutant data with sequence and structural features of the protein, nucleic acid and its complex, experimental conditions, thermodynamic parameters such as the dissociation constant, the free energy of binding, and the change in free energy upon mutation. Each entry in the database is also cross-linked with other databases such as GenBank, UniProt, PDB, ProThermDB, PROSITE, DisProt, and PubMed. The database has options for uploading and downloading the data. The database is available at https://web.iitm.ac.in/bioinfo2/pronab/. The database also includes details of the secondary structure and accessible surface area (ASA) of the mutated residue. An example entry in ProNAB is shown in **Table 6.1**.

Table 6.1 An example of entries in the ProNAB database

Description	Example
Description	Example
Entry ID	156
Protein Name	Early growth response protein 1
Synonyms	EGR-1; Nerve growth factor-induced protein A; NGFI-A; Transcription factor Zif268; Zinc finger protein Krox-24
EC number	—
Protein Source	Mus musculus (Mouse)
Sequence	MAAAKAEMQLMSPLQISDPFGSFPHSPTMDNYPKLEEM MLLSNGAPQFLGAAGTPEGSGGNSSSSTSSGGGGGG SNSGSSAFNPQGEPSEQPYEHLTTESFSDIALNNEKAM VETSYPSQTTRLPPITYTGRFSLEPAPNSGNTLWP

Table 6.1 *(Continued)*

Description	Example
Length	533
Mass (Da)	56590
UniProt ID	P08046
PROSITE ID	PS00028; PS50157;
DisProt ID	—
PDB of Free Protein	—
ASA of Free protein (Å2)	—
ProTherm Id	—
Mutation in protein	R349A, D351A
Nucleic acid name	Zinc finger binding site
Nucleic acid source	Synthetic
Type of nucleic acid	DNA
Sequence	agcagctgagcgtgggcgtagtgagct
Mutation in nucleic acid	G 18 C
Genbank ID	2098366; 2098367
PDB complex	1aay
NDB complex	pdt039
ASA of Complex (Å2)	147.0, 78.0
Sec str	Coil, Helix
pH	7.8
Temperature (K)	298
Buffer	15 mM Hepes
Additives	glycerol(v/v)(5%), acetylated BSA (0.1 mg/mL), Igepal-CA630 (1 mg/mL)
Method	Gel shift
Kd wild (M)	2.00E-08
Kd mutant (M)	5.10E-09
Ka wild (M−1)	50000000
Ka mutant (M−1)	196000000

(Continued)

Table 6.1 (*Continued*)

Description	Example
ΔG wild (kcal/mol)	−10.5
ΔG mutant (kcal/mol)	−11.31
ΔΔG (kcal/mol)	−0.81
ΔH wild (kcal/mol)	—
ΔH mutant (kcal/mol)	—
Stoichiometry	—
Reference	*The Journal of Biological Chemistry, 274*(27), 19281–19285.
Title	Binding studies with mutants of Zif268. Contribution of individual side chains to binding affinity and specificity in the Zif268 zinc finger-DNA complex.
Authors	Elrod-Erickson M, Pabo CO
Keywords	Zif268, zinc-finger-DNA complex, specificity
PubMed	10383437
DOI	http://dx.doi.org/10.1074/jbc.274.27.19281
Location of data	Table 1; Page No: 19282
Remarks	—

Other thermodynamic databases provide binding affinity, but not the change in affinity on mutation. PDBbind (Liu *et al.*, 2015) contains the experimentally determined binding affinity of biomolecules, including protein–ligand, protein–DNA/RNA, and protein–protein complexes, which is specific for the complexes with the experimental known structures. Another database, the Protein–Nucleic Acid Thermodynamics Database (PNATDB) has about 12,000 binding affinity values, along with details of molecular interactions in the complexes (Mei *et al.*, 2023). It is available at http://chemyang.ccnu.edu.cn/ccb/database/PNAT/.

6.4 Generation of mutant protein–nucleic acid complexes

Experimentally determined structures of protein-nucleic acid complexes are limited, and it is necessary to model the wild-type and mutant structures

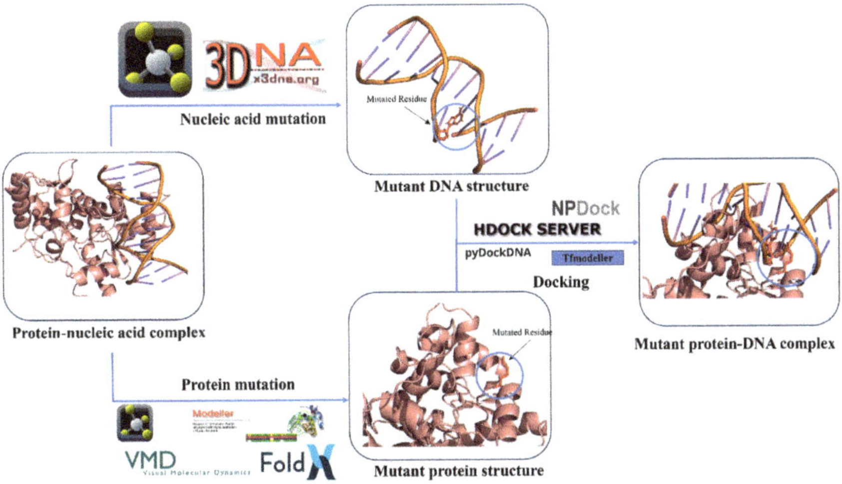

Figure 6.1 Preparing mutant protein–nucleic acid complex structures. It has three steps: (a) consider the protein–nucleic acid complex structure, (b) construct the mutant protein and DNA, and (c) obtain the mutant protein–DNA complex structure

computationally. **Figure 6.1** shows the tools for preparing the mutant protein-nucleic acid complexes.

6.4.1 *Methods to generate mutant protein and nucleic acid structures*

Numerous software are available to incorporate mutations in proteins, such as FoldX (Schymkowitz *et al.*, 2005), Modeller (Webb & Sali, 2014), VMD (Humphrey *et al.*, 1996), and PyMOL (DeLano, 2002). FoldX has the option to generate mutant structures using the "Build Model." Modeller generates 3D structures using the protein sequence by aligning them to the known three-dimensional structures of similar proteins. The mutant structures are generated using wild-type proteins as templates. In addition, the mutator plugin in VMD can be used for generating mutant structures using the CHARMM force field. Once the mutant structures are generated, the structure has to be energy minimized for relaxing conformations using the AMBER program, and a short period of MD simulation is performed to refine the orientation of residue side chains. Similarly,

PyMOL generates mutant structures by simple substitution of single amino acid in the wild-type conformation. On the other hand, nucleic acid mutations are performed using 3DNA (Li *et al.*, 2022) and PyMOL.

6.4.2 *Prediction of protein–nucleic acid complex structures*

This section outlines the methods for modeling protein-nucleic acid complex structures. Nucleic Acid–Protein Dock (NPDock) performs a series of tasks during modeling (Tuszynska *et al.*, 2015), such as modeling protein–nucleic acid complexes, scoring poses with statistical potentials, clustering the best models, and refinement of the selected best models. The web server is available at https://genesilico.pl/NPDock/.

HDOCK utilizes homology-based searching and template-based modeling for structure prediction, with the option to incorporate biological information (Li *et al.*, 2022). The server is available at http://hdock.phys.hust.edu.cn/. TFmodeller is developed based on the information about protein–DNA interfaces in the Protein Data Bank (PDB). The software is available at http://www.ccg.unam.mx/tfmodeller (Contreras-Moreira *et al.*, 2007). pyDockDNA takes the apo-protein and DNA molecules as input, utilizes FTDOCK to generate protein–DNA orientations and DNA-based scoring functions and outputs the best 10 or 100 docking models along with options for visualizations (Rodriguez-Lumbreras *et al.*, 2022). It is available at https://model3dbio.csic.es/pydockdna. MELD-DNA has additional options to identify preferred binding modes from multiple poses and provide qualitative binding preferences between DNA sequences (Esmaeeli *et al.*, 2023). The software can be downloaded from the GitHub repository, github.com/maccallumlab/meld.

6.5 Tools for extracting features from the structures of proteins, nucleic acids, and complexes

The tools available for extracting features from the structures of proteins, nucleic acids, or protein–nucleic acid complexes are listed in **Tables 6.2** and **6.3**.

Table 6.2 Tools for extracting features from protein and nucleic acid structures

Tool	Description	Link	References
Protein structures			
AAindex	Physicochemical properties of amino acid residue in the proteins	https://www.genome.jp/aaindex/	Kawashima *et al.* (1999)
DSSP	Defines the secondary structure, geometrical features, and solvent exposure of proteins	https://www3.cmbi.umcn.nl/xssp/	Kabsch and Sander (1983)
DynaMut	Impact of mutation on protein stability and dynamics	https://biosig.lab.uq.edu.au/dynamut/	Rodrigues *et al.* (2018)
ENDES	Identifies near-native structure using scoring functions	http://sparks.informatics.iupui.edu.	Liang *et al.* (2009)
NAPS	Analysis of residue interaction networks in proteins	http://bioinf.iiit.ac.in/NAPS/	Chakrabarty and Parekh (2016)
PSAIA	Solvent properties in protein structures	http://complex.zesoi.fer.hr/tools/PSAIA.html	Mihel *et al.* (2008)
Nucleic acid structures			
Curves+	Analyzing nucleic acid steps and helical parameters	http://gbio-pbil.ibcp.fr/cgi/Curves_plus/.	Blanchet *et al.* (2011)
w3DNA 2.0	Analysis, visualization, rebuilding, and mutation in nucleic acid structures	http://web.x3dna.org/.	Li *et al.* (2019)

6.5.1 *Extraction of features from protein structures*

The AAindex database is a reservoir of physicochemical and biochemical properties of amino acids (Kawashima *et al.*, 1999), which are obtained from protein structures and/or sequences. It contains 437 amino acid indices, each with numerical values for each of the 20 amino acids. Moreover, AAindex2 contains numerous amino acid substitution matrices, which can be symmetrical or non-symmetrical. In addition, it contains a

Table 6.3 Tools for extracting the interaction features in protein–DNA complexes

Tool	Description	Link	References
Interaction features			
HBPLUS	Details of hydrogen bonds	https://www.ebi.ac.uk/thornton-srv/software/HBPLUS/	McDonald *et al.* (1994)
Naccess	Atomic and residue-wise solvent accessibility	http://www.bioinf.manchester.ac.uk/naccess/	Hubbard and Thornton (1993)
Arpeggio	Interatomic interactions in protein structures	http://structure.bioc.cam.ac.uk/arpeggio/	Jubb *et al.* (2017)
PDBsum1	Secondary structure, interacting residues, hydrogen bonds, motifs, and cavities in a complex	https://www.ebi.ac.uk/thornton-srv/software/PDBsum1/	Laskowski (2022)
DNAproDB	Functional, biophysical, and structural features	https://dnaprodb.usc.edu/	Sagendorf *et al.* (2020)
ProDFace	Hydrogen bond donors and conservation of residues	http://structbioinfo.iitj.ac.in/resources/bioinfo/pd_interface/	Pal *et al.* (2022)
FoldX	Interaction energies of the complex, stability of a protein, etc.	http://foldx.embl.de/	Schymkowitz *et al.* (2005)

list of amino acid pairwise-contact potentials. DSSP is a secondary structure assignment program that determines the secondary structure of each residue by calculating the hydrogen bond energy between all atoms (Kabsch & Sander, 1983). The program provides geometrical features and solvent exposure for proteins.

DynaMut (Rodrigues *et al.*, 2018) is a web server that can be used to visualize and analyze protein dynamics and assess the impact of mutations on protein dynamics and stability. The stability of the protein, in turn, has an impact on the binding affinity. Empirical Near-native Docking-decoy Enrichment *Score (ENDES)* (Liang *et al.*, 2009) is a scoring that can calculate different scoring functions using energy, conservation, and interface propensity. In addition, residue energy, side-chain energy, conservation, accessibility, and interface propensity can be obtained. Network Analysis of Protein Structures (NAPS) (Chakrabarty & Parekh, 2016) is a web server for quantitative and qualitative analysis of residue interaction networks in the protein structure. The server provides numerous interaction features of the residues, like degree, closeness, clustering coefficient, betweenness, and eccentricity. PSAIA (Mihel *et al.*, 2008) is a server that can calculate several solvent-based geometric parameters such as ASA, protrusion index, and depth index of residue in proteins.

6.5.2 *Extraction of features from nucleic acid structures*

The features based on DNA are also reported to be important for understanding the interactions between protein and DNA as well as the binding affinity. Several tools are currently available in the literature to extract these features. Curves+ (Blanchet *et al.*, 2011) is software for analyzing the conformation of nucleic acid structures, which takes the nucleic acid structure files as input and provides helical, backbone, and groove parameters, inter-base pair and intra-base pair parameters, and sugar puckers of nucleic acids. The software is freely available at http://gbio-pbil.ibcp.fr/cgi/Curves_plus/.

w3DNA 2.0 (Li *et al.*, 2019) has six main modules for analysis, visualization, rebuilding, composite, fiber model, and mutation. The "analysis module" provides information about the conformational features of DNA/RNA, such as base step parameters (shift, slide, rise, tilt, roll, twist), local base-pair parameters (shear, stretch, stagger, buckle, propeller, opening), groove width, and torsional angles, whereas the "visualization module" is used to create high-resolution base-block schematic images. The last four modules focus on modeling atomic-level simulations of proteins or DNA/RNA structures. The w3DNA is available at http://web.x3dna.org/.

6.5.3 *Extraction of interaction features from protein–DNA complex structures*

HBPLUS provides a detailed list of hydrogen bonding residues and their geometries, positions, donor, and acceptor atoms of the residues, and conformations (McDonald *et al.*, 1994). Naccess gives the atomic and residue-wise solvent-ASA of the complex by rolling the probe around the van der Waals surface of the given structure (Hubbard & Thornton, 1993). It provides the ASA and relative ASA (RSA) of each residue and changes in ASA upon complex formation.

The Arpeggio web server (Jubb *et al.*, 2017) computes different types of interactions, such as van der Waals, ionic, metal, hydrophilic, halogen bond, hydrogen bonds, and aromatic. PDBsum1 is a standalone program that provides the pictorial representation of a 3D structure along with possible interactions in the complex (Laskowski, 2022). The software can be downloaded and installed from https://www.ebi.ac.uk/thornton-srv/software/PDBsum1. DNAproDB is a visualization tool that maps the interactions between the protein and DNA in a complex and is available at https://dnaprodb.usc.edu/ (Sagendorf *et al.*, 2020). ProDFace is a web server to understand protein–DNA interface structures (Pal *et al.*, 2022). It takes the protein-DNA complex in PDB format as input. It provides physicochemical parameters of the interface, details of hydrogen bond donors, water-mediated hydrogen bonds, and conservation of residues in the interface core and rim of the protein as output. The web tool is available at http://structbioinfo.iitj.ac.in/resources/bioinfo/pd_interface/.

FoldX is an empirical force field developed for calculating the free energy of a macromolecule based on a given 3D structure (Schymkowitz *et al.*, 2005). It is also helpful for calculating the stability of a protein, repairing PDB files, analyzing the effect of mutations, and calculating the interaction energies of protein–protein, protein–DNA, and protein–RNA complexes. FoldX is available for downloading at foldxsuite.crg.eu.

6.6 Functional and disease-causing effects of mutations in protein–DNA complexes

Mutations of amino acid residues in DNA-binding proteins affect function and may cause diseases (Shiroma *et al.*, 2020). The Wilms tumor protein is a transcriptional factor consisting of a DNA-binding domain, which has four zinc finger repeats that determine sequence-specific binding to DNA (Hamilton *et al.*, 1995). While two of the zinc fingers bind to the DNA, others are essential for recognizing the cognate nucleotide base.

One of these zinc fingers that is responsible for recognizing the cognate nucleotide base undergoes a mutation (M342R) that enhances the affinity for a different nucleotide base leading to errors in transcription (Wang *et al.*, 2018).

Missense mutations in the MH1 domain of Smad tumor suppressor proteins are identified in cancer patients, specifically the R133C mutation in Smad2 in colon carcinoma (Eppert *et al.*, 1996) and the R100T mutation in Smad4 in pancreatic carcinoma (Schutte *et al.*, 1996). These mutations increase the affinity of MH1 for the MH2 domain in the Smad proteins. This prevents the Smad2 and Smad4 interactions and further affects signaling, keeping them inactive (Hata *et al.*, 1998).

The S392E mutant of the human p53 protein is considered an "activated" form of p53. This mutation does not affect affinity in the absence of a non-specific DNA sequence. However, the affinity of the S392E mutant for the nonspecific DNA is increased with the proportion of non-specific sequences (Nichols & Matthews, 2002). The R249S mutation in the p53 protein is associated with liver cancer. The mutation affects the S1-S2 turn and extends to a beta-strand at the DNA binding interface, which decreases the affinity of interactions (Liu *et al.*, 2020).

The nuclear factor of activated T cells 5 (NFAT5) is a dimeric transcriptional activator protein that is stable upon binding to DNA, whereas the absence of DNA confers high flexibility to the protein. Mutation in NFAT5 either increases the flexibility of the protein complex by compromising the DNA binding (R217A/E223A/R226A mutation) or increases the structural stability (T222D mutation) (Li *et al.*, 2013). Stat5b is a transcriptional regulator of a variety of cytokines. The L327M mutation in the DNA binding domain affects the regulatory pathways in nonobese diabetic mice by reducing its affinity to the DNA (Davoodi-Semiromi *et al.*, 2004).

Transcription factor, PrfA, regulates the pathogenicity factors in *Listeria monocytogenes.* Mutation G145S in the helix-turn-helix (HTH) of the protein rearranges itself and stabilizes the HTH motif, thereby increasing the DNA binding affinity (Eiting *et al.*, 2005). GT-1 is a plant transcription factor that acts as a molecular switch in response to light. The phosphorylated DNA binding domain of the T133D mutant showed enhanced binding affinity to the cis-acting element BoxII (Nagata *et al.*, 2010).

6.7 Prediction of protein–DNA binding affinity change upon mutations

This section outlines different methods developed to predict the binding affinity change upon mutation. **Table 6.4** lists the available online tools along with datasets.

mCSM-NA is based on the mutational cut-off scanning matrices (mCSM) and graph-based signatures (Pires *et al.*, 2017). It utilizes two signature vectors (Pires *et al.*, 2014), such as the distance matrix from the geometric center of wild-type residues and the "pharmacophore count" vector, which captures the change in the atom types due to mutations. Each atom in a residue is classified into eight possible categories: hydrophobic, positive, negative, neutral, aromatic, sulfur, hydrogen acceptor, and donor, represented using PMapper. It also includes nucleic acid base signatures, where the atoms can be classified based on the nature of the nucleotide (purines/pyrimidines) or based on phosphate, sugar, and base. In addition, the distance to nucleic acids and predicted protein stability change is also considered for prediction. The server is available at http://structure.bioc.cam.ac.uk/mcsm_na.

SAMPDI (Peng *et al.*, 2018) uses a modified Molecular Mechanics Poisson–Boltzmann Surface Area (MM/PBSA) approach to predict the effect of single protein mutations on protein–DNA binding affinity. The user interface is implemented using the HTML http://compbio.clemson.edu/SAMPDI/.

PremPDI (Zhang *et al.*, 2018) is a multiple linear regression-based model developed to predict the effect of mutations using molecular mechanics force fields and fast side-chain optimization algorithms. It generates the mutant structures using FoldX (Schymkowitz *et al.*, 2005) and calculates different binding energy terms along with differences in the number of

Table 6.4 Methods to predict the change in binding affinities of protein-DNA complexes

Tool	Methods and features	Dataset	Link	References
mCSM-NA	Graph-based signatures to predict the effect of single and tested for multiple mutations	264 mutations from six complexes	structure.bioc.cam.ac.uk/mcsm_na	Pires *et al.* (2017)
SAMPDI	MM/PBSA-based energy and knowledge-based energy for prediction	105 mutations from 13 complexes	http://compbio.clemson.edu/SAMPDI-3D/	Peng *et al.* (2018)
PremPDI	Using multiple linear regression, molecular mechanics force fields, and fast side-chain optimization	219 mutations from 49 complexes	https://lilab.jysw.suda.edu.cn/research/PremPDI/	Zhang *et al.* (2018)
PEMPNI	Random forest regressor-based model using geometric partition-based energy and interface structural features	324 mutations from 73 complexes	http://liulab.hzau.edu.cn/PEMPNI	Jiang *et al.* (2021)
SAMPDI-3D	Physicochemical properties and structural properties of mutations to predict the effect of a protein or DNA mutation using a gradient-boosting decision tree algorithm	419 protein mutations from 96 complexes; 463 DNA mutations from 30 complexes	http://compbio.clemson.edu/SAMPDI-3D/	Li *et al.* (2021)

hydrogen bonds, the ratio of the ASA of the complex to DNA, and pairwise statistical potential from AAindex for prediction. PremPDI is available at http://lilab.jysw.suda.edu.cn/research/PremPDI/.

PEMPNI (Jiang *et al.*, 2021) is a random forest regressor-based model developed using energy and non-energy-based interaction features to predict the change in binding affinity upon mutation in protein–nucleic acid complexes. It generates the mutant structures using MODELLER (Webb & Sali, 2014). It utilizes the information on interface, and non-interface regions and the parameters residue–residue pairs, residue–nucleotide pairs, nucleotide–nucleotide pairs, solvent accessibilities, hydrogen bonds, contacts, and evolutionary information for predicting the binding affinity. The server is available at http://liulab.hzau.edu.cn/PEMPNI.

SAMPDI-3D (Li *et al.*, 2021) is based on the gradient-boosting decision tree algorithm to predict the effect of a single-point protein or DNA mutation. Different structural and interaction features, such as secondary structure propensity, solvent accessibility, and backbone torsion angles, contacts, hydrogen bonds, and stacking interactions, and other knowledge-based terms are used for prediction. The server is freely available at http://compbio.clemson.edu/SAMPDI-3D/.

6.8 Identification of hotspot residues in protein–DNA complexes

There are numerous prediction methods available to identify hotspot residues at the interface of protein–DNA complexes that contribute to the binding free energy of the complexes. The residue is considered as hotspot if the binding free energy change upon mutation ($\Delta\Delta G$) $\geq$1.0 kcal/mol. **Table 6.5** lists the available methods for identifying hotspot residues. Most of the methods utilize machine learning techniques and structure-based features.

PrPDH (Zhang *et al.*, 2020) is a SVM-based classifier that is based on solvent ASA, position-specific scoring matrix, secondary structure, disorder regions, and contact energies, along with structural and network

Table 6.5 Methods to classify hotspot residues in protein–DNA complexes

Tool	Methods and features	Dataset	Link	References
PrPDH	Sequence, structure, and network-based features using random forests and SVM classifier	40 complexes, 62 hot spots, and 88 non-hot spots	http://bioinfo.ahu.edu.cn:8080/PrPDH	Zhang *et al.* (2020)
inpPDH	Structural and interface-adjacent property features using SVM	62 and 88 hotspots and non-hotspots from 40 complexes	http://bioinfo.ahu.edu.cn/inpPDH	Zhang *et al.* (2021)
sxPDH	Supervised isometric feature mapping and extreme gradient boosting (XGBoost)	62 hot spots and 88 non-hot spots from 40 DNA-binding proteins	https://github.com/xialab-ahu/sxPDH	Li *et al.* (2020)
PreHots	Target residue attributes and network information using an ensemble stacking classifier	123 hot spots and 137 non-hot spots from 89 protein–DNA complexes	http://dmb.tongji.edu.cn/tools/PreHots/	Pan *et al.* (2020)
WTL-PDH	Discrete wavelet transform (DWT) and wavelet packet transform (WPT)-based feature selection and light gradient boosting machine	339 mutations in 117 protein–DNA complexes	https://github.com/chase2555/WTL-PDH	Sun *et al.* (2023)
HISNAPI	Flexibility of protein–nucleic acid complexes by sampling conformations using molecular dynamics simulation	299 mutants from 40 protein–nucleic acid complexes	http://agroda.gzu.edu.cn:9999/ccb/server/HISNAPI/	Mei *et al.* (2021)

features for identifying hotspot residues (http://bioinfo.ahu.edu.cn:8080/PrPDH). This method has been improved as inpPDH (Zhang *et al.*, 2021) using SVM classifier and two-step feature selection, such as SVM-based recursive feature elimination and the elimination of highly correlated features (http://bioinfo.ahu.edu.cn/inp PDH). sxPDH (Li *et al.*, 2020) utilizes extreme gradient boosting (XGBoost) and supervised isometric feature mapping to reduce the feature dimensionality for the prediction. The source code is available at https://github.com/xialab-ahu/sxPDH. PreHots (Pan *et al.*, 2020) is an ensemble stacking classifier to identify hotspot residues based on solvent exposure, interaction networks, hydrogen bonds, consensus scores, secondary structure, ASA, and physiochemical properties (http://dmb.tongji.edu.cn/tools/PreHots/).

Sun *et al.* (2023) developed WTL-PDH based on a gradient-boosting decision tree using discrete wavelet transform (DWT) and wavelet packet transform (WPT) for feature processing. It also utilizes conventional features such as SASA, secondary structure, hydrogen bonds, and depth features to build the model. The dataset and source code are available at https://github.com/chase2555/WTL-PDH. Mei *et al.* (2021) proposed a dynamic method, HISNAPI using Molecular Dynamics (MD) simulations in which the binding energy of the conformational ensemble of wild-type and mutants is calculated using FoldX. The server is available at http://agroda.gzu.edu.cn:9999/ccb/server/HISNAPI/.

6.9 Conclusions

The development of computational tools for analyzing protein–nucleic acid complexes is important for understanding the recognition mechanism and the development of therapeutic strategies. We have reviewed various databases available for binding affinity upon mutations in protein–nucleic acid complex structures. Further, computational tools reported in the literature for extracting residue-wise features in proteins, DNA, and protein–DNA complexes have been outlined. Moreover, various algorithms for predicting the binding affinity change upon mutation of protein–DNA complexes have been discussed, along with identifying hotspot residues. The performance of currently available prediction methods could be improved upon the availability of a large number of experimental data for protein–DNA complexes and their mutants.

Acknowledgments

The authors acknowledge the Indian Institute of Technology Madras and the High-Performance Computing Environment (HPCE) for computational facilities. We thank the members of the Protein Bioinformatics Lab for providing valuable suggestions. The work is partially supported by the Science and Engineering Research Board (SERB), Ministry of Science and Technology, Government of India to MMG (No. CRG/2020/000314).

References

Arai, T., Hasegawa, M., Akiyama, H., Ikeda, K., Nonaka, T., Mori, H., Mann, D., Tsuchiya, K., Yoshida, M., Hashizume, Y., & Oda, T. (2006). TDP-43 is a component of ubiquitin-positive tau-negative inclusions in frontotemporal lobar degeneration and amyotrophic lateral sclerosis. *Biochemical and Biophysical Research Communications, 351*(3), 602–611.

Bejarano, L., Schuhmacher, A. J., Méndez, M., Megías, D., Blanco-Aparicio, C., Martínez, S., Pastor, J., Squatrito, M., & Blasco, M. A. (2017). Inhibition of TRF1 telomere protein impairs tumor initiation and progression in glioblastoma mouse models and patient-derived Xenografts. *Cancer Cell, 32*(5), 590–607.e4.

Blanchet, C., Pasi, M., Zakrzewska, K., & Lavery, R. (2011). CURVES+ web server for analyzing and visualizing the helical, backbone and groove parameters of nucleic acid structures. *Nucleic Acids Research, 39*(Web Server issue), W68–73.

Bott, A. J., Peng, I. C., Fan, Y., Faubert, B., Zhao, L., Li, J., Neidler, S., Sun, Y., Jaber, N., Krokowski, D., Lu, W., Pan, J. A., Powers, S., Rabinowitz, J., Hatzoglou, M., Murphy, D. J., Jones, R., Wu, S., Girnun, G., & Zong, W. X. (2015). Oncogenic Myc induces expression of glutamine synthetase through promoter demethylation. *Cell Metabolism, 22*(6), 1068–1077.

Brázda, V., & Fojta, M. (2019). The rich world of p53 DNA binding targets: The role of DNA Structure. *International Journal of Molecular Sciences, 20*(22), 5605.

Chakrabarty, B., & Parekh, N. (2016). NAPS: Network analysis of protein structures. *Nucleic Acids Research, 44*(W1), W375–W382.

Contreras-Moreira, B., Branger, P. A., & Collado-Vides, J. (2007). TFmodeller: Comparative modelling of protein-DNA complexes. *Bioinformatics, 23*(13), 1694–1696.

Crocker, J., Noon, E. P., & Stern, D. L. (2016). The soft touch: Low-affinity transcription factor binding sites in development and evolution. *Current Topics in Developmental Biology, 117*, 455–469.

Davoodi-Semiromi, A., Laloraya, M., Kumar, G. P., Purohit, S., Jha, R. K., & She, J. X. (2004). A mutant Stat5b with weaker DNA binding affinity defines a key defective pathway in nonobese diabetic mice. *The Journal of Biological Chemistry, 279*(12), 11553–11561.

DeLano, W. L. (2002). *The PyMOL molecular graphics system.* DeLano Scientific. http://www.pymol.org

Eiting, M., Hagelüken, G., Schubert, W. D., & Heinz, D. W. (2005). The mutation G145S in PrfA, a key virulence regulator of Listeria monocytogenes, increases DNA-binding affinity by stabilizing the HTH motif. *Molecular Microbiology, 56*(2), 433–446.

Eppert, K., Scherer, S. W., Ozcelik, H., Pirone, R., Hoodless, P., Kim, H., Tsui, L.-C., Bapat, B., Gallinger, S., Andrulis, I. L., Thomsen, G. H., Wrana, J. L., & Attisano, L. (1996). MADR2 maps to 18q21 and encodes a TGFb-regulated MAD-related protein that is functionally mutated in colorectal carcinoma. *Cell, 86*, 543–552.

Esmaeeli, R., Bauza, A., & Perez, A. (2023). Structural predictions of protein-DNA binding: MELD-DNA. *Nucleic Acids Research, 51*(4), 1625–1636.

Galaktionov, K., Chen, X., & Beach, D. (1996). Cdc25 cell-cycle phosphatase as a target of c-myc. *Nature, 382*(6591), 511–517.

Gromiha, M. M. (2020). *Protein Interactions: Computational methods, analysis and applications*. World Scientific.

Gromiha, M. M., & Nagarajan, R. (2013). Computational approaches for predicting the binding sites and understanding the recognition mechanism of protein-DNA complexes. *Advances in Protein Chemistry and Structural Biology, 91*, 65–99.

Hamilton, T. B., Barilla, K. C., & Romaniuk, P. J. (1995). High affinity binding sites for the Wilms' tumour suppressor protein WT1. *Nucleic acids research, 23*(2), 277–284.

Harini, K., Kihara, D., & Gromiha, M. M. (2023). PDA-Pred: Predicting the binding affinity of protein-DNA complexes using machine learning techniques and structural features. *Methods, 213*, 10–17.

Harini, K., Srivastava, A., Kulandaisamy, A., & Gromiha, M. M. (2022). ProNAB: Database for binding affinities of protein-nucleic acid complexes and their mutants. *Nucleic Acids Research, 50*(D1), D1528–D1534.

Hata, A., Shi, Y., & Massagué, J. (1998). TGF-beta signaling and cancer: structural and functional consequences of mutations in Smads. *Molecular Medicine Today, 4*(6), 257–262.

Herman, L., Todeschini, A. L., & Veitia, R. A. (2021). Forkhead transcription factors in health and disease. *Trends in Genetics: TIG, 37*(5), 460–475.

Hubbard, S. J., & Thornton, J. M. (1993). *NACCESS. Department of Biochemistry and Molecular Biology*. University College London.

Humphrey, W., Dalke, A., & Schulten, K. (1996). VMD: visual molecular dynamics. *Journal of Molecular Graphics, 14*(1), 33–28.

Jiang, Y., Liu, H. F., & Liu, R. (2021). Systematic comparison and prediction of the effects of missense mutations on protein-DNA and protein-RNA interactions. *PLoS Computational Biology, 17*(4), e1008951.

Jubb, H. C., Higueruelo, A. P., Ochoa-Montaño, B., Pitt, W. R., Ascher, D. B., & Blundell, T. L. (2017). Arpeggio: A Web Server for Calculating and Visualising Interatomic Interactions in Protein Structures. *Journal of molecular biology*, *429*(3), 365–371.

Kabsch, W., & Sander, C. (1983). Dictionary of protein secondary structure: pattern recognition of hydrogen-bonded and geometrical features. *Biopolymers, 22*(12), 2577–2637.

Kawashima, S., Ogata, H., & Kanehisa, M. (1999). AAindex: Amino acid index database. *Nucleic Acids Research, 27*(1), 368–369.

Ladbury, J. E. (1995). Counting the calories to stay in the groove. *Structure, 3*(7), 635–639.

Laskowski, R. A. (2022). PDBsum1: A standalone program for generating PDBsum analyses. *Protein Science, 31*(12), e4473.

Li, G., Panday, S. K., Peng, Y., & Alexov, E. (2021). SAMPDI-3D: Predicting the effects of protein and DNA mutations on protein-DNA interactions. *Bioinformatics (Oxford, England), 37*(21), 3760–3765.

Li, H., Huang, E., Zhang, Y., Huang, S. Y., & Xiao, Y. (2022). HDOCK update for modeling protein-RNA/DNA complex structures. *Protein Science, 31*(11), e4441.

Li, K., Zhang, S., Yan, D., Bin, Y., & Xia, J. (2020). Prediction of hot spots in protein-DNA binding interfaces based on supervised isometric feature mapping and extreme gradient boosting. *BMC Bioinformatics, 21*(Suppl 13), 381.

Li, M., Shoemaker, B. A., Thangudu, R. R., Ferraris, J. D., Burg, M. B., & Panchenko, A. R. (2013). Mutations in DNA-binding loop of NFAT5 transcription factor produce unique outcomes on protein-DNA binding and dynamics. *The Journal of Physical Chemistry B, 117*(42), 13226–13234.

Li, S., Olson, W. K., & Lu, X. J. (2019). Web 3DNA 2.0 for the analysis, visualization, and modeling of 3D nucleic acid structures. *Nucleic Acids Research, 47*(W1), W26–W34.

Liang, S., Meroueh, S. O., Wang, G., Qiu, C., & Zhou, Y. (2009). Consensus scoring for enriching near-native structures from protein-protein docking decoys. *Proteins, 75*(2), 397–403.

Liu, L., Xiong, Y., Gao, H., Wei, D. Q., Mitchell, J. C., & Zhu, X. (2018). dbAMEPNI: A database of alanine mutagenic effects for protein-nucleic acid interactions. *Database (Oxford), 2018*, bay034.

Liu, X., Tian, W., Cheng, J., Li, D., Liu, T., & Zhang, L. (2020). Microsecond molecular dynamics simulations reveal the allosteric regulatory mechanism of p53 R249S mutation in p53-associated liver cancer. *Computational biology and chemistry*, 84, 107194.

Liu, Z., Li, Y., Han, L., Li, J., Liu, J., Zhao, Z., Nie, W., Liu, Y., & Wang, R. (2015). PDB-wide collection of binding data: current status of the PDBbind database. *Bioinformatics, 31*(3), 405–412.

McDonald, I. K., & Thornton, J. M. (1994). Satisfying hydrogen bonding potential in proteins. *Journal of Molecular Biology, 238*(5), 777–793.

Mei, L. C., Hao, G. F., & Yang, G. F. (2023). Thermodynamic database supports deciphering protein-nucleic acid interactions. *Trends in Biotechnology, 41*(2), 140–143.

Mei, L. C., Wang, Y. L., Wu, F. X., Wang, F., Hao, G. F., & Yang, G. F. (2021). HISNAPI: A bioinformatic tool for dynamic hot spot analysis in nucleic acid-protein interface with a case study. *Briefings in Bioinformatics, 22*(5), bbaa373.

Mihel, J., Sikić, M., Tomić, S., Jeren, B., & Vlahovicek, K. (2008). PSAIA - protein structure and interaction analyzer. *BMC Structural Biology, 8*, 21.

Mullen, J., Kato, S., Sicklick, J. K., & Kurzrock, R. (2021). Targeting ARID1A mutations in cancer. *Cancer Treatment Reviews, 100*, 102287.

Nagarajan, R., Ahmad, S., & Gromiha, M. M. (2013). Novel approach for selecting the best predictor for identifying the binding sites in DNA binding proteins. *Nucleic Acids Research, 41*(16), 7606–7614.

Nagata, T., Niyada, E., Fujimoto, N., Nagasaki, Y., Noto, K., Miyanoiri, Y., Murata, J., Hiratsuka, K., & Katahira, M. (2010). Solution structures of the trihelix DNA-binding domains of the wild-type and a phosphomimetic mutant of Arabidopsis GT-1: Mechanism for an increase in DNA-binding affinity through phosphorylation. *Proteins, 78*(14), 3033–3047.

Neumann, M., Sampathu, D. M., Kwong, L. K., Truax, A. C., Micsenyi, M. C., Chou, T. T., Bruce, J., Schuck, T., Grossman, M., Clark, C. M., McCluskey, L. F., Miller, B. L., Masliah, E., Mackenzie, I. R., Feldman, H., Feiden, W., Kretzschmar, H. A., Trojanowski, J. Q., & Lee, V. M. (2006). Ubiquitinated TDP-43 in frontotemporal lobar degeneration and amyotrophic lateral sclerosis. *Science (New York, N.Y.), 314*(5796), 130–133.

Nichols, N. M., & Matthews, K. S. (2002). Human p53 phosphorylation mimic, S392E, increases nonspecific DNA affinity and thermal stability. *Biochemistry, 41*(1), 170–178.

Novak, D., Hüser, L., Elton, J. J., Umansky, V., Altevogt, P., & Utikal, J. (2020). SOX2 in development and cancer biology. *Seminars in Cancer Biology, 67*(Pt 1), 74–82.

Pal, A., Chakrabarti, P., & Dey, S. (2022). ProDFace: A web-tool for the dissection of protein-DNA interfaces. *Frontiers in Molecular Biosciences, 9*, 978310.

Pan, Y., Zhou, S., & Guan, J. (2020). Computationally identifying hot spots in protein-DNA binding interfaces using an ensemble approach. *BMC Bioinformatics, 21*(Suppl 13), 384.

Peng, Y., Sun, L., Jia, Z., Li, L., & Alexov, E. (2018). Predicting protein-DNA binding free energy change upon missense mutations using modified MM/PBSA approach: SAMPDI webserver. *Bioinformatics (Oxford, England), 34*(5), 779–786.

Pires, D. E., Ascher, D. B., & Blundell, T. L. (2014). mCSM: Predicting the effects of mutations in proteins using graph-based signatures. *Bioinformatics (Oxford, England), 30*(3), 335–342.

Pires, D. E. V., & Ascher, D. B. (2017). mCSM-NA: predicting the effects of mutations on protein-nucleic acids interactions. *Nucleic Acids Research, 45*(W1), W241–W246.

Prabakaran, P., An, J., Gromiha, M. M., Selvaraj, S., Uedaira, H., Kono, H., & Sarai, A. (2001). Thermodynamic database for protein-nucleic acid interactions (ProNIT). *Bioinformatics (Oxford, England), 17*(11), 1027–1034.

Pulice, J. L., & Kadoch, C. (2016). Composition and function of mammalian SWI/SNF chromatin remodeling complexes in human disease. *Cold Spring Harbor Symposia on Quantitative Biology, 81*, 53–60.

Rastogi, C., Rube, H. T., Kribelbauer, J. F., Crocker, J., Loker, R. E., Martini, G. D., Laptenko, O., Freed-Pastor, W. A., Prives, C., Stern, D. L., Mann, R. S., & Bussemaker, H. J. (2018). Accurate and sensitive quantification of protein-DNA binding affinity. *Proceedings of the National Academy of Sciences of the United States of America, 115*(16), E3692–E3701.

Rodrigues, C. H., Pires, D. E., & Ascher, D. B. (2018). DynaMut: predicting the impact of mutations on protein conformation, flexibility and stability. *Nucleic Acids Research, 46*(W1), W350–W355.

Rodriguez-Lumbreras, L. A., Jimenez-Garcia, B., Gimenez-Santamarina, S., Fernandez-Recio, J. (2022). pyDockDNA: A new web server for energy-based protein-DNA docking and scoring. *Frontiers in Molecular Biosciences, 9*, 988996.

Sagendorf, J. M., Markarian, N., Berman, H. M., & Rohs, R. (2020). DNAproDB: An expanded database and web-based tool for structural analysis of DNA-protein complexes. *Nucleic Acids Research, 48*(D1), D277–D287.

Sangermano, F., Delicato, A., & Calabrò, V. (2020). Y box binding protein 1 (YB-1) oncoprotein at the hub of DNA proliferation, damage and cancer progression. *Biochimie, 179*, 205–216.

Santoro, A., Vlachou, T., Luzi, L., Melloni, G., Mazzarella, L., D'Elia, E., Aobuli, X., Pasi, C. E., Reavie, L., Bonetti, P., Punzi, S., Casoli, L., Sabò, A., Moroni, M. C., Dellino, G. I., Amati, B., Nicassio, F., Lanfrancone, L., & Pelicci, P. G. (2019). p53 Loss in breast cancer leads to Myc activation, increased cell plasticity, and expression of a mitotic signature with prognostic value. *Cell Reports, 26*(3), 624–638.e8.

Schutte, M., Hruban, R. H., Hedrick, L., Cho, K. R., Nadasdy, G. M., Weinstein, C. L., Bova, G. S., Isaacs, W. B., Cairns, P., Nawroz, H., Sidransky, D., Casero Jr, R. A., Meltzer, P. S., Hahn, S. A., & Kern, S. E. (1996). DPC4 gene in various tumor types. *Cancer Research, 56*, 2527–253.

Schymkowitz, J., Borg, J., Stricher, F., Nys, R., Rousseau, F., & Serrano, L. (2005). The FoldX web server: An online force field. *Nucleic Acids Research, 33*(Web Server issue), W382–388.

Semenza G. L. (2003). Targeting HIF-1 for cancer therapy. Nature reviews. *Cancer, 3*(10), 721–732.

Shen, J., Peng, Y., Wei, L., Zhang, W., Yang, L., Lan, L., Kapoor, P., Ju, Z., Mo, Q., Shih, I.eM., Uray, I. P., Wu, X., Brown, P. H., Shen, X., Mills, G. B., & Peng, G. (2015). ARID1A Deficiency impairs the DNA damage checkpoint and sensitizes cells to PARP inhibitors. *Cancer Discovery, 5*(7), 752–767.

Shiroma, Y., Takahashi, R. U., Yamamoto, Y., & Tahara, H. (2020). Targeting DNA binding proteins for cancer therapy. *Cancer Science, 111*(4), 1058–1064.

Sun, Y., Wu, H., Xu, Z., Yue, Z., & Li, K. (2023). Prediction of hot spots in protein-DNA binding interfaces based on discrete wavelet transform and wavelet packet transform. *BMC Bioinformatics, 24*(1), 129.

Taniguchi, H., Fujimoto, A., Kono, H., Furuta, M., Fujita, M., & Nakagawa, H. (2018). Loss-of-function mutations in Zn-finger DNA-binding domain of HNF4A cause aberrant transcriptional regulation in liver cancer. *Oncotarget, 9*(40), 26144–26156.

Tuszynska, I., Magnus, M., Jonak, K., Dawson, W., & Bujnicki, J. M. (2015). NPDock: A web server for protein-nucleic acid docking. *Nucleic Acids Research, 43*(w1), w425–430.

Wang, D., Horton, J. R., Zheng, Y., Blumenthal, R. M., Zhang, X., & Cheng, X. (2018). Role for first zinc finger of WT1 in DNA sequence specificity: Denys-Drash syndrome-associated WT1 mutant in ZF1 enhances affinity for a subset of WT1 binding sites. *Nucleic acids research, 46*(8), 3864–3877.

Webb, B., & Sali A. (2014). Protein structure modeling with MODELLER. *Methods in Molecular Biology, 1137*, 1–15.

Wu, K. J., Grandori, C., Amacker, M., Simon-Vermot, N., Polack, A., Lingner, J., & Dalla-Favera, R. (1999). Direct activation of TERT transcription by c-MYC. *Nature Genetics, 21*(2), 220–224.

Wu, Y., Lu, J., & Kang, T. (2016). Human single-stranded DNA binding proteins: Guardians of genome stability. *Acta Biochimica et Biophysica Sinica, 48*(7), 671–677.

Yang, W., & Deng, L. (2020). PreDBA: A heterogeneous ensemble approach for predicting protein-DNA binding affinity. *Scientific Reports, 10*(1), 1278.

Yasen, M., Kajino, K., Kano, S., Tobita, H., Yamamoto, J., Uchiumi, T., Kon, S., Maeda, M., Obulhasim, G., Arii, S., & Hino, O. (2005). The up-regulation of Y-box binding proteins (DNA binding protein A and Y-box binding protein-1) as prognostic markers of hepatocellular carcinoma. *Clinical Cancer Research: An Official Journal of the American Association for Cancer Research, 11*(20), 7354–7361.

Zhang, N., Chen, Y., Zhao, F., Yang, Q., Simonetti, F. L., & Li, M. (2018). PremPDI estimates and interprets the effects of missense mutations on protein-DNA interactions. *PLoS Computational Biology, 14*(12), e1006615.

Zhang, S., Wang, L., Zhao, L., Li, M., Liu, M., Li, K., Bin, Y., & Xia, J. (2021). An improved DNA-binding hot spot residues prediction method by exploring interfacial neighbor properties. *BMC Bioinformatics, 22*(Suppl 3), 253.

Zhang, S., Zhao, L., Zheng, C. H., & Xia, J. (2020). A feature-based approach to predict hot spots in protein-DNA binding interfaces. *Briefings in Bioinformatics, 21*(3), 1038–1046.

Zhang, Y., Bao, W., Cao, Y., Cong, H., Chen, B., & Chen, Y. (2022). A survey on protein-DNA-binding sites in computational biology. *Briefings in Functional Genomics, 21*(5), 357–375.

https://doi.org/10.1142/9789811293269_0007

Chapter 7

Computational resources for understanding the effect of mutations in binding affinities of protein–RNA complexes

K. Harini[1], Sowmya Ramaswamy Krishnan[1], M. Sekijima[2], and M. Michael Gromiha[1,3,*]

[1]*Department of Biotechnology, Bhupat and Jyoti Mehta School of Biosciences, Indian Institute of Technology Madras, Chennai 600036, India*

[2]*Department of Computer Science, Tokyo Institute of Technology, 4259-J3-23, Nagatsuta-cho, Midori-ku, Yokohama 226-8501, Japan*

[3]*International Research Frontiers Initiative, Department of Computer Science, Tokyo Institute of Technology, Yokohama 226-8501, Japan*

Abstract

Protein–nucleic acid interactions play a crucial role in maintaining cellular homeostasis. Quantitatively, these interactions are described in terms of the dissociation constant or free energy change observed during

*Corresponding author
Tel: +914422574138
Fax: +914422574102
MMG: gromiha@iitm.ac.in

protein–nucleic acid complex formation. These interactions are impaired in the presence of mutations affecting the binding affinity of the complexes, in turn leading to numerous diseases. Therefore, understanding how binding affinity changes due to mutations in protein-nucleic acid complexes is vital. While experimental techniques are highly accurate, they are also time-consuming and labor-intensive. On the other hand, computational techniques are emerging as valuable alternatives, with numerous databases and computational tools to study protein–RNA complexes. In this chapter, we discuss various databases that provide information on binding affinities and their changes upon mutation in protein–RNA complexes. Additionally, tools available to extract different structural and interaction features from the complexes are given in detail. Furthermore, we offer insights into prediction methods reported to predict the change in binding affinity upon mutation of protein–RNA complexes. We also cover the existing methods for hotspot residue identification in these complexes.

Keywords: Protein–RNA complexes; binding affinity change upon mutation; prediction methods; computational tools; machine learning; hotspots

7.1 Introduction

Proteins and nucleic acids are the primary biomolecules of the cell. Among the nucleic acids, ribonucleic acids (RNAs) form a highly diverse and versatile ensemble of functional elements with essential roles in the regulation of every stage of the central dogma (Quake, 2021). A predominant fraction of RNA subtypes has been primarily shown to function through its association with a plethora of different proteins, resulting in the formation of protein–RNA complexes. They are known to involve in the control of transcriptional, post-transcriptional, translational, and post-translational processes in a cell (Mukherjee *et al.*, 2019; Jeon *et al.*, 2022). Some of the notable examples of protein–RNA complexes in the cell are the ribosome (Balcerak *et al.*, 2019), spliceosome (Will & Lührmann, 2011), chromatin modification complex (Hendrickson *et al.*, 2016; Statello *et al.*, 2021), and nuclear sub-compartmental elements such as paraspeckles (Bond & Fox, 2009; Fox & Lamond, 2010). Owing to their significance in cell survival, malformations and mutations in protein–RNA complexes have been implicated in several human diseases (Cooper *et al.*, 2009; Zhang *et al.*, 2020). Several *cis*- and *trans*-acting mutations have been identified in RNA-binding proteins, which have a role in diseases including Prader–Willi syndrome (Sledziowska *et al.*, 2023), spinal muscular atrophy (SMA) (Keinath *et al.*, 2021), prostate cancer (Annala *et al.*, 2018), Huntington's disease (Bogomazova *et al.*, 2019), and fragile X syndrome (Malecki *et al.*, 2020). These mutations prevent the formation of the cognate protein–RNA complex by affecting RNA-binding domains, leading to the loss of RNA-mediated regulatory function.

In order to understand the effect of protein–RNA complex formation and mutations in diseases, it is necessary to identify the factors governing protein–RNA interactions. In this regard, measurement of the binding affinity of protein–RNA complexes and the change in affinity upon mutations in RNA-binding protein domains can provide useful insights (Sun *et al.*, 2021). Structure-based features such as the interacting residue–nucleotide pair preference, physicochemical nature of the protein–RNA binding interface, binding hotspots, pathways involved, localization of the complex, and so on can provide a cellular view of the complex and its centrality in the regulatory network. The binding affinity of protein–RNA complexes is usually quantified in terms of the dissociation constant (K_d) and/or the binding free energy (ΔG) (Jarmoskaite *et al.*, 2020). There exist several well-established experimental methods to determine the binding affinity of protein–RNA complexes, including electrophoretic mobility shift assays, isothermal titration calorimetry (ITC), surface plasmon resonance (SPR), biolayer interferometry (BLI), and fluorescence-based methods (Ramanathan *et al.*, 2019; Chopra *et al.*, 2022; Harini *et al.*, 2022; Vega-Badillo *et al.*, 2023). The resultant binding affinities for over 20,000 protein–nucleic acid complexes and their mutants reported in the literature have been curated and made available through several databases, including the ProNAB database, which exists as the currently largest repository of binding affinity information for protein–RNA complexes (Harini *et al.*, 2022).

Although experimental measurement of binding affinity is necessary to validate every protein–RNA interaction conclusively, it is time- and resource-intensive to screen through the vast number of possible interactions through such experiments (Harini *et al.*, 2023). Consequently, the available data on protein–RNA binding affinity should be efficiently leveraged for the development of computational methods to screen, identify, and prioritize potentially novel protein–RNA interactions at scale. The emergence of such *in silico* models will also help accelerate the time to screening and save additional resources invested in the elimination of false positive protein-RNA interactions. Computational methods can also facilitate rapid and exhaustive site-saturation mutagenesis studies, which investigate the effects of mutations on the binding affinity of protein–RNA complexes (Pires *et al.*, 2016; Rizvanovic *et al.*, 2021). This chapter will focus on the development of such computational resources to study protein–RNA interactions and the effects of mutations on protein–RNA complexes, with experimental binding affinities collected from the literature as the primary knowledge base. The overall workflow for explaining the different sections of this chapter is provided in **Figure 7.1.**

7.2 Importance of protein–RNA complexes in diseases

Protein–RNA complexes are involved in the regulation of cellular organization and various stages of the central dogma through association with a multitude of protein partners. Consequently, experimental evidence shows the involvement of almost every protein–RNA interaction in a disease pathway due to the essentiality of the gene expression network for cell survival (Cooper *et al.*, 2009; Barta & Jantsch, 2017). There are several instances of long non-coding RNAs (lncRNAs) and large intergenic non-coding RNAs (lincRNAs) being involved in epigenetic regulation of gene expression, specifically in cancer cells. Due to the ability of protein–RNA complexes to silence gene expression by altering the methylation status of targeted genomic loci, tumorigenesis can be induced in several cancer types, such as breast, gastric, colorectal, and cervical cancers. While antisense RNA-mediated chromatin modifications can operate only in *cis*,

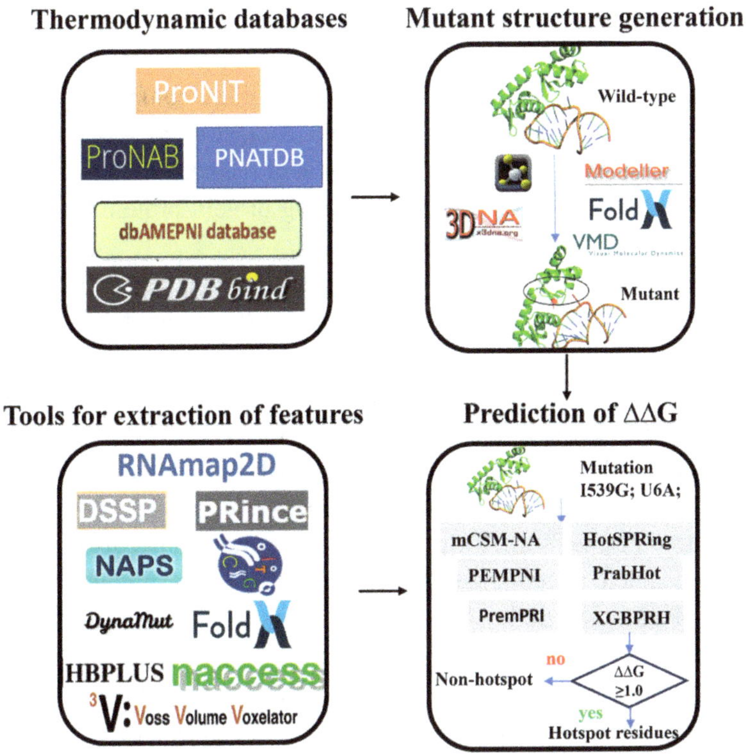

Figure 7.1 Overall workflow for understanding mutational effects in protein–RNA complexes

lincRNA-mediated modifications can affect genes in both *cis* and *trans* fashion (Khalil & Rinn, 2011). For example, the growth arrest-specific transcript 5 (GAS5) is a lncRNA with multi-faceted regulatory roles in the cell involving proliferation, apoptosis, invasion, and metastasis in more than eight human cancers (Lambrou *et al.*, 2020).

In case of gender-biased diseases such as autoimmune diseases, gonadotrophic cancers, and neurodegenerative diseases, the primary cause is the dysregulation of gene expression dosage levels. In humans, the gene expression dosage is equalized between males and females through the X-inactive specific transcript (XIST), a non-coding RNA that controls the random inactivation of one of the copies of the X chromosome in female placental mammals. Similarly, several protein–RNA complexes play a role in the regulation of gene expression in several cancers, autoimmune diseases, and neurodegenerative diseases. Hence, it is highly necessary to understand the nature of protein–RNA complexes and their dynamic rearrangements from a structural context to be able to identify therapeutic opportunities through disruption of the protein–RNA complex formation, similar to interventions involving protein-protein interactions in therapy.

Recently, therapeutics such as Sotorasib (Lumakras) and adagrasib (Krazati) have been designed to target the G12C mutation of the *KRAS* gene, resulting in a protein isoform unique to non-small cell lung cancer patients (Lee, 2022; Jänne *et al.*, 2022). Similarly, Osimertinib (Tagrisso) is an inhibitor targeting the T790M mutation of the *EGFR* gene in cancers (Wu *et al.*, 2020), while Mobocertinib (Exkivity) is a drug that targets an exon 20 mutation in the same gene (Campelo *et al.*, 2022). Entire monoclonal antibodies such as Amivantamab (Rybrevant) have also been specifically designed to recognize and target mutated *EGFR* proteins in cancer subtypes (Park *et al.*, 2021). The development and FDA approval of more such drugs with specificity toward mutated proteins involved in diseases, provides further emphasis on the necessity to map all possible mutations in disease-associated biomolecules and identify their phenotypes.

7.3 Analysis of the effect of mutations in protein-RNA complexes

7.3.1 *Mutations extend the binding site with new interactions*

The MS2 protein of the bacteriophage is a structural coat protein and a translational repressor that interacts with the RNA stem-loop region to inhibit translation of the viral replicase. The V29I mutation in the MS2 coat protein enhanced the affinity of the protein, compared to the wild-type with the $\Delta\Delta G$ of −1.1. kcal/mol and it did not affect its ability to form capsid. In contrast, it extended the binding sites by introducing weak additional interactions with RNA, thereby increasing the affinity without affecting the interactions with other proteins (Lim & Peabody, 1994).

7.3.2 *Protein mutations impair RNA binding function leading to protein aggregation*

The P112H mutation in the RNA recognition motif (RRM) domain of the TDP-43 protein contributes to frontotemporal dementia (Agrawal *et al.*, 2021). The mutation causes conformational changes that affect the stacking interactions between the W113 residue and the RNA partner. Another pair of mutations namely, K181E and K263E, located close to the RRM domain, impairs their interactions with RNA, accelerating protein aggregation and leading to ALS/FTD disease pathogenesis (Chen *et al.*, 2019).

7.3.3 *Mutation enhances ribosomal resistance to erythromycin*

Thermus thermophilus ribosomal protein L22 interacts with all domains of the 23S rRNA. Deletion of three consecutive residues (L82, K83, R84) in the L22 protein confers resistance to erythromycin (Davydova *et al.*, 2002). Deletion of these residues affects the interaction between the L22 protein and 23S rRNA, as the β-hairpin bends inward the ribosome tunnel, modifying the shape of its narrowest part. The mutation in the L22 β-hairpin affects the orientation and distance between the interacting nucleotides. This destabilizes the erythromycin-binding "pocket" formed by 23S rRNA nucleotides exposed at the tunnel surface, leading to drug resistance.

7.3.4 *Mutation in RRM reduces the binding to RNA*

The poly(A)-binding protein, pab1, of *Saccharomyces cerevisiae* contains four tandem RRM domains with proline-rich linkers. Single point mutations in one of the RRM domains, expected to disturb the RNA binding ability, were suppressed by the redundant function of the other three RRM domains in the protein. Mutation F170V in the RRM2 domain reduced the protein binding to a poly(A) RNA by greater than 97%, and the K166Q mutation in the RRM2 domain, along with similar mutation in the other three RRMs, reduced binding to the poly(A) RNA by greater than 70% compared to the wild-type protein (Melamed *et al.*, 2013).

7.3.5 *Mutation misregulates splicing by altering RNA binding affinities*

Serine/arginine-rich splicing factor 2 (SRSF2) is important in mRNA precursor splicing. Mutations in this protein are found mostly in myelodysplastic syndromes and certain subtypes of leukemia (Zhang *et al.*, 2015). The P95H mutation in the SRSF2 protein is reported to misregulate 548 splicing events, including exon inclusion (with the UCCA/UG motif) and exclusion (with the UGGA/UG motif), correlated with stronger or weaker RNA binding, respectively. The mutated residue forms a hydrogen bond with the second cytosine in the UCCA/UG motif of one RNA monomer, whereas in the other monomer, the second guanine is in *syn* conformation. Hence, the histidine cannot form hydrogen bonds, thus reducing the affinity to SRSF2.

Similarly, mutations in protein–RNA complexes have also been implicated in targeted inhibition of oncogenic miRNA and interaction between viral RNA and protein components. From the above cases, it is clear that mutations can affect the binding affinity of protein–RNA complexes drastically, leading to numerous disease conditions. Hence, it is important to understand and analyze the effect of mutations on the binding affinity of protein-RNA complexes. **Figure 7.2** illustrates the influence of mutations on the binding of protein–RNA complexes.

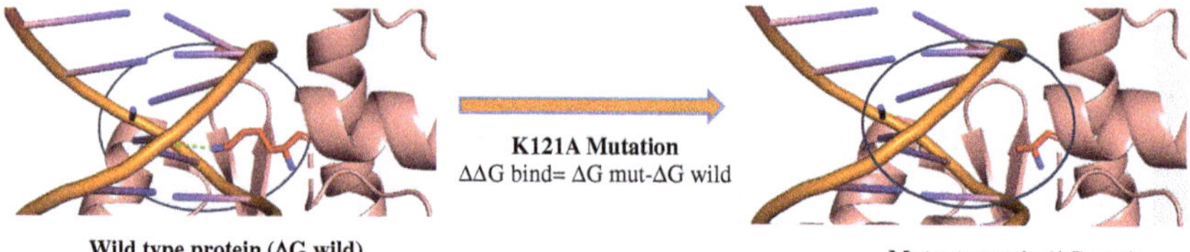

Figure 7.2 Representation of the effect of mutation in Protein-RNA complexes

7.4 Databases for protein–RNA binding affinity change upon mutation

The thermodynamic databases available for studying protein–nucleic acid complexes are listed in **Table 7.1**. These databases provide the experimentally determined binding affinity (ΔG) and changes in binding affinity upon mutations (ΔΔG) of protein-nucleic acid complexes. ProNIT (Prabakaran *et al.*, 2001) is the first developed thermodynamic database with protein/RNA sequences, structures, and mutational details. Thermodynamic parameters such as binding affinity and free energy are also provided for every entry. dbAMEPNI (Liu *et al.*, 2018) is only specific for alanine mutations, while the PDBbind database (Liu *et al.*, 2015)

Table 7.1 Thermodynamic databases of protein-nucleic acid complexes

Database	Description	Number of data	Link	References
ProNIT	Experimentally determined thermodynamic interaction data	12,174	http://dna00.bio.kyutech.ac.jp/pronit/	Prabakaran *et al.* (2001)
PDBbind	Binding affinity of biomolecular complexes with known three-dimensional structures	1052	http://www.pdbbind.org.cn/	Liu *et al.* (2015)
dbAMEPNI	Alanine mutagenic effects on protein–nucleic acid interactions with affinity change upon mutation	578	http://zhulab.ahu.edu.cn/dbAMEPNI	Liu *et al.* (2018)
ProNAB	Binding affinities of protein–nucleic acid complexes and their mutants with binding affinity change upon mutation	20,219	https://web.iitm.ac.in/bioinfo2/pronab/	Harini *et al.* (2022)
PNATDB	Binding affinities of protein–nucleic acid interactions with molecular interactions	12,635	http://chemyang.ccnu.edu.cn/ccb/database/PNAT/	Mei *et al.* (2023)

has affinity values only if PDB structures of the protein–RNA complex are known. ProNAB (Harini *et al.*, 2022) is currently the largest available database with protein and nucleic acid sequence, and structural information mapped to numerous databases such as UniProt, PDB, NDB, Genbank, Prosite, and ProTherm. PNATDB (Mei *et al.*, 2023) contains thermodynamic values and additional details about molecular interactions in complexes. These databases are great resources for extracting structural features of protein–RNA complexes and for classifying and training predictive models to estimate the binding affinity of protein–RNA complexes.

7.5 Tools for extracting features from protein–RNA complex structures

Several tools, such as "BuildModel" of FoldX (Schymkowitz *et al.*, 2005), Modeller (Webb & Sali, 2014), VMD (Humphrey *et al.*, 1996), and PyMOL (DeLano, 2002) are available for generating mutant protein structures. Further, protein–RNA complex structures can be modeled using docking software such as NPDock (Tuszynska *et al.*, 2015), and Tfmodeller (Contreras-Moreira *et al.*, 2007).

Several tools are reported in the literature for calculating features from protein–RNA complexes. **Table 7.2** lists the tools available for extracting features from protein–RNA complexes. DSSP (Kabsch & Sandor, 1983), NACCESS (Hubbard & Thornton, 1993), and HBPLUS (McDonald & Thornton, 1994) provide details on the secondary structure of residues, atom/residue-wise accessible surface area, and hydrogen bonds in the complex. The residue interaction network parameters are extracted using NAPS (Chakrabarty *et al.*, 2016) and FoldX (Schymkowitz *et al.*, 2005), which provide the predicted free energy of the complex along with protein stability and other energy-based parameters. Curves+ (Blanchet *et al.*, 2011) and w3DNA (Li *et al.*, 2019) provide structural parameters of the nucleic acids.

Focusing on tools specific to protein–RNA complexes, RNAmap2D provides information about interactions between proteins and RNA along with contact maps (Pietal *et al.*, 2012). It is available at http://iimcb.genesilico.pl/rnamap2d.html. 3V is a cavity, channel, cleft volume calculator,

Table 7.2 Tools for extracting residue-wise features in protein–RNA complexes

Tool	Description	Link	References
DSSP	Secondary structure, geometrical features, and solvent exposure of proteins	https://www3.cmbi.umcn.nl/xssp/	Kabsch and Sander (1983)
Naccess	Atomic and residue-wise solvent accessibility	http://www.bioinf.manchester.ac.uk/naccess/	Hubbard and Thornton (1993)
HBPLUS	Details of hydrogen bonds	https://www.ebi.ac.uk/thornton-srv/software/HBPLUS/	McDonald and Thornton (1994)
NAPS	Analysis of residue interaction network in proteins	http://bioinf.iiit.ac.in/NAPS/	Chakrabarty *et al.* (2016)
FoldX	Interaction energies of the complex, stability of a protein, etc.	http://foldx.embl.de/	Schymkowitz *et al.* (2005)
RNAmap2D	Stand-alone program for calculation, visualization, and analysis of contact and distance maps for RNA and protein–RNA complex structures	https://genesilico.pl/software/stand-alone/rnamap2d	Pietal *et al.* (2012)
3V: webserver	Internal volumes from RNA and protein structures	http://3vee.molmovdb.org/	Voss *et al.* (2010)
PRince	Solvent accessible surface area, hydration, and structural and physicochemical properties of the protein–RNA interface.	http://www.facweb.iitkgp.ac.in/~rbahadur/prince/home.html	Barik *et al.* (2012)

and extractor (Voss *et al.*, 2010), which automatically extracts and analyzes all the internal volumes of RNA and protein structures by taking the difference between two rolling probes. The server is available at http://3vee.molmovdb.org/. PRince is a web server for analyzing the interface of protein–RNA complexes, which provides solvent-accessible surface area and structural, physicochemical, and hydration properties of the interface (Barik *et al.*, 2012). It is available at http://www.facweb.iitkgp.ernet.in/~rbahadur/prince/home.html.

7.6 Prediction of protein–RNA binding affinity change upon mutations

This section outlines the methods available for the prediction of changes in binding affinity resulting from mutations. **Table 7.3** enumerates the binding affinity change prediction methods applicable to protein–RNA complexes.

mCSM-NA (Pires & Ascher, 2017) employs graph-based signatures extracted from the wild-type protein and nucleic acid structures, eliminating the need for mutant structure modeling. The method is available at https://structure.bioc.cam.ac.uk/mcsm_na. **PEMPNI** (Jiang *et al.*, 2021) is a random forest-based prediction technique that utilizes energy-based descriptors and various non-energy-based features, including solvent accessibilities, interactions, and evolutionary data. The server can be accessed at http://liulab.hzau.edu.cn/PEMPNI.

PremPRI (Zhang *et al.*, 2020) is a multiple linear regression-based model using structure and sequence features specific to protein-RNA complexes. The model uses various energy-based features, such as differences in van

Table 7.3 Methods to predict the effect of mutation on protein–RNA binding affinity

Tool	Features	Link	References
mCSM-NA	Graph-based signatures	structure.bioc.cam.ac.uk/mcsm_na	Pires and Ascher (2017)
PEMPNI	Geometric partition-based energy and interface structural features	http://liulab.hzau.edu.cn/PEMPNI	Jiang *et al.* (2021)
PremPRI	Interface interactions and graph-based features	https://lilab.jysw.suda.edu.cn/research/PremPRI/.	Zhang *et al.* (2020)

der Waals interaction energies, repulsive energies, and electrostatic interaction energies between the wild-type and mutant proteins, computed using the CHARMM program (MacKerell *et al.*, 1998). In addition, amino acids at the interface, ASA features, hydrophobicity, and the closeness of the nodes in the residue interaction network are also used for prediction. PremPRI is freely available at https://lilab.jysw.suda.edu.cn/research/PremPRI/.

7.7 Prediction of hotspot residues in protein–RNA complexes

This section outlines the methods that can identify the hotspot residues at the interface of protein–RNA complexes. A residue is considered as a hotspot, if the binding free energy change upon mutation (ΔΔG) is more than 1.0 kcal/mol. **Table 7.4** lists the available methods for identifying hotspot residues.

HotSPRing (Barik *et al.*, 2016) is a web server for predicting the effect of mutations in five ranges of ΔΔG (<−1.0, −1 to 0.2, 0.2 to 1, 1 to 2, >2 kcal/mol.) using a random forest-based classifier. It includes structure-based and physicochemical features such as solvent-accessible surface area and conservation for the classification. The web server is available at http://www.csb.iitkgp.ernet.in/applications/HotSPRing/main.

Table 7.4 Methods to identify hotspot residues in protein–RNA complexes

Tool	Method and features	Link	References
HotSPRing	Random forest-based classifier using structural and physicochemical features	http://www.csb.iitkgp.ernet.in/applications/HotSPRing/main.	Barik *et al.* (2016)
PrabHot	Network, solvent, sequence, and structure-based features using an ensemble voting classifier	http://denglab.org/PrabHot/.	Pan *et al.* (2018)
XGBPRH	XGBoost-based classification using interface features.	https://github.com/SupermanVip/XGBPRH.	Deng *et al.* (2019)

PrabHot (Pan *et al.*, 2018) predicts hotspots in protein–RNA complex structures using sequence, structure, solvent, and network-based features and an ensemble voting classifier. The web server is accessible at http://denglab.org/PrabHot/.

XGBPRH (Deng *et al.*, 2019) uses network-based centralities, solvent-based properties, protrusion, depth index, and hydrophobicity of the residues using the extreme gradient boosting (XGBoost) algorithm for identifying hotspot residues. The method is available at https://github.com/SupermanVip/XGBPRH.

7.8 Conclusions

The development of computational tools for analyzing protein–nucleic acid interactions is crucial for gaining insights into the mechanisms underlying these interactions, which in turn can aid in the development of therapeutic strategies. In this chapter, we have examined several databases cataloging the binding affinity upon mutations in protein-RNA complex structures. Further, various computational tools that enable the extraction of residue-specific features from proteins, RNA, or features related to their interactions are discussed. Moreover, we have provided an overview of the algorithms currently accessible for predicting the alterations in binding affinity due to mutations in protein–RNA complexes. Additionally, we have explored different classification methods that can be employed to identify hotspot residues in RNA-binding proteins.

Acknowledgments

The authors acknowledge the Indian Institute of Technology Madras and the High-Performance Computing Environment (HPCE) for computational facilities. We thank the members of the Protein Bioinformatics Lab for providing valuable suggestions. The work is partially supported by the Science and Engineering Research Board (SERB), Ministry of Science and Technology, Government of India to MMG (No. CRG/2020/000314).

References

Agrawal, S., Jain, M., Yang, W. Z., & Yuan, H. S. (2021). Frontotemporal dementia-linked P112H mutation of TDP-43 induces protein structural change and impairs its RNA binding function. *Protein Science: A Publication of the Protein Society, 30*(2), 350–365.

Annala, M., Taavitsainen, S., Vandekerkhove, G., Bacon, J. V. W., Beja, K., Chi, K. N., Nykter, M., & Wyatt, A. W. (2018). Frequent mutation of the *FOXA1* untranslated region in prostate cancer. *Communications Biology, 1*, 122.

Balcerak, A., Trebinska-Stryjewska, A., Konopinski, R., Wakula, M., & Grzybowska, E. A. (2019). RNA-protein interactions: Disorder, moonlighting and junk contribute to eukaryotic complexity. *Open Biology, 9*(6), 190096.

Barik, A., Mishra, A., & Bahadur, R. P. (2012). PRince: A web server for structural and physico-chemical analysis of protein-RNA interface. *Nucleic Acids Research, 40*, W440–444.

Barik, A., Nithin, C., Karampudi, N. B., Mukherjee, S., & Bahadur, R. P. (2016). Probing binding hot spots at protein-RNA recognition sites. *Nucleic Acids Research, 44*(2), e9.

Barta, A., & Jantsch, M. F. (2017). RNA in disease and development. *RNA Biology, 14*(5), 457–459.

Blanchet, C., Pasi, M., Zakrzewska, K., & Lavery, R. (2011). CURVES+ web server for analyzing and visualizing the helical, backbone and groove parameters of nucleic acid structures. *Nucleic Acids Research,* W68–73.

Bogomazova, A. N., Eremeev, A. V., Pozmogova, G. E., & Lagarkova, M. A. (2019). The Role of mutant RNA in the pathogenesis of Huntington's disease and other polyglutamine diseases. *Molekuliarnaia biologiia, 53*(6), 954–967.

Bond, C. S., & Fox, A. H. (2009). Paraspeckles: Nuclear bodies built on long noncoding RNA. *Journal of Cell Biology, 186*(5), 637–644.

Campelo, M. R.G, Zhou, C., Ramalingam, S. S., Lin, H. M., Kim, T. M., Riely, G. J., Mekhail, T., Nguyen, D., Goodman, E., Mehta, M., Popat, S., & Jänne, P. A. (2022). Mobocertinib (TAK-788) in EGFR Exon 20 Insertion+ Metastatic NSCLC: Patient-Reported Outcomes from EXCLAIM Extension Cohort. *Journal of clinical medicine, 12*(1), 112.

Chakrabarty, B., & Parekh, N. (2016). NAPS: Network analysis of protein structures. *Nucleic Acids Research, 44*(W1), W375–W382.

Chen, H. J., Topp, S. D., Hui, H. S., Zacco, E., Katarya, M., McLoughlin, C., King, A., Smith, B. N., Troakes, C., Pastore, A., & Shaw, C. E. (2019). RRM adjacent TARDBP mutations disrupt RNA binding and enhance TDP-43 proteinopathy. *Brain: A Journal of Neurology, 142*(12), 3753–3770.

Chopra, A., Balbous, F., & Biggar, K. K. (2022). Assessing the in vitro binding affinity of protein-RNA interactions using an RNA pull-down technique. *Bio-Protocol, 12*(23), e4560.

Contreras-Moreira, B., Branger, P. A., & Collado-Vides, J. (2007). TFmodeller: Comparative modelling of protein-DNA complexes. *Bioinformatics, 23*(13), 1694–1696.

Cooper, T. A., Wan, L., & Dreyfuss, G. (2009). RNA and disease. *Cell, 136*(4), 777–793.

Davydova, N., Streltsov, V., Wilce, M., Liljas, A., & Garber, M. (2002). L22 ribosomal protein and effect of its mutation on ribosome resistance to erythromycin. *Journal of Molecular Biology, 322*(3), 635–644.

DeLano, W. L. (2002). *The PyMOL Molecular Graphics System.* DeLano Scientific. https://www.pymol.org

Deng, L., Sui, Y., & Zhang, J. (2019). XGBPRH: Prediction of binding hot spots at protein RNA interfaces utilizing extreme gradient boosting. *Genes, 10*(3), 242.

Fox, A. H., & Lamond, A. I. (2010). Paraspeckles. *Cold Spring Harbor Perspectives in Biology, 2*(7), a000687.

Harini, K., Kihara, D., & Michael Gromiha, M. (2023). PDA-Pred: Predicting the binding affinity of protein-DNA complexes using machine learning techniques and structural features. *Methods (San Diego, Calif.), 213*, 10–17.

Harini, K., Srivastava, A., Kulandaisamy, A., & Gromiha, M. M. (2022). ProNAB: Database for binding affinities of protein-nucleic acid complexes and their mutants. *Nucleic Acids Research, 50*(D1), D1528–D1534.

Hendrickson, D., Kelley, D. R., Tenen, D., Bernstein, B., & Rinn, J. L. (2016). Widespread RNA binding by chromatin-associated proteins. *Genome Biology, 17*, 28.

Hubbard, S. J., & Thornton, J. M. (1993). *NACCESS.* Department of Biochemistry and Molecular Biology, University College London.

Humphrey, W., Dalke, A., & Schulten, K. (1996). VMD: Visual molecular dynamics. *Journal of Molecular Graphics, 14*(1), 33–28.

Jänne, P. A., Riely, G. J., Gadgeel, S. M., Heist, R. S., Ou, S. I., Pacheco, J. M., Johnson, M. L., Sabari, J. K., Leventakos, K., Yau, E., Bazhenova, L., Negrao, M. V., Pennell, N. A., Zhang, J., Anderes, K., Der-Torossian, H., Kheoh, T., Velastegui, K., Yan, X., Christensen, J. G., . . . Spira, A. I. (2022). Adagrasib in non-small-cll lung cancer harboring a *KRASG12C* mutation. *New England Journal of Medicine, 387*(2), 120–131.

Jarmoskaite, I., AlSadhan, I., Vaidyanathan, P. P., & Herschlag, D. (2020). How to measure and evaluate binding affinities. *eLife, 9*, e57264.

Jeon, P., Ham, H. J., Park, S., & Lee, J. A. (2022). Regulation of cellular ribonucleoprotein granules: From assembly to degradation via post-translational modification. *Cells, 11*(13), 2063.

Jiang, Y., Liu, H. F., & Liu, R. (2021). Systematic comparison and prediction of the effects of missense mutations on protein-DNA and protein-RNA interactions. *PLoS Computational Biology, 17*(4), e1008951.

Kabsch, W., & Sander, C. (1983). Dictionary of protein secondary structure: Pattern recognition of hydrogen-bonded and geometrical features. *Biopolymers, 22*(12), 2577–2637.

Keinath, M. C., Prior, D. E., & Prior, T. W. (2021). Spinal muscular atrophy: Mutations, testing, and clinical relevance. *Application of Clinical Genetics, 14*, 11–25.

Khalil, A. M., & Rinn, J. L. (2011). RNA-protein interactions in human health and disease. *Seminars in Cell & Developmental Biology, 22*(4), 359–365.

Lambrou, G. I., Hatziagapiou, K., & Zaravinos, A. (2020). The non-coding RNA *GAS5* and its role in tumor therapy-induced resistance. *International Journal of Molecular Sciences, 21*(20), 7633.

Lee, A. (2022). Sotorasib: A review in KRAS G12C mutation-positive non-small cell lung cancer. *Targeted Oncology, 17*(6), 727–733.

Li, S., Olson, W. K., & Lu, X. J. (2019). Web 3DNA 2.0 for the analysis, visualization, and modeling of 3D nucleic acid structures. *Nucleic Acids Research, 47*(W1), W26–W34.

Lim, F., & Peabody, D. S. (1994). Mutations that increase the affinity of a translational repressor for RNA. *Nucleic Acids Research, 22*(18), 3748–3752.

Liu, L., Xiong, Y., Gao, H., Wei, D. Q., Mitchell, J. C., & Zhu, X. (2018). dbAMEPNI: A database of alanine mutagenic effects for protein-nucleic acid interactions. *Database (Oxford), 2018.*

Liu, Z., Li, Y., Han, L., Li, J., Liu, J., Zhao, Z., Nie, W., Liu, Y., & Wang, R. (2015). PDB-wide collection of binding data: Current status of the PDBbind database. *Bioinformatics, 31*(3), 405–412.

MacKerell, A. D., Bashford, D., Bellott, M., Dunbrack, R. L., Evanseck, J. D., Field, M. J., Fischer, S., Gao, J., Guo, H., Ha, S., Joseph-McCarthy, D., Kuchnir, L., Kuczera, K., Lau, F. T., Mattos, C., Michnick, S., Ngo, T., Nguyen, D. T., Prodhom, B., Reiher, W. E., . . . Karplus, M. (1998). All-atom empirical potential for molecular modeling and dynamics studies of proteins. *Journal of Physical Chemistry B, 102*(18), 3586–3616.

Malecki, C., Hambly, B. D., Jeremy, R. W., & Robertson, E. N. (2020). The RNA-binding fragile-X mental retardation protein and its role beyond the brain. *Biophysical Reviews, 12*(4), 903–916.

McDonald, I. K., & Thornton, J. M. (1994). Satisfying hydrogen bonding potential in proteins. *Journal of Molecular Biology, 238*(5), 777–793.

Mei, L. C., Hao, G. F., & Yang, G. F. (2023). Thermodynamic database supports deciphering protein-nucleic acid interactions. *Trends Biotechnology, 41*(2), 140–143.

Melamed, D., Young, D. L., Gamble, C. E., Miller, C. R., & Fields, S. (2013). Deep mutational scanning of an RRM domain of the Saccharomyces cerevisiae poly(A)-binding protein. *RNA (New York, N.Y.), 19*(11), 1537–1551.

Mukherjee, N., Wessels, H. H., Lebedeva, S., Sajek, M., Ghanbari, M., Garzia, A., Munteanu, A., Yusuf, D., Farazi, T., Hoell, J. I., Akat, K. M., Akalin, A., Tuschl, T., & Ohler, U. (2019). Deciphering human ribonucleoprotein regulatory networks. *Nucleic Acids Research, 47*(2), 570–581.

Pan, Y., Wang, Z., Zhan, W., & Deng, L. (2018). Computational identification of binding energy hot spots in protein-RNA complexes using an ensemble approach. *Bioinformatics (Oxford, England), 34*(9), 1473–1480.

Park, K., Haura, E. B., Leighl, N. B., Mitchell, P., Shu, C. A., Girard, N., Viteri, S., Han, J. Y., Kim, S. W., Lee, C. K., Sabari, J. K., Spira, A. I., Yang, T. Y., Kim, D. W., Lee, K. H., Sanborn, R. E., Trigo, J., Goto, K., Lee, J. S., Yang, J. C., … Cho, B. C. (2021). Amivantamab in EGFR Exon 20 insertion-mutated non-small-cell lung cancer progressing on platinum chemotherapy: Initial results from the CHRYSALIS phase I study. *Journal of Clinical Oncology: Official Journal of the American Society of Clinical Oncology, 39*(30), 3391–3402.

Pietal, M. J., Szostak, N., Rother, K. M., & Bujnicki, J. M. (2012). RNAmap2D — calculation, visualization and analysis of contact and distance maps for RNA and protein-RNA complex structures. *BMC Bioinformatics, 13*, 333.

Pires, D. E., Chen, J., Blundell, T. L., & Ascher, D. B. (2016). In silico functional dissection of saturation mutagenesis: Interpreting the relationship between phenotypes and changes in protein stability, interactions and activity. *Scientific Reports, 6*, 19848.

Pires, D. E. V., & Ascher, D. B. (2017). mCSM-NA: Predicting the effects of mutations on protein-nucleic acids interactions. *Nucleic Acids Research, 45*(W1), W241–W246.

Prabakaran, P., An, J., Gromiha, M. M., Selvaraj, S., Uedaira, H., Kono, H., & Sarai, A. (2001). Thermodynamic database for protein-nucleic acid interactions (ProNIT). *Bioinformatics, 17*(11), 1027–1034.

Quake, S. R. (2021). The cell as a bag of RNA. *Trends in Genetics, 37*(12), 1064–1068.

Ramanathan, M., Porter, D. F., & Khavari, P. A. (2019). Methods to study protein-RNA interactions. *Nature Methods, 16*(3), 225–234.

Rizvanovic, A., Kjellin, J., Söderbom, F., & Holmqvist, E. (2021). Saturation mutagenesis charts the functional landscape of Salmonella ProQ and reveals a gene regulatory function of its C-terminal domain. *Nucleic Acids Research, 49*(17), 9992–10006.

Schymkowitz, J., Borg, J., Stricher, F., Nys, R., Rousseau, F., & Serrano, L. (2005). The FoldX web server: An online force field. *Nucleic Acids Research, 33* (Web Server issue), W382–388.

Sledziowska, M., Winczura, K., Jones, M., Almaghrabi, R., Mischo, H., Hebenstreit, D., Garcia, P., & Grzechnik, P. (2023). Non-coding RNAs associated with Prader-Willi syndrome regulate transcription of neurodevelopmental genes in human induced pluripotent stem cells. *Human Molecular Genetics, 32*(4), 608–620.

Statello, L., Guo, C. J., Chen, L. L., & Huarte, M. (2021). Gene regulation by long non-coding RNAs and its biological functions. *Nature Reviews. Molecular Cell Biology, 22*(2), 96–118.

Sun, L., Xu, K., Huang, W., Yang, Y. T., Li, P., Tang, L., Xiong, T., & Zhang, Q. C. (2021). Predicting dynamic cellular protein-RNA interactions by deep learning using in vivo RNA structures. *Cell Research, 31*(5), 495–516.

Tuszynska, I., Magnus, M., Jonak, K., Dawson, W., & Bujnicki, J. M. (2015). NPDock: A web server for protein-nucleic acid docking. *Nucleic Acids Research, 43*(W1), W425–430.

Vega-Badillo, J., Zamore, P. D., & Jouravleva, K. (2023). Protocol to measure protein-RNA binding using double filter-binding assays followed by phosphorimaging or high-throughput sequencing. *STAR Protocols, 4*(2), 102336. Advance online publication.

Voss, N. R., & Gerstein, M. (2010). 3V: Cavity, channel and cleft volume calculator and extractor. *Nucleic Acids Research, 38*(Web Server issue), W555–562.

Webb, B., & Sali, A. (2014). Protein structure modeling with MODELLER. *Methods Molecular Biology, 1137*, 1–15.

Will, C. L., & Lührmann, R. (2011). Spliceosome structure and function. *Cold Spring Harbor Perspectives in Biology, 3*(7), a003707.

Wu, Y. L., Tsuboi, M., He, J., John, T., Grohe, C., Majem, M., Goldman, J. W., Laktionov, K., Kim, S. W., Kato, T., Vu, H. V., Lu, S., Lee, K. Y., Akewanlop, C., Yu, C. J., de Marinis, F., Bonanno, L., Domine, M., Shepherd, F. A.,

Zeng, L., … ADAURA Investigators. (2020). Osimertinib in resected *EGFR*-mutated non-small-cell lung cancer. *New England Journal of Medicine, 383*(18), 1711–1723.
Zhang, J., Lieu, Y. K., Ali, A. M., Penson, A., Reggio, K. S., Rabadan, R., Raza, A., Mukherjee, S., & Manley, J. L. (2015). Disease-associated mutation in SRSF2 misregulates splicing by altering RNA-binding affinities. *Proceedings of the National Academy of Sciences of the United States of America, 112*(34), E4726–E4734.
Zhang, N., Lu, H., Chen, Y., Zhu, Z., Yang, Q., Wang, S., & Li, M. (2020). PremPRI: Predicting the effects of missense mutations on protein-RNA interactions. *International Journal of Molecular Sciences, 21*(15), 5560.

https://doi.org/10.1142/9789811293269_0008

Chapter 8

Computational analysis on the effect of mutations for the binding affinity of protein–carbohydrate complexes

N. R. Siva Shanmugam[1,*], S. Lekshmi[1], and M. Michael Gromiha[1,*]

[1]*Department of Biotechnology, Bhupat and Jyoti Mehta School of Biosciences, Indian Institute of Technology Madras, Chennai, Tamil Nadu 600036, India*

Abstract

Carbohydrates are known as sugars or saccharides, which range from simple sugars (glucose) to complex polysaccharides (cellulose). Proteins that interact with carbohydrates are called carbohydrate-binding proteins, and the interaction between proteins and carbohydrates is important for several biological processes, such as cell–cell adhesion, immune response, pathogen recognition, and enzyme catalysis. Mutations in these proteins may impact their structure, binding affinity and function, and understanding the mutational effects on protein–carbohydrate interactions is essential for elucidating their molecular mechanisms. In this chapter, we provide a comprehensive overview of (i) databases available

*Corresponding authors
NRSS: nrsivashanmugam@gmail.com
MMG: gromiha@iitm.ac.in

for binding affinity of protein–carbohydrate complexes and their mutants, (ii) analysis and prediction of mutational effects on binding affinity of protein–carbohydrate complexes, (iii) effects of mutation at the interaction sites based on diseases, and (iv) potential applications. The information provided in this chapter will provide insights to understand mutational effects on binding affinity and disease-causing mutations in protein–carbohydrate complexes and design therapeutic strategies.

8.1 Introduction

The biological realm is profoundly shaped by a quartet of molecules: proteins, carbohydrates, nucleic acids, and lipids (Kanie & Kanie, 2017). Within this ensemble, carbohydrates emerge as wonders, existing in various forms as monosaccharides, oligosaccharides, and polysaccharides and interacting or binding with various other biological compounds. They engage in molecular interactions with proteins as glycoproteins, with lipids as glycolipids, and with nucleic acids as ribose or deoxyribose sugars. The binding between protein and carbohydrate is mainly governed by noncovalent interactions featuring electrostatic, van der Waals forces, hydrogen bonds, and hydrophobic interactions. Shanmugam *et al.* (2018) analyzed interactions between proteins and carbohydrates and identified key residues that are involved in both the binding and stability of protein–carbohydrate complexes.

Functionally, carbohydrates act as receptors for toxins and viruses, guide cell adhesion and trafficking, adapt the intricate interactions of egg and sperm, facilitate the adhesion of bacteria, fungi, and parasites, and cell–matrix interaction. Carbohydrates also play the essential role of clearing away damaged glycoconjugates and cells, acting as messengers in signal transduction, orchestrating the recognition between hosts and pathogens, and inflammation (del Carmen Fernandez-Alonso *et al.*, 2012; Varki, 2017). Nature has inherently designed sugar molecules to facilitate a multitude of complex biological recognition processes, leveraging the remarkable ability to generate diverse structures through the strategic manipulation of anomers, glycosidic linkages, and branching, even with a limited number of sugar molecules.

Carbohydrates are omnipresent in biological systems, and they play a profoundly intriguing and multifaceted role both intracellularly and on the cellular periphery. Carbohydrate moieties, when bound to proteins on the cell surface, facilitate intricate interactions with toxins, viruses, and neighboring cells. Particularly, glycoproteins are pivotal in performing crucial biological recognition events through protein–carbohydrate interactions. These encompass a spectrum of processes, such as cell–cell recognition and developmental pathways (Murrey & Hsieh-Wilson, 2008). Recent research over the past decade has unveiled the multifaceted involvement of specific proteins that possess a keen ability to recognize and bind carbohydrates. These proteins extend their participation in an array of biological processes, including both pathological and physiological functions within the cell. In numerous instances, sugars play an integral role in conferring stability to proteins, with a notable example being their role in maintaining the structural integrity of plant lectins (Kumar *et al.*, 2012).

Interactions between proteins and carbohydrates serve as the initial foothold for infections, where parasites, fungi, bacteria, and viruses adhere to carbohydrate moieties on host cells, a pivotal aspect of cell adhesion. These protein–carbohydrate interactions are often straightforward, representing the inaugural stage in complex events, frequently culminating in intricate signaling cascades. Carbohydrate-binding proteins encompass various categories, including enzymes, anti-carbohydrate antibodies, sugar transporters, and lectins. They exhibit a wide range of structural diversity, spanning differences in size, tertiary, and quaternary structures. Consequently, the binding pockets within these proteins also exhibit significant variability (Sharon, 2006). A comprehensive understanding of the critical structural intricacies at the atomic and molecular levels is paramount. Such knowledge is indispensable for effectively designing molecules intended for therapeutic applications.

Mutations at the interface of protein–carbohydrate complexes profoundly affect vital biological functions. These mutations often lead to changes in the conformation, electrostatic properties, and hydrogen bonding patterns within the complexes, subsequently impacting their stability and functionality. In some cases, mutations may enhance the binding affinity, rendering the interaction more robust, while in others, they can weaken or disrupt it altogether. Understanding the consequences of mutations in protein–carbohydrate interactions is pivotal not only for unraveling the molecular underpinnings of various biological processes but also for guiding the development of therapeutic strategies and targeted interventions in diseases where these interactions are crucial.

Mutations of amino acid residues at the protein–carbohydrate interface change the binding affinity. The ProCaff database has data on experimentally determined binding affinity change upon mutations using frontal affinity chromatography (FAC), isothermal titration calorimetry (ITC), surface plasmon resonance (SPR), and fluorescence spectroscopy, which is a useful resource for the binding affinity of protein–carbohydrate complexes and their mutants. Experimentally determining binding affinities is resource-intensive and time-consuming, and hence computational methods are necessary to predict the binding affinity change of protein–carbohydrate complexes upon mutation. On the other hand, binding affinity change upon mutation is reported to be important for understanding the effect of disease-causing mutations (Salomonsson *et al.*, 2010; Ni *et al.*, 2014; Ruiz *et al.*, 2014).

This review aims to overview current knowledge and recent updates in protein–carbohydrate interactions. Specifically, we survey the databases available for mutations in protein–carbohydrate complexes and carbohydrate-binding proteins. In addition, specific interactions dominating the interface and factors influencing the binding affinity of protein–carbohydrate complexes will be outlined, along with prediction methods for identifying the binding sites and free energy of binding. Further, applications of molecular dynamics (MD) simulations for understanding the recognition of protein–carbohydrate complexes will be described.

8.2 Experimental studies on the binding interface and affinity of protein–carbohydrate complexes

Experimental studies are crucial in understanding the binding interface and affinity of protein–carbohydrate complexes. A range of biophysical techniques can be used to study protein–carbohydrate interactions, which are SPR, nuclear magnetic resonance (NMR) spectroscopy in conjunction with ITC and fluorescence spectroscopy, FAC, dual polarization interferometry, microarray techniques, quartz crystal microbalance, and enzyme-linked lectin assays (Linman *et al.*, 2008). These techniques allow researchers to quantify the binding affinity and obtain information on the kinetics and thermodynamics of the binding process.

SPR is a powerful technique that provides real-time information on protein–carbohydrate interactions. In SPR, a protein–carbohydrate complex is formed on a sensor surface, and changes in the refractive index of the surface are monitored in response to binding. The binding affinity can then be calculated from the changes in the signal. Linman *et al.* (2008) fabricated a novel sensing surface composed of biotinylated sialosides for the real-time quantification of lectin–carbohydrate binding affinities. The biotinylated sialoside surface was designed to facilitate multiple experiments using glycine-stripping buffers. The inert and hydrophilic hexaethylene glycol spacer introduced between the biotin and the sialic acid molecules allows conformational flexibility of the carbohydrate making this surface robust for better estimation of binding affinity between lectins and carbohydrates using the principle of SPR.

ITC is another technique that is widely used for the measurement of protein–carbohydrate binding affinity (Morgan *et al.*, 2015; Mobbs *et al.*, 2015). ITC measures the heat generated or absorbed during the binding process, which provides information of protein-carbohydrate complexes on the energetics of complex formation. The binding affinity can be calculated from the energetics data, allowing for a detailed analysis of the thermodynamics of the binding process. ITC can be used where the heat changes on titration of substrate into the enzyme solution are measured using a titration calorimeter.

NMR spectroscopy is a valuable tool for studying the structure and dynamics of protein–carbohydrate complexes. NMR spectroscopy can provide information on the structure of the complex, including the location of the carbohydrate component within the complex and the interactions between the protein and carbohydrate components. This information is essential for understanding the molecular basis of binding affinity (Lete *et al.*, 2023).

FAC has been widely used for understanding lectin–carbohydrate interactions, as it is useful for analyzing weak interactions like protein–carbohydrate interactions (Kasai, 2021). In this technique, an immobilized lectin column is used, and labeled glycan molecules in solution are passed through the column. From the elution volume and retarded volume, properties like the binding affinity of the lectin to carbohydrates can be accurately quantified. Lund *et al.* (2014) used heparin affinity chromatography to understand the differential binding of insulin-like growth factor binding protein 2 (IGFBP2) to heparin when in complex with IGF2 as well as in free states. In cases where a single binding site has alternate modes of binding that facilitate different monosaccharides to bind to the protein, fluorescence titration spectroscopy is shown to be more sensitive in capturing the differences in binding energies (Agostino *et al.*, 2015).

8.3 Databases for the binding affinity of protein–carbohydrate complexes

Only few databases are available in the literature that provide information on the binding affinity of protein–carbohydrate complexes and their mutants (**Table 8.1**).

The ProCaff database (Shanmugam *et al.*, 2020) contains data on experimental binding affinity of protein–carbohydrate complexes and their mutants, collected from the literature. It includes sequence and

Table 8.1 Databases for the binding affinity of protein–carbohydrate complexes*

Name	Contents	URL	References
ProCaff	Dissociation constant (K_d), Gibbs free energy change (ΔΔG), experimental conditions, sequence, structure, and literature information	https://web.iitm.ac.in/bioinfo2/procaff/	Shanmugam *et al.* (2020)
ProCarbDB[#]	3D structures of protein–carbohydrate complexes	http://www.procarbdb.science/procarb	Copoiu *et al.* (2020)
PDBBind	Binding affinity data for biomolecular complexes	http://www.pdbbind.org.cn/	Wang *et al.* (2004)
BindingDB	Binding affinities of protein–ligands	https://www.bindingdb.org/bind/index.jsp	Liu *et al.* (2007)

*Last accessed on December 18, 2023; [#]Not available.

structure information, thermodynamic data such as binding free energy and dissociation constant for wild-type as well as change in free energy and dissociation constant for mutants, experimental conditions, and literature information. ProCaff is a valuable resource to gain insights for understanding the importance of specific interactions at the interface of protein–carbohydrate complexes and the recognition mechanism of protein–carbohydrate complexes.

The ProCarbDB database (Copoiu *et al.*, 2020) encompasses three-dimensional structures of protein–carbohydrate complexes. It also provides information on the interactions between the protein and the ligand in a complex along with the binding affinity and effects of mutations. The PDBbind database is a valuable resource, offering both experimentally measured binding affinity information and structural insights across various biomolecular interactions (Wang *et al.*, 2004), including protein–carbohydrate complexes. BindingDB is a publicly accessible binding affinity database that is focused on the interactions between proteins (particularly potential drug targets) and small drug-like molecules (Liu *et al.*, 2007).

8.4 Analysis of the influence of mutations

The influence of mutations in protein–carbohydrate complexes has been extensively studied in recent years, focusing on the effect of mutations on binding affinity and specificity. Mutations in the protein component of a complex can have a profound impact on the binding interface, altering the shape and charge distribution of the protein and thereby affecting the interactions with the carbohydrate component, which may lead to diseases.

Vermersch *et al.* (1990) reported that the mutation P254G in the hinge region of the L-arabinose-binding protein (ABP) of *Escherichia coli* enhances binding to galactose up to 20-fold and alters the specificity. The enhancement of binding is due to stabilizing contacts between the strands of the hinge region (G254) when bound to the galactose. Labrie *et al.* (2015) reported the modulation of apoptosis by galectin-7 through its carbohydrate-recognition domain (CRD)-defective mutant form of galectin-7 (R74S). Su *et al.* (2010) studied the influence of mutation on two closely related Family 16 carbohydrate binding modules (CBMs). The mutation Q121E increased affinity for all substrates (glucose-, mannose-, and glucose/mannose-), whereas Q21G and N97R exhibited decreased substrate affinity. Bakkers *et al.* (2016) explained the change in the specificity of the Hemagglutinin-esterases (HEs), which occurred due to the mutation A74S.

On the other hand, computational methods, such as MD simulations, molecular docking, and empirical methods have been widely used to study the influence of mutations in protein–carbohydrate complexes. Parasuraman *et al.* (2014) carried out mutational studies on wheat germ agglutinin (WGA) to identify the binding specificity toward Neu5Acα(2-3)Gal and reported that a single point mutation, S114N, and two double point mutations (S114N, G118K, and S114N, G118R) increased the affinity toward the sialyl galactose compared to the wild-type, which is due to the direct and water-mediated hydrogen bonds and hydrophobic interactions with sialyl galactose. Wang *et al.* (2022) reported that E92A and E107A are required in E-selectin to stabilize the interaction with sulfated glycan 6′-sulfo-sialyl Lewis X (6′-sulfo-sLe^x), which mimics the specificity of the unrelated protein Siglec-8, for which sulfated glycan is the preferred ligand.

Mishra *et al.* (2012) generated 114 mutations at six positions (Arg17, Glu28, Gly39, Ala40, Trp76, and Trp81) of *Ralstonia solanacearum* lectin and found that polar residues Arg17 and Glu28 and nonpolar residues Trp76 and Trp81 are essential for binding with Me-α-L-fucoside. Amon *et al.* (2018) used computational and experimental methods to define the structural origin of antibody recognition using mutational analysis in antibody-sialyl-Tn (carbohydrate antigen).

Shanmugam *et al.* (2020) collected data on experimentally determined binding affinity change upon mutation, and the distribution of the K_d and ΔG values of wild type and mutations on protein–carbohydrate complexes is shown in **Figure 8.1**. We noticed that single mutations moderately increased the binding affinity in terms of K_d and ΔG.

8.5 Prediction of binding affinity and change in binding affinity upon mutation in protein–carbohydrate complexes

In this section, we will discuss the role of computational methods in predicting the binding affinity and change in binding affinity upon

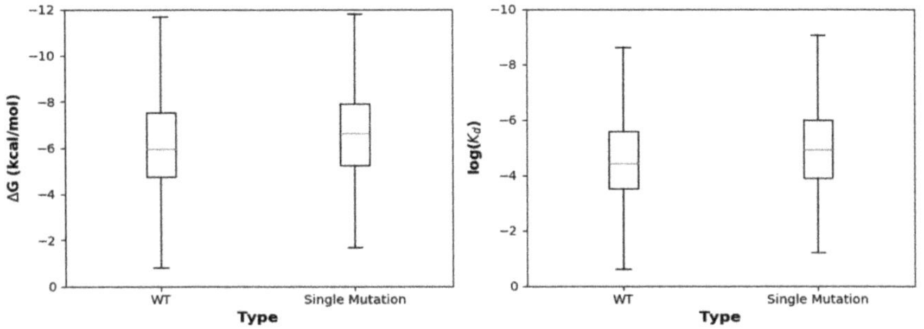

Figure 8.1 Distribution of the dissociation constant (K_d) and binding free energy (ΔG) values of wild-type and single mutations in protein–carbohydrate complexes

mutations in protein–carbohydrate complexes. Shanmugam *et al.* (2021) derived several structure-based features such as contact potentials, interaction energy, number of binding residues, and contacts between different types of atoms and developed a web server PCA-Pred (Protein–Carbohydrate Affinity Predictor) for predicting the binding affinity of protein–carbohydrate complexes. It is available at https://web.iitm.ac.in/bioinfo2/pcapred/. Nguyen *et al.* (2022) used information on both protein and carbohydrate complementarity in terms of shape and chemistry, which was captured using graph-based structural signatures, and developed a method, cutoff scanning matrix (CSM)-carbohydrate for predicting the binding affinity. The web server is available at https://biosig.lab.uq.edu.au/csm_carbohydrate/.

Kazan *et al.* (2022) studied the impact of mutations (I34K, I34L, I34Y, A71S, and A71T) in cyanovirin-N, which binds with the D-mannose using MD simulations and reported that mutations I34K/L/Y had a diverse effect on glycan binding, either improving twofold or abolishing completely, due to the dynamics of residues in the binding pocket. Zhao *et al.* (2014) developed a structure-based function-prediction technique called SPOT-Struc, which identifies carbohydrate-recognizing proteins and their binding amino acid residues by alignment program (SPalign) and binding affinity scoring according to a knowledge-based statistical potential (DFIRE).

Shanmugam *et al.* (2022) related the binding affinity change of mutations with sequence and structural features of amino acid residues at the

Table 8.2 Tools for predicting the binding affinity and change in binding affinity upon mutation in protein–carbohydrate complexes*

Method	Features	URL	References
PCA-Pred	Binding site residues, accessible surface area, interactions between atoms, and interaction energy	https://web.iitm.ac.in/bioinfo2/pcapred/	Shanmugam *et al.* (2021)
PCA-MutPred	Accessible surface area, secondary structure, mutation preference, conservation score, hydrophobicity, and contact energies	https://web.iitm.ac.in/bioinfo2/pcamutpred	Shanmugam *et al.* (2022)
CSM-carbohydrate	Physicochemical properties, inter- and intramolecular interactions, molecular surface area of interaction	http://biosig.unimelb.edu.au/csm_carbohydrate/	Nguyen *et al.* (2022)
SPOT-Struc	Knowledge-based statistical potential	NA	Zhao *et al.* (2014)

*Last accessed on December 18, 2023.

mutant sites. They found that the accessible surface area, secondary structure, mutation preference, conservation score, hydrophobicity, and contact energies are important to understand the binding affinity change upon mutation. A web server (PCA-MutPred) has been developed for predicting the change in binding affinity, and it is available at https://web.iitm.ac.in/bioinfo2/pcamutpred/. Table 8.2 summarizes the available tools for predicting protein-carbohydrate binding affinities along with sequence and structural features.

8.6 Disease-causing mutations at the interface of protein–carbohydrate complexes

Mutations at the interface of protein–carbohydrate complexes are involved in various diseases, including cancer, autoimmune diseases, and infectious diseases. Polfus *et al.* (2019) reported that mutations associated with E-selectin may disrupt the binding to ligands expressed by immune cells, leading to increased leukocyte recruitment as well as the loss of function due to the missense variant E274K in Fucosyltransferase 6. Ruiz *et al.* (2014) investigated the mutation of F19 to Y in human galectin-8 with lactose, which is involved in rheumatoid arthritis, and it altered the conformation of the complex and binding affinity.

Infectious diseases are another area where disease-causing mutations at the interface of protein–carbohydrate complexes can significantly affect human health. Many infectious diseases, including viral, bacterial, and fungal infections, are associated with changes in the binding affinity and stability of protein–carbohydrate complexes. For example, mutations in the surface proteins of viruses, such as the human immunodeficiency virus (HIV), can alter the binding affinity and stability of the protein–carbohydrate complex, leading to the development of resistance to antiviral drugs (Van Duyne *et al.*, 2019; Spillings *et al.*, 2022). Ni *et al.* (2014) studied the effect of the F95Y mutation using glycan microarray analysis on influenza B virus hemagglutinin. They showed that this mutation changed the binding affinity by 4.08 kcal/mol and enhanced the pathogenicity.

The accumulation of data on disease-causing mutations helps to understand the relationship among sequence, structure, and diseases. Currently, available databases (HUMSAVAR (http://www.uniprot.org/docs/Humsavar), SwissVar (Mottaz *et al.*, 2010), 1000 Genomes Project, and COSMIC (Tate *et al.*, 2019)) are generic to different types of proteins and their complexes (1000 Genomes Project Consortium, 2015).

8.7 Prediction of binding sites in carbohydrate-binding proteins

Several computational tools have been developed to predict carbohydrate binding sites in proteins and are listed in **Table 8.3**.

The CBS-Pred module available in the PROCARB database (Malik *et al.*, 2010) predicts carbohydrate binding sites in proteins using position-specific scoring matrices and solvent accessibility of residues. Taherzadeh *et al.* (2016) used similar features and developed Sequence-based Prediction of Residue-level INTeraction–CarBoHydrates (SPRINT-CBH) based on support vector machines (SVMs). On the other hand, computational tools have been developed specifically for mannose-interacting proteins due to

Table 8.3 List of webservers for predicting carbohydrate binding sites in proteins

Server/software	Types of binding sites	URL	References
Sequence based			
CBS-Pred	Carbohydrate	https://www.procarb.org/procarbdb/cbs-pred.html	Malik *et al.* (2010)
SPRINT-CBH	Carbohydrate	http://sparks-lab.org/server/SPRINT-CBH	Taherzadeh *et al.* (2016)
PreMieR	Mannose	http://www.imtech.res.in/raghava/premier/	Agarwal *et al.* (2011)
MOWGLI	Mannose	NA	Pai and Mondal (2016)
GlycoPP	Glycosite	http://www.imtech.res.in/raghava/glycopp	Chauhan *et al.* (2012)
GlycoEP	Glycosite	http://www.imtech.res.in/raghava/glycoep/	Chauhan *et al.* (2013)
SBRP	Acidic and non-acidic sugars	https://zenodo.org/records/61513	Banno *et al.* (2017)
SPRINT-Gly	N- and O-linked glycosylation	https://sparks-lab.org/server/sprint-gly/	Taherzadeh *et al.* (2019)
Structure-based			
CAPSIF	Carbohydrate	https://github.com/Graylab/CAPSIF	Canner *et al.* (2023)

#Last accessed on December 18, 2023.

their importance in the elimination of pathogens by binding to mannose present on the surface of pathogens. These tools include PreMieR (Agarwal *et al.*, 2011) and MOWGLI (Pai & Mondal, 2016), which use machine learning techniques and sequence-based features, position-specific scoring matrix (PSSM), and amino acid composition.

Glycosylations are the most diverse post-translational modifications, which lead to the formation of advanced glycation end products that play important functions in pathophysiological conditions (Fournet *et al.*, 2018). Hence, computational methods have been developed to predict N- and O-linked glycosylation sites in prokaryotic (GlycoPP) and eukaryotic (GlycoEP) sequences (Chauhan *et al.*, 2012; 2013). These methods are SVM-based models that use binary profile of patterns, composition profile of patterns, PSSM profile of patterns, predicted secondary structure, and surface accessibility as features.

SBRP or Sugar Binding Residue Predictor (Banno *et al.*, 2017) is a SVM-based method for identifying the binding sites for acidic and non-acidic sugars using PSSM and other sequence-based features. Taherzadeh *et al.* (2019) developed SPRINT-Gly based on deep neural networks and SVMs for predicting N-linked and O-linked glycosylation sites, respectively. It uses evolutionary information, physicochemical properties, and predicted structure-based features to detect glycosylation sites. Canner *et al.* (2023) proposed a deep neural network-based predictor (CAPSIF) for identifying carbohydrate binding sites in proteins using solvent-accessible surface area, backbone orientation, hydrophobicity index, aromatophilicity index, hydrogen bond donor capability, and hydrogen bond acceptor capability.

8.8 Potential applications

Proteins interacting with carbohydrates can be potential targets for developing antiviral and anti-cancer drugs. A novel glucose-conjugated aza-BODIPY dye analog containing iodine (AZB-Glc-I) has been reported to be useful to induce cytotoxicity in cancer cells as the uptake of this glycoconjugate was higher in breast cancer cells (Treekoon *et al.*, 2021). Liu *et al.* (2020) showed that Flt3 Receptor Interacting Lectin (FRIL) derived from hyacinth beans has been found to be effective against influenza and SARS-CoV-2 viral infections. The lectin binds to the heavily glycosylated glycoproteins hemagglutinin (HA) and spike protein (S) in human influenza virus and the Sars-Cov-2 virus, respectively, aiding the therapeutic potential.

An alteration in expression or an incomplete expression of glycans is observed as a hallmark of cancer as reported (Bellis *et al.*, 2022). The ability of lectins to specifically and selectively bind to glycosylated molecules with good sensitivity is leveraged currently for its therapeutic potential against cancer (Gupta, 2020). CancerLectinDB encompasses information on lectins specifically associated with cancer (Damodaran *et al.*, 2008), which has sequence, structural, and functional annotations of lectins associated with cancer. Utilizing this database, Butt and Khan

(2019) developed a method for predicting cancer-associated lectins (CanLect-Pred) using statistical moment features and a random forest classifier. Ali *et al.* (2022) proposed a deep-learning model for the prediction of cancer lectins (Deep-PCL) using pseudo-amino acid composition and PSSM. Liao *et al.* (2023) found that increased levels of mannose-binding lectin2 (MBL2) are positively correlated with the inhibition of cell growth in hepatocellular carcinoma (HCC) and reduced tumor progression. Therefore, cancer lectins have therapeutic potential against various types of cancer and studies on interactions between lectins and carbohydrates in glycosylated proteins have important applications in diseases including cancer.

Further, multivalency in binding capacity, high density of functional groups, diversity in composition, and biocompatibility of carbohydrates make them interesting candidates for drug delivery applications. Carbohydrates are used in various forms like pure carbohydrate drugs, carbohydrate conjugates, as scaffolds or in glyconanomaterials (Wang *et al.*, 2021). The potential applications of carbohydrates for drug delivery demand a comprehensive understanding of the diverse ways in which proteins and carbohydrates interact.

8.9 Conclusions

In conclusion, computational analysis has played a crucial role in understanding the effects of mutations on the binding affinity of protein–carbohydrate complexes. The results obtained from various computational methods have shown that mutations can have a significant impact on the binding affinity of these complexes. These findings have important implications for the design of new drugs and therapies, as they can be used to identify the key residues that are critical for binding and to optimize the interactions between proteins and carbohydrates. Further studies are necessary to validate these computational results experimentally and to extend these analyses to other types of protein–carbohydrate interactions. Nevertheless, computational analysis remains a valuable tool for understanding the molecular basis of protein–carbohydrate interactions and for designing new treatments for diseases.

Acknowledgments

The authors thank the Department of Biotechnology, IIT Madras, for the computational resources. The work is partially supported by the Department of Biotechnology, Government of India to MMG (BT/PR39164/BID/7/965/2020).

References

1000 Genomes Project Consortium. (2015). A global reference for human genetic variation. *Nature, 526*(7571), 68.

Agarwal, S., Mishra, N. K., Singh, H., & Raghava, G. P. (2011). Identification of mannose interacting residues using local composition. *PLoS One, 6*(9), e24039.

Agostino, M., Velkov, T., Dingjan, T., Williams, S. J., Yuriev, E., & Ramsland, P. A. (2015). The carbohydrate-binding promiscuity of Euonymus europaeus lectin is predicted to involve a single binding site. *Glycobiology, 25*(1), 101–114.

Ali, F., Ghulam, A., Maher, Z. A., Asif Khan, M., Afzal Khan, S., & Hongya, W. (2022). Deep-PCL: A deep learning model for prediction of cancerlectins and non cancerlectins using optimized integrated features. *Chemometrics and Intelligent Laboratory Systems, 221*, 104484.

Amon, R., Grant, O. C., Leviatan Ben-Arye, S., Makeneni, S., Nivedha, A. K., Marshanski, T., Norn, C., Yu, H., Glushka, J. N., Fleishman, S. J., Chen, X., Woods, R. J., & Padler-Karavani, V. (2018). A combined computational-experimental approach to define the structural origin of antibody recognition of sialyl-Tn, a tumor-associated carbohydrate antigen. *Scientific Reports, 8*(1), 10786.

Bakkers, M. J., Zeng, Q., Feitsma, L. J., Hulswit, R. J., Li, Z., Westerbeke, A., van Kuppeveld, F. J., Boons, G. J., Langereis, M. A., Huizinga, E. G., & de Groot, R. J. (2016). Coronavirus receptor switch explained from the stereochemistry of protein–carbohydrate interactions and a single mutation. *Proceedings of the National Academy of Sciences, 113*(22), E3111–E3119.

Banno, M., Komiyama, Y., Cao, W., Oku, Y., Ueki, K., Sumikoshi, K., Nakamura, S., Terada, T., & Shimizu, K. (2017). Development of a sugar-binding residue prediction system from protein sequences using support vector machine. *Computational Biology and Chemistry, 66*, 36–43.

Bellis, S. L., Reis, C. A., Varki, A., Kannagi, R., & Stanley, P (2022). Glycosylation changes in cancer.

Butt, A. H., & Khan, Y. D. (2019). CanLect-Pred: A cancer therapeutics tool for prediction of target cancerlectins using experiential annotated proteomic sequences. *IEEE Access, 8*, 9520–9531.

Canner, S. W., Shanker, S., & Gray, J. J. (2023). Structure-based neural network protein–carbohydrate interaction predictions at the residue level. *Frontiers in Bioinformatics, 3*, 1186531.

Chauhan, J. S., Bhat, A. H., Raghava, G. P., & Rao, A. (2012). GlycoPP: A webserver for prediction of N-and O-glycosites in prokaryotic protein sequences. *PloS One, 7*(7), e40155.

Chauhan, J. S., Rao, A., & Raghava, G. P. (2013). In silico platform for prediction of N-, O- and C-glycosites in eukaryotic protein sequences. *PloS One, 8*(6), e67008.

Copoiu, L., Torres, P. H., Ascher, D. B., Blundell, T. L., & Malhotra, S. (2020). ProCarbDB: A database of carbohydrate-binding proteins. *Nucleic Acids Research, 48*(D1), D368–D375.

Damodaran, D., Jeyakani, J., Chauhan, A., Kumar, N., Chandra, N. R., & Surolia, A. (2008). CancerLectinDB: A database of lectins relevant to cancer. *Glycoconjugate Journal, 25*, 191–198.

del Carmen Fernandez-Alonso, M., Díaz, D., Alvaro Berbis, M., Marcelo, F., Cañada, J., & Jiménez-Barbero, J. (2012). Protein-carbohydrate interactions studied by NMR: Fom molecular recognition to drug design. *Current Protein and Peptide Science, 13*(8), 816–830.

Fournet, M., Bonté, F., & Desmoulière, A. (2018). Glycation damage: A possible hub for major pathophysiological disorders and aging. *Aging and Disease, 9*(5), 880.

Gupta, A. (2020). Emerging applications of lectins in cancer detection and biomedicine. *Materials Today: Proceedings, 31*, 651–661.

Kanie, Y., & Kanie, O. (2017). Addressing the glycan complexity by using mass spectrometry: In the pursuit of decoding glycologic. *Digestion, 3*, 5.

Kasai, K. (2021). Frontal affinity chromatography: An excellent method of analyzing weak biomolecular interactions based on a unique principle. *Biochimica et Biophysica Acta (BBA)-General Subjects, 1865*(1), 129761.

Kazan, I. C., Sharma, P., Rahman, M. I., Bobkov, A., Fromme, R., Ghirlanda, G., & Ozkan, S. B. (2022). Design of novel cyanovirin-N variants by modulation of binding dynamics through distal mutations. *Elife, 11*, e67474.

Kumar, K. K., Chandra, L. P. K., Sumanthi, J., Reddy, G. S., Shekar, P. C., & Reddy, B. (2012). Biological role of lectins: A review. *Journal of Orofacial Sciences, 4*(1), 20–25.

Labrie, M., Vladoiu, M., Leclerc, B. G., Grosset, A. A., Gaboury, L., Stagg, J., & St-Pierre, Y. (2015). A mutation in the carbohydrate recognition domain drives a phenotypic switch in the role of galectin-7 in prostate cancer. *PloS One, 10*(7), e0131307.

Lete, M. G., Franconetti, A., Bertuzzi, S., Delgado, S., Azkargorta, M., Elortza, F., Millet, O., Jiménez-Osés, G., Arda, A., & Jiménez-Barbero, J. (2023). NMR investigation of protein-carbohydrate interactions: The recognition of glycans by galectins engineered with fluorotryptophan residues. *Chemistry — A European Journal, 29*(5), e202202208.

Liao, H., Yang, J., Xu, Y., Xie, J., Li, K., Chen, K., Pei, J., Luo, Q., & Pan, M. (2023). Mannose-binding lectin 2 as a potential therapeutic target for hepatocellular carcinoma: Multi-omics analysis and experimental validation. *Cancers, 15*(19), 4900.

Linman, M. J., Taylor, J. D., Yu, H., Chen, X., & Cheng, Q. (2008). Surface plasmon resonance study of protein–carbohydrate interactions using biotinylated sialosides. *Analytical Cchemistry, 80*(11), 4007–4013.

Liu, T., Lin, Y., Wen, X., Jorissen, R. N., & Gilson, M. K. (2007). BindingDB: A web-accessible database of experimentally determined protein–ligand binding affinities. *Nucleic Acids Research, 35*(suppl_1), D198–D201.

Liu, Y. M., Shahed-Al-Mahmud, M., Chen, X., Chen, T. H., Liao, K. S., Lo, J. M., Wu, Y. M., Ho, M. C., Wu, C. Y., Wong, C. H., Jan, J. T., & Ma, C. (2020). A carbohydrate-binding protein from the edible lablab beans effectively blocks the infections of influenza viruses and SARS-CoV-2. *Cell Reports, 32*(6).

Lund, J., Søndergaard, M. T., Conover, C. A., & Overgaard, M. T. (2014). Heparin-binding mechanism of the IGF2/IGF-binding protein 2 complex. *Journal of Molecular Endocrinology, 52*(3), 345–355.

Malik, A., Firoz, A., Jha, V., & Ahmad, S. (2010). PROCARB: A database of known and modelled carbohydrate-binding protein structures with sequence-based prediction tools. *Advances in Bioinformatics, 2010.*

Mishra, S. K., Adam, J., Wimmerová, M., & Koča, J. (2012). In silico mutagenesis and docking study of Ralstonia solanacearum RSL lectin: Performance of docking software to predict saccharide binding. *Journal of Chemical Information and Modeling, 52*(5), 1250–1261.

Mobbs, J. I., Koay, A., Di Paolo, A., Bieri, M., Petrie, E. J., Gorman, M. A., Doughty, L., Parker, M. W., Stapleton, D. I., Griffin, M. D., & Gooley, P. R. (2015). Determinants of oligosaccharide specificity of the carbohydrate-binding modules of AMP-activated protein kinase. *Biochemical Journal, 468*(2), 245–257.

Morgan, A., Sepuru, K. M., Feng, W., Rajarathnam, K., & Wang, X. (2015). Flexible linker modulates glycosaminoglycan affinity of decorin binding protein A. *Biochemistry, 54*(32), 5113–5119.

Mottaz, A., David, F. P., Veuthey, A. L., & Yip, Y. L. (2010). Easy retrieval of single amino-acid polymorphisms and phenotype information using SwissVar. *Bioinformatics, 26*(6), 851–852.

Murrey, H. E., & Hsieh-Wilson, L. C. (2008). The chemical neurobiology of carbohydrates. *Chemical Reviews, 108*(5), 1708–1731.

Nguyen, T. B., Pires, D. E., & Ascher, D. B. (2022). CSM-carbohydrate: Protein-carbohydrate binding affinity prediction and docking scoring function. *Briefings in Bioinformatics, 23*(1), bbab512.

Ni, F., Mbawuike, I. N., Kondrashkina, E., & Wang Q. (2014). The roles of hemagglutinin Phe-95 in receptor binding and pathogenicity of influenza B virus. *Virology, 450*, 71–83.

Pai, P. P., & Mondal, S. (2016). MOWGLI: Prediction of protein–MannOse interacting residues with ensemble classifiers usinG evoLutionary Information. *Journal of Biomolecular Structure and Dynamics, 34*(10), 2069–2083.

Parasuraman, P., Murugan, V., Selvin, J. F., Gromiha, M. M., Fukui, K., & Veluraja, K. (2014). Insights into the binding specificity of wild type and mutated wheat germ agglutinin towards Neu5Acα (2-3) Gal: A study by in silico mutations and molecular dynamics simulations. *Journal of Molecular Recognition, 27*(8), 482–492.

Ruiz, F. M., Scholz, B. A., Buzamet, E., Kopitz, J., André, S., Menéndez, M., Romero, A., Solís, D., & Gabius, H. J. (2014). Natural single amino acid polymorphism (F19Y) in human galectin-8: Detection of structural alterations and increased growth-regulatory activity on tumor cells. *The FEBS Journal, 281*(5), 1446–1464.

Salomonsson, E., Carlsson, M. C., Osla, V., Hendus-Altenburger, R., Kahl-Knutson, B., Oberg, C. T., Sundin, A., Nilsson, R., Nordberg-Karlsson, E., Nilsson, U. J., Karlsson, A., Rini, J. M., & Leffler, H. (2010). Mutational tuning of galectin-3 specificity and biological function. *Journal of Biological Chemistry, 285*(45), 35079–35091.

Shanmugam, N. R. S., Blessy, J. J., Veluraja, K., & Michael Gromiha, M. (2020). ProCaff: Protein–carbohydrate complex binding affinity database. *Bioinformatics, 36*(11), 3615–3617.

Shanmugam, N. R. S., Blessy, J. J., Veluraja, K., & Michael Gromiha, M. (2021). Prediction of protein–carbohydrate complex binding affinity using structural features. *Briefings in Bioinformatics*, *22*(4), bbaa319.

Shanmugam, N. R. S., Selvin, J. F. A., Veluraja, K., & Michael Gromiha, M. (2018). Identification and analysis of key residues involved in folding and binding of protein-carbohydrate complexes. *Protein and Peptide Letters, 25*(4), 379–389.

Shanmugam, N. R. S., Veluraja, K., & Gromiha, M. M. (2022). PCA-MutPred: Prediction of binding free energy change upon missense mutation in protein-carbohydrate Complexes. *Journal of Molecular Biology*, *434*(11), 167526.

Sharon, N. (2006). Carbohydrates as future anti-adhesion drugs for infectious diseases. *Biochimica et Biophysica Acta (BBA)-General Subjects, 1760*(4), 527–537.

Spillings, B. L., Day, C. J., Garcia-Minambres, A., Aggarwal, A., Condon, N. D., Haselhorst, T., Purcell, D. F. J., Turville, S. G., Stow, J. L., Jennings, M. P., & Mak, J. (2022). Host glycocalyx captures HIV proximal to the cell surface via oligomannose-GlcNAc glycan-glycan interactions to support viral entry. *Cell Reports, 38*(5).

Su, X., Agarwal, V., Dodd, D., Bae, B., Mackie, R. I., Nair, S. K., & Cann, I. K. (2010). Mutational insights into the roles of amino acid residues in ligand binding for two closely related family 16 carbohydrate binding modules. *Journal of Biological Chemistry, 285*(45), 34665–34676.

Taherzadeh, G., Dehzangi, A., Golchin, M., Zhou, Y., & Campbell, M. P. (2019). SPRINT-Gly: Predicting N-and O-linked glycosylation sites of human and mouse proteins by using sequence and predicted structural properties. *Bioinformatics, 35*(20), 4140–4146.

Taherzadeh, G., Zhou, Y., Liew, A. W. C., & Yang, Y. (2016). Sequence-based prediction of protein–carbohydrate binding sites using support vector machines. *Journal of Chemical Information and Modeling, 56*(10), 2115–2122.

Tate, J. G., Bamford, S., Jubb, H. C., Sondka, Z., Beare, D. M., Bindal, N., Boutselakis, H., Cole, C. G., Creatore, C., Dawson, E., Fish, P., Harsha, B., Hathaway, C., Jupe, S. C., Kok, C. Y., Noble, K., Ponting, L., Ramshaw, C. C., Rye, C. E., Speedy, H. E., … Forbes, S. A. (2019). COSMIC: The catalogue of somatic mutations in cancer. *Nucleic Acids Research, 47*(D1), D941–D947.

Treekoon, J., Pewklang, T., Chansaenpak, K., Gorantla, J. N., Pengthaisong, S., Lai, R. Y., Ketudat-Cairns, J. R., & Kamkaew, A. (2021). Glucose conjugated aza-BODIPY for enhanced photodynamic cancer therapy. *Organic & Biomolecular Chemistry, 19*(26), 5867–5875.

Van Duyne, R., Kuo, L. S., Pham, P., Fujii, K., & Freed, E. O. (2019). Mutations in the HIV-1 envelope glycoprotein can broadly rescue blocks at multiple steps in the virus replication cycle. *Proceedings of the National Academy of Sciences, 116*(18), 9040–9049.

Varki, A. (2017). Biological roles of glycans. *Glycobiology, 27*(1), 3–49.

Vermersch, P. S., Tesmer, J. J., Lemon, D. D., & Quiocho, F. A. (1990). A Pro to Gly mutation in the hinge of the arabinose-binding protein enhances binding and alters specificity. Sugar-binding and crystallographic studies. *Journal of Biological Chemistry, 265*(27), 16592–16603.

Wang, J., Zhang, Y., Lu, Q., Xing, D., & Zhang, R. (2021). Exploring carbohydrates for therapeutics: A review on future directions. *Frontiers in Pharmacology, 12*, 756724.

Wang, R., Fang, X., Lu, Y., & Wang, S. (2004). The PDBbind database: Collection of binding affinities for protein–ligand complexes with known three-dimensional structures. *Journal of Medicinal Chemistry, 47*(12), 2977–2980.

Wang, X., Hanes, M. S., Cummings, R. D., & Woods, R. J. (2022). Computationally guided conversion of the specificity of E-selectin to mimic that of Siglec-8. *Proceedings of the National Academy of Sciences, 119*(41), e2117743119.

Zhao, H., Yang, Y., von Itzstein, M., & Zhou, Y. (2014). Carbohydrate-binding protein identification by coupling structural similarity searching with binding affinity prediction. *Journal of Computational Chemistry, 35*(30), 2177–2183.

https://doi.org/10.1142/9789811293269_0009

Chapter 9

Elucidating the effects of mutations on protein function

Govindarajan Sudha[1,*] and M. Michael Gromiha[1]

[1]*Department of Biotechnology, Bhupat and Jyoti Mehta School of Biosciences, Indian Institute of Technology Madras, Chennai, Tamil Nadu 600036, India*

Abstract

Elucidating the consequences of mutations on protein function is important for identifying the functionally critical residues and designing proteins with enhanced function. Currently, several databases accumulate information on protein functions for experimentally known variants. On the other hand, prediction tools have been developed to reveal the effects of mutations on protein function. In this chapter, the advancements in experimental methods from traditional site-directed mutagenesis to recent deep mutational scanning experiments are discussed. In addition, we explain variant effect prediction methods ranging from simple amino acid substitution probabilities to state-of-the-art zero-shot language models, along with potential applications. Further, recent efforts such as the "Atlas of variant effects alliance" to accelerate research in variant effect

*Corresponding author
GS: sgovindarajan6@gmail.com

prediction and the Critical Assessment of Genome Interpretation (CAGI) experiment for assessing the performance of the different variant effect predictors are outlined.

9.1 Background

Understanding the function of a protein is critical to explain the biological processes driven in the cell (Morris *et al.*, 2022). Protein function can be classified at different levels, such as molecular function, cellular component, or biological function (Gene Ontology Consortium *et al.*, 2023). Molecular functions refer to the activities of the protein at a molecular level, which include stability, binding affinity, and catalytic activity (Morris *et al.*, 2022). Cellular components deal with locations relative to cellular structures where the molecular function is performed. Biological function is focused on a large biological program that utilizes the molecular function of a protein.

Understanding the sequence–structure–function relationships (Koehler Leman *et al.*, 2023) elucidates the molecular basis of biological phenomena and design proteins with enhanced function. Further, amino acid mutations may alter protein structure, stability, binding affinity, and function (Morris *et al.*, 2022). Function annotation is elucidated using various experimental methods such as enzyme assays, immunoprecipitation, fluorescence microscopy, mass spectrometry, X-ray diffraction, cryo-electron microscopy, phage display, fluorescent resonance energy transfer, surface plasmon resonance, and isothermal titration calorimetry (Shoemaker & Panchenko, 2007; Zhou *et al.*, 2016; Morris *et al.*, 2022).

Several mutagenesis experiments have dissected the roles of functional residues by mutation and investigated their effects of mutations on their physiological function. Catalytic residues D52 and E35 have been identified by site-directed mutagenesis in chicken lysozyme, and the mutations D52N and E35Q show only 5% of wild-type lytic activity against bacterial cell walls (Malcolm *et al.*, 1989). Mutation of D262 on estrogen receptor alpha has been shown to abrogate the estradiol signaling pathway (Xu *et al.*, 2023), which has been shown to affect uterine and follicular development. Yun *et al.* (2007) reported that the mutations L858R and G719S in the kinase domain of epidermal growth factor receptor (EGFR)

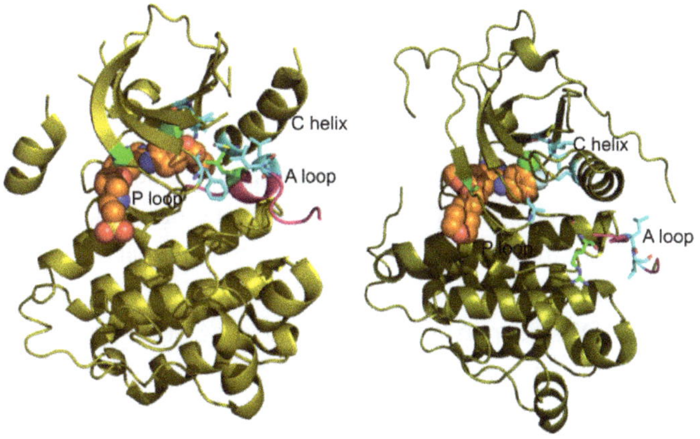

Figure 9.1 Structures of (a) wild-type inactive (PDB ID: 1xkk) and (b) L858R mutant active kinases (PDB ID: 2itz). Hydrophobic residues, mutated residues, and activation loops are shown in cyan, green, and magenta, respectively

activate the kinase by disrupting autoinhibitory interactions, which accelerate the catalysis up to 50-fold. In the wild-type, clusters of hydrophobic residues stabilize the inactive conformation (**Figure 9.1a**), whereas in the mutant (L858R) structure (**Figure 9.1b**), Arg is not favorably accommodated in the hydrophobic pocket. Similarly, substitution of G719S destabilizes the inactive conformation of P-loop and activates the kinase. The activation loop (A loop), P loop, and C helix are reorganized in mutant structures to switch from inactive to active kinase conformations.

In this chapter, we outline the experimental methods for characterizing the effect of protein function upon mutation along with existing databases. Further, tools available for predicting the effect of mutations have been discussed with potential applications. Moreover, we explain the recent progress in assessing the performance of variant prediction algorithms.

9.2 Experimental methods

Several experimental techniques are available in the literature to explore the effect of mutations on protein function. Site-directed mutagenesis (Carter, 1986) involves the mutation of one amino acid to another, whereas combinatorial alanine scanning (Morrison & Weiss, 2001) considers mutations of several amino acids to alanine. The most recent and popular experimental method is the deep mutational scanning (DMS), where mutations are carried out systematically from one residue to all other residues with high-throughput functional assays and deep sequencing (Fowler & Fields, 2014). Next generation sequencing is used for the high-throughput sequencing of the diverse library, which enables simultaneous analysis of thousands or millions of variants. Some of the functional assays tested in DMS experiments include testing yeast growth rate, viral replication assays, survival assays for antibiotic resistance, two hybrid assays for protein–protein interactions, fluorescence for protein stability, enzyme activity, and expression level assays to quantify the amount of protein (Fowler & Fields, 2014). The limitations of DMS are its suitability for only a few classes of proteins, generating functional assays associated with one particular function, and large proteins are not suitable due to the high cost of generating variant libraries. These limitations are being addressed by computational methods independently or in conjunction with experimental methods (Horne & Shukla, 2022).

Directed evolution mimics the natural selection process by several iterations of mutagenesis and selection, followed by screening the functional impact of the mutation (Packer & Liu, 2015). However, exhaustive exploration of combinatorial libraries becomes experimentally impractical, and a probable solution is to couple directed evolution with machine learning. This approach is cost- and time-effective with promising results (Yang *et al.*, 2019).

9.3 Databases for protein functions

Several databases have been developed to accumulate experimental data on protein functions. MaveDB (Esposito *et al.*, 2019) is a repository that has variant effect data generated by DMS experiments. ProtaBank (Wang *et al.*, 2019) stores both experimental and computational data related to various functional properties such as activity, binding, stability, folding, and solubility. It includes mutation data from experiments such as DMS and directed evolution, as well as computational approaches. ProteinGym (Notin *et al.*, 2022) is a large curated collection of DMS assays for benchmarking fitness predictors. It includes 94 functional assays, which is two times the size of previous benchmarks, as well as viral datasets and data on indel benchmark.

Apart from these databases, data for different functional properties can be collected from a variety of databases, such as changes in binding affinity upon mutation from Skempi2 (Jankauskaite *et al.*, 2019) and Proximate (Jemimah *et al.*, 2017), changes in stability upon mutation from ProThermDB (Nikam *et al.*, 2021) and MPTherm (Kulandaisamy *et al.*, 2021), changes in solubility upon mutations from SoluProtMutDB (Velecký *et al.*, 2022), changes in aggregation rate upon mutation from Curated Protein Aggregation Database (CPAD) (Thangakani *et al.*, 2016), and so on.

Studies about the effects of variants have been intensified with efforts such as the Atlas of variant effects alliance. It is an international collaboration comprising hundreds of researchers, technologists, and clinicians to develop comprehensive variant effect maps for important regions of human and pathogen genomes. This will assist in the diagnosis, prognosis, and treatment of diseases (Fowler *et al.*, 2023). The Critical Assessment of Genome Interpretation (CAGI) experiment (Hoskins *et al.*, 2017) involves the prediction of phenotypes from the given genotypes, and the predictors are assessed against novel unseen datasets, including DMS experiments.

9.4 Computational methods for predicting protein functions

9.4.1 *Prediction of variant effects*

Predicting the effect of mutations on protein function are popularly called "variant effect predictors" (VEPs) (Yang *et al.*, 2019). With the advent of machine learning, VEPs have improved tremendously. VEP programs are primarily designed for disease predictions (Cheng *et al.*, 2023) due to the fact that mutations generally cause diseases and affect several functions of proteins (Backwell & Marsh, 2022). We have focused on computational tools that generally predict the effects of mutations on protein function and are listed in **Table 9.1.** However, in general, a VEP can be adapted and

Table 9.1 A list of unsupervised, supervised, and meta variant effect predictor programs

No.	Predictor name	Method	Features
(a)	**Unsupervised Methods**		
1.	SIFT — Sorting Intolerant from Tolerant	Empirical	MSA
2.	PROVEAN — PROtein Variant Effect ANalyzer	Empirical	MSA
3.	MutationAssessor	Combinatorial entropy	MSA
4.	LIST_S2	Empirical	MSA
5.	COSMIS — COntact Set MISsense tolerance	Empirical	Structure
6.	EVmutation	Empirical	MSA
7.	GEMME — Global Epistatic Model for predicting Mutational Effects	Empirical	MSA
8.	DeepSequence	Variational Autoencoder	MSA
9.	EVE — Evolutionary model of Variant Effects	Variational Autoencoder	MSA
10.	DeMask	Empirical	MSA
11.	ESM-1v	Language model	Sequence
12.	SeqDesign	Autoregressive generative model	Sequence
(b)	**Supervised Methods**		
13.	EvoRator2	Multi-layer perceptron	Structure
14.	Envision	Gradient-boosting regression	Sequence, Structure
15.	Sequence UNET	Convolutional Neural Network	Sequence, Structure
16.	nn4dms (Neural networks for deep mutational scanning data)	GCN (Graph Convolutional Networks)	Sequence, function
(c)	**Supervised — Meta Predictor**		
17.	Condel — Consensus deleteriousness score	Consensus (FATHMM and MutationAssessor)	

customized for identifying and predicting function-enhancing and function-disruptive mutations.

Large variant effect data from several DMS studies are useful to benchmark and assess the performance of VEPs. VEPs showed a strong correlation with DMS data, which shows that variant effects on protein functions can be reliably predicted with computational methods (Livesey & Marsh, 2020, 2023). VEPs generally use features from protein sequence, structure, and evolution, and machine learning techniques. The unsupervised and supervised methods for VEP will be discussed below.

9.4.2 *Unsupervised predictors*

Unsupervised learning is based on the idea that biological properties can be read directly from sequences without supervision from experimental measurements. One of the first unsupervised VEPs is sorting intolerant from tolerant (SIFT), which relies on simple statistical methods such as collecting homologous sequences and constructing a multiple sequence alignment (MSA) followed by computing amino acid substitution probability (Ng & Henikoff, 2003). PROVEAN uses an alignment-based score for single and multiple substitutions, which measures the change in sequence similarity of a query to homolog before and after introducing the mutation to the query (Choi & Chan, 2015). MutationAssessor distinguishes conservation patterns within families and subfamilies of homologues and determines functional specificity (Reva *et al.*, 2007). LIST-S2 incorporates local sequence identity and taxonomy distances. It determines whether the mutated residue occurs in a close or distant homologue and assesses the likelihood of changing the reference to mutant amino acid (Malhis *et al.*, 2020). COSMIS uses the information of neighboring amino acid sites in 3D space or contact sets to understand the mutational and functional constraints of each site. It quantifies the observed versus expected missense variations in 3D structures (Li *et al.*, 2022).

There is also a set of programs that not only focuses on independent conservation but also dependent co-conservation. EVmutation is based on statistical models capturing dependencies across pairs of residues and also account for epistasis between positions, and are also capable of predicting the effect of multiple mutations (Hopf *et al.*, 2017). Gemme (Laine *et al.*, 2019) predicts mutational effects by modeling the evolutionary history of natural sequences. These methods are improvised by the program "Deep sequence" considering a nonlinear latent variable model to implicitly capture high-order interaction between positions in a sequence (Riesselman *et al.*, 2018). Using MSA as an input, the evolutionary model of variant effect (EVE) uses a variational autoencoder (James *et al.*, 2023). Simple approaches such as DeMaSk derive the amino acid substitution matrix from DMS datasets, substitution scores with evolutionary conservation scores, and frequencies of the variant amino acids across homologs (Munro *et al.*, 2021).

Large language models trained on massive quantities of non-aligned protein sequences from diverse families address the problems of shallow MSA for a few protein families, such as ESM-1V (Mansoor *et al.*, 2023). Tranception (Notin *et al.*, 2022) uses autoregressive transformers that are trained on an unaligned sequence, overcoming the limitations of using MSA. It augments the autoregressive prediction with predictions based on the empirical distribution of amino acids observed at a position in homologous sequences. ECNet (Luo *et al.*, 2021) combines language embeddings and evolutionary-based direct coupling models. Further, a protein design approach called SeqDesign is an unsupervised and alignment-free method that makes use of autoregressive generative models (Shin *et al.*, 2021).

9.4.3 *Supervised predictors*

Supervised VEP methods rely on fitting models to experimental measurements of sequence-function relationships from DMS experiments. EvoRator2 (Nagar *et al.*, 2023) exploits deep learning to predict site-specific tolerated amino acids, and it performs well for orphan or de novo designed proteins also. It incorporates protein structural signatures composed of physicochemical, geometrical, and graph-based features, as well as features from MSA that capture evolutionary constraints. Envision is based on gradient-boosting regression that uses biological, structural, and physicochemical features (Gray *et al.*, 2018). Sequence UNET

(Dunham *et al.*, 2023) is a highly scalable program that uses a fully convolutional neural network (CNN) that incorporates both Protein Data Bank structures (Burley *et al.*, 2023) and MSA information.

Methods are also designed to generate variants with enhanced properties using labeled information from public datasets that include globular and membrane proteins as well as enzymes. CNN models built with amino acids and structure-based descriptors are shown to perform better than other machine learning methods (Xu *et al.*, 2020). Gelman *et al.* González-Pérez and López-Bigas (2011) used neural networks for DMS data and graph convolutional networks to predict new uncharacterized designed variants. Condel (González-Pérez & López-Bigas, 2011) is a supervised meta predictor that integrates the output of five predictive tools, such as SIFT, Logre, MAPP, PPH2, and Massessor into a unified classification and considered the weighted average of the normalized scores of the individual methods.

9.4.4 *Other methods for understanding the effect of mutations on protein functions*

Apart from VEPs, other methods such as (i) machine learning-guided protein evolution and (ii) ancestral protein reconstruction, which combine both computational and experimental workflows are reported in the literature to understand the effects of mutations on protein functions and design mutant proteins.

(i) Machine learning-guided directed protein evolution
Directed protein evolution accumulates beneficial mutations via sequence diversification to generate a library of variants followed by screening to identify variants with improved properties. Machine learning methods (Yang *et al.*, 2019) are used to navigate the fitness landscape efficiently, and it accelerates the directed evolution by learning from a training dataset and using the information to select sequences that show enhanced properties.

(ii) Ancestral protein reconstruction
Phylogenetic-based methods such as ancestral protein reconstruction (Merkl & Sterner, 2016) are used to understand the effects of mutation on

protein function by comparing the ancestral and extant sequences. It identifies specific amino acid changes that have occurred over time, and the mutations influence the structure and function of the protein. This method generally relies on the experimental characterization and validation of designed sequences.

9.5 Potential applications for studying sequence-function landscapes

Four different applications of protein engineering using variant effect prediction methods and data-driven protein design are highlighted in **Figure 9.2**.

9.5.1 *Variant effect prediction methods*

9.5.1.1 *Sequence design of nanobody libraries*

Deep generative models adapted from natural language processing have been used to predict missense, indel effects, design, and test diverse nanobody libraries with better expression (Shin *et al.*, 2021). Single-domain

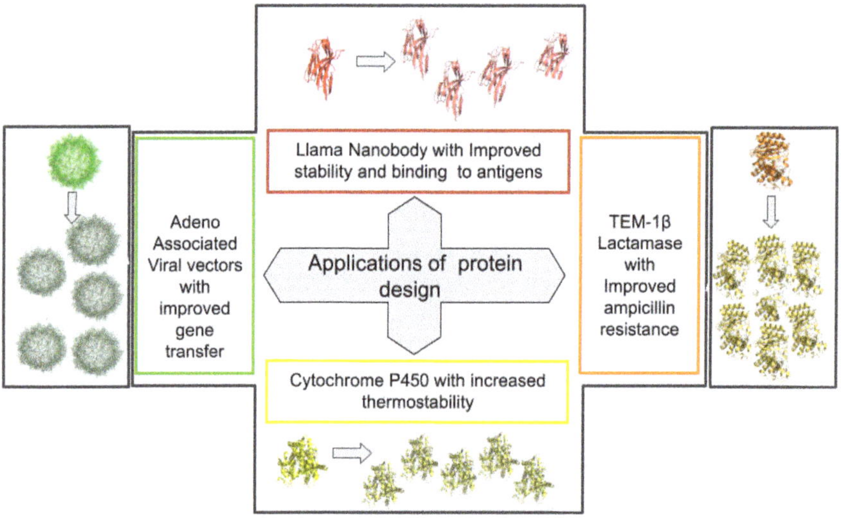

Figure 9.2 Examples for the applications of protein design

antibodies or nanobodies are composed solely of the variable domain of the canonical antibody heavy chain. The advantage of this method is being alignment-free thereby finding applications for disordered proteins and the design of antibodies due to the highly variable complementary determining regions (Malcolm *et al.*, 1989). It includes the following steps: (i) natural sequences are used to learn functional constraints by predicting the likelihood of each residue in the sequence, (ii) constraints are used to generate millions of novel designed nanobody sequences, and (iii) designed sequences are screened for expression and binding target antigens (Shin *et al.*, 2021).

9.5.1.2 *Engineering of TEM-1 β-lactamase and variants with improved ampicillin resistance*

ECNet (Luo *et al.*, 2021) (Evolutionary context-integrated neural network) is a method for engineering proteins with improved ampicillin resistance (Yun *et al.*, 2007) involves the following steps: (i) a global evolutionary context is obtained by sequence representations learned from language models from protein sequence databases such as UniProt and Pfam, (ii) local evolutionary contexts are obtained by dependencies between residues that are captured using a direct coupling analysis model, and (iii) these representations are provided as input for a deep learning model that predicts protein fitness.

Quantitative fitness data obtained from DMS for all point mutation variants and 12% of double mutation variants of TEM-1 are used to train the deep learning model. ECNet has been used to engineer TEM-1 β-lactamase and identify variants with improved ampicillin resistance. It revealed 37 new TEM-1 variants, which contain high-order mutants ranging from two to six mutations and did not overlap with any other reported variants. Experimental validation of the variants obtained from ECNet demonstrated improved fitness as compared to the wild type.

9.5.2 *Data-driven protein design*

Data-driven protein design is based on generating thousands of variants with single or multiple mutations and mapping their phenotypes using a functional assay. This will create a "Fitness landscape" to map the sequence–function relationship, and the landscape will be expanded using machine learning. The model is used for predicting the function-enhancing variants, which are not observed experimentally (Yang *et al.*, 2019). A few examples are discussed below.

9.5.2.1 *Increased thermostability in cytochrome P450*

Increased thermostability in cytochrome p450 (Romero *et al.*, 2013) was achieved by recombining sequence fragments from the heme domains of the bacterial cytochrome P450 proteins CYP102A1, CYP102A2, and CYP102A3. The sequence fragments are selected to minimize the number of lost contacts, which are amino acids within 4.5 Å. They trained Gaussian process models for T50 (the temperature at which an enzyme loses 50% of its activity) and the presence or absence of function on 242 chimeric P450s. They evaluated the model performance on a test set of chimeric P450s, which were generated with different boundaries. With sufficient training data, Bayesian optimization was used to identify thermostable variants (Romero *et al.*, 2013).

9.5.2.2 *Improved adeno-associated viral vectors for gene transfer*

Bryant *et al.* (2021) applied recent deep learning to design highly diverse AAV2 capsid protein variants that remain viable for packaging the DNA. The 28 amino acid segment that overlaps both known heparin and antibody binding sites has been focused to generate thousands of viable engineered capsids with 12 to 29 mutations. On the other hand, experimental measurement of production of virus was carried out using a high-throughput viability assay. The experimental mutant data generated by complete, random, or additive strategies are considered training datasets and developed as a deep learning model to accurately predict adeno-associated viral (AAV) capsid viability (Bryant *et al.*, 2021).

9.6 Conclusions

Understanding the effects of mutation on protein function has seen tremendous progress in terms of methodological advancements over the years, from a simple amino acid substitution probability to the most recent state-of-the-art language models. These programs serve to understand the functional landscape of the protein and to explore the unexplored functional landscape that can generate protein-designed sequences with enhanced function. The increasing wealth of data on protein sequences, experimental and modeled structures, and DMS datasets are major reasons to expedite the improved prediction performance of the VEP programs. Also, the technical advancements of DMS and sequencing experiments have accelerated the growth of variant effect prediction. The various applications of protein design discussed in this chapter using both supervised and unsupervised machine learning approaches have highlighted their immense scope and applications to generate designed proteins with enhanced function.

Acknowledgments

This research was supported by the Department of Biotechnology as a research associate fellowship with SG.

References

Backwell, L., & Marsh, J. A. (2022). Diverse molecular mechanisms underlying pathogenic protein mutations: Beyond the loss-of-function paradigm. *Annual Review of Genomics and Human Genetics, 23*, 475–498.

Bryant, D. H., Bashir, A., Sinai, S., Jain, N. K., Ogden, P. J., Riley, P. F., Church, G. M., Colwell, L. J., & Kelsic, E. D. (2021). Deep diversification of an AAV capsid protein by machine learning. *Nature Biotechnology, 39*, 691–696.

Burley, S. K., Bhikadiya, C., Bi, C., Bittrich, S., Chao, H., Chen, L., Craig, P. A., Crichlow, G. V., Dalenberg, K., Duarte, J. M., Dutta, S., Fayazi, M., Feng, Z., Flatt, J. W., Ganesan, S., Ghosh, S., Goodsell, D. S., Green, R. K., Guranovic, V.,

Henry, J., … Zardecki, C. (2023). RCSB Protein Data Bank (RCSB.org): Delivery of experimentally-determined PDB structures alongside one million computed structure models of proteins from artificial intelligence/machine learning. *Nucleic Acids Research, 51*, D488–D508.

Carter, P. (1986). Biochemical Society (Londres). Site-directed Mutagenesis.

Cheng, J., Novati, G., Pan, J., Bycroft, C., Žemgulytė, A., Applebaum, T., Pritzel, A., Wong, L. H., Zielinski, M., Sargeant, T., Schneider, R. G., Senior, A. W., Jumper, J., Hassabis, D., Kohli, P., & Avsec, Ž. (2023). Accurate proteome-wide missense variant effect prediction with AlphaMissense. *Science, 381*, eadg7492.

Choi, Y., & Chan, A. P. (2015). PROVEAN web server: A tool to predict the functional effect of amino acid substitutions and indels. *Bioinformatics, 31*, 2745–2747.

Dunham, A. S., Beltrao, P., & AlQuraishi, M. (2023). High-throughput deep learning variant effect prediction with Sequence UNET. *Genome Biology, 24*, 110.

Esposito, D., Weile, J., Shendure, J., Starita, L. M., Papenfuss, A. T., Roth, F. P., Fowler, D. M., & Rubin, A. F. (2019). MaveDB: An open-source platform to distribute and interpret data from multiplexed assays of variant effect. *Genome Biology, 20*, 223.

Fowler, D. M., Adams, D. J., Gloyn, A. L., Hahn, W. C., Marks, D. S., Muffley, L. A., Neal, J. T., Roth, F. P., Rubin, A. F., Starita, L. M., & Hurles, M. E. (2023). An atlas of variant effects to understand the genome at nucleotide resolution. *Genome Biology, 24*, 147.

Fowler, D. M., & Fields, S. (2014). Deep mutational scanning: A new style of protein science. *Nature Methods, 11*, 801–807.

Gene Ontology Consortium, Aleksander, S. A., Balhoff, J., Carbon, S., Cherry, J. M., Drabkin, H. J., Ebert, D., Feuermann, M., Gaudet, P., Harris, N. L., Hill, D. P., Lee, R., Mi, H., Moxon, S., Mungall, C. J., Muruganugan, A., Mushayahama, T., Sternberg, P. W., Thomas, P. D., Van Auken, K., … Westerfield, M. (2023). The gene ontology knowledgebase in 2023. *Genetics, 224*. Doi: 10.1093/genetics/iyad031

González-Pérez, A., & López-Bigas, N. (2011). Improving the assessment of the outcome of nonsynonymous SNVs with a consensus deleteriousness score, Condel. *American Journal of Human Genetics, 88*, 440–449.

Gray, V. E., Hause, R. J., Luebeck, J., Shendure, J., & Fowler, D. M. (2018). Quantitative missense variant effect prediction using large-scale mutagenesis data. *Cell System, 6*, 116–124.e3.

Hopf, T. A., Ingraham, J. B., Poelwijk, F. J., Schärfe, C. P. I., Springer, M., Sander, C., & Marks, D. S. (2017). Mutation effects predicted from sequence co-variation. *Nature Biotechnology*, *35*, 128–135.

Horne, J., & Shukla, D. (2022). Recent advances in machine learning variant effect prediction tools for protein engineering. *Industrial England Chemical Research, 61*, 6235–6245.

Hoskins, R. A., Repo, S., Barsky, D., Andreoletti, G., Moult, J., & Brenner, S. E. (2017). Reports from CAGI: The critical assessment of genome interpretation. *Human Mutation, 38*, 1039–1041.

James, J. K., Norland, K., Johar, A. S., & Kullo, I. J. (2023). Deep generative models of LDLR protein structure to predict variant pathogenicity. *Journal of Lipid Research,* 100455.

Jankauskaite, J., Jiménez-García, B., Dapkunas, J., Fernández-Recio, J., & Moal, I. H. (2019). SKEMPI 2.0: An updated benchmark of changes in protein–protein binding energy, kinetics and thermodynamics upon mutation. *Bioinformatics*, *35*, 462–469.

Jemimah, S., Yugandhar, K., & Michael Gromiha, M. (2017). PROXiMATE: A database of mutant protein-protein complex thermodynamics and kinetics. *Bioinformatics, 33*, 2787–2788.

Koehler Leman, J., Szczerbiak, P., Renfrew, P. D., Gligorijevic, V., Berenberg, D., Vatanen, T., Taylor, B. C., Chandler, C., Janssen, S., Pataki, A., Carriero, N., Fisk, I., Xavier, R. J., Knight, R., Bonneau, R., & Kosciolek, T. (2023). Sequence-structure-function relationships in the microbial protein universe. *Nature Communication, 14*, 2351.

Kulandaisamy, A., Sakthivel, R., & Gromiha, M. M. (2021). MPTherm: Database for membrane protein thermodynamics for understanding folding and stability. *Brief Bioinformation, 22*, 2119–2125.

Laine, E., Karami, Y., & Carbone, A. (2019). GEMME: A simple and fast global epistatic model predicting mutational effects. *Molecular Biology and Evolution, 36*, 2604–2619.

Li, B., Roden, D. M., & Capra, J. A. (2022). The 3D mutational constraint on amino acid sites in the human proteome. *Nature Communications, 13*, 3273.

Livesey, B. J., & Marsh, J. A. (2020). Using deep mutational scanning to benchmark variant effect predictors and identify disease mutations. *Molecular Systems Biology, 16*, e9380.

Livesey, B. J., & Marsh, J. A. (2023). Updated benchmarking of variant effect predictors using deep mutational scanning. *Molecular Systems Biology, 19*, e11474.

Luo, Y., Jiang, G., Yu, T., Liu, Y., Vo, L., Ding, H., Su, Y., Qian, W. W., Zhao, H., & Peng, J. ECNet is an evolutionary context-integrated deep learning framework for protein engineering. *Nature Communication, 12*, 5743.

Malcolm, B. A., Rosenberg, S., Corey, M. J., Allen, J. S., de Baetselier, A., & Kirsch, J. F. (1989). Site-directed mutagenesis of the catalytic residues Asp-52 and Glu-35 of chicken egg white lysozyme. *Proceedings National Academy of Science USA, 86*, 133–137.

Malhis, N., Jacobson, M., Jones, S. J. M., & Gsponer, J. (2020). LIST-S2: Taxonomy based sorting of deleterious missense mutations across species. *Nucleic Acids Research, 48*, W154–W161.

Mansoor, S., Baek, M., Juergens, D., Watson, J. L., & Baker, D. (2023). Zero-shot mutation effect prediction on protein stability and function using RoseTTAFold. *Protein Sci*ence, e4780.

Merkl, R., & Sterner, R. (2016). Ancestral protein reconstruction: Techniques and applications. *Journal of Biology Chemistry, 397*, 1–21.

Morris, R., Black, K. A., & Stollar, E. J. (2022). Uncovering protein function: From classification to complexes. *Essays Biochemistry, 66*, 255–285.

Morrison, K. L., & Weiss, G. A. (2001). Combinatorial alanine-scanning. *Current Opinion Chemical Biology, 5*, 302–307.

Munro, D., & Singh, M. (2021). DeMaSk: A deep mutational scanning substitution matrix and its use for variant impact prediction. *Bioinformatics, 36*, 5322–5329.

Nagar, N., Tubiana, J., Loewenthal, G., Wolfson, H. J., Ben Tal, N., & Pupko, T. (2023). EvoRator2: Predicting site-specific amino acid substitutions based on protein structural information using deep learning. *Journal of Molecular Biology, 435*, 168155.

Ng, P. C., & Henikoff, S. (2003). SIFT: Predicting amino acid changes that affect protein function. *Nucleic Acids Research, 31*, 3812–3814.

Nikam, R., Kulandaisamy, A., Harini, K., Sharma, D., & Gromiha, M. M. (2021). ProThermDB: Thermodynamic database for proteins and mutants revisited after 15 years. *Nucleic Acids Research, 49*, D420–D424.

Notin, P., Dias, M., Frazer, J., Marchena-Hurtado, J., Gomez, A., Marks, D. S., & Gal, Y. (2022). Tranception: Protein fitness prediction with autoregressive transformers and inference-time retrieval. *Machine Learning*. doi:10.48550/ARXIV.2205.13760

Packer, M. S., & Liu, D. R. (2015). Methods for the directed evolution of proteins. *Nature Reviews Genetics, 16*, 379–394.

Reva, B., Antipin, Y., & Sander, C. (2007). Determinants of protein function revealed by combinatorial entropy optimization. *Genome Biology, 8*, R232.

Riesselman, A. J., Ingraham, J. B., & Marks, D. S. (2018). Deep generative models of genetic variation capture the effects of mutations. *Nature Methods, 15*, 816–822.

Romero, P. A., Krause, A., & Arnold, F. H. (2013). Navigating the protein fitness landscape with Gaussian processes. *Proceedings of the National Academy Science USA, 110*, E193–201.

Shin, J.-E., Riesselman, A. J., Kollasch, A. W., McMahon, C., Simon, E., Sander, C., Manglik, A., Kruse, A. C., & Marks, D. S. (2021). Protein design and variant prediction using autoregressive generative models. *Nature Communication, 12*, 2403.

Shoemaker, B. A., & Panchenko, A. R. (2007). Deciphering protein–protein interactions. Part I. Experimental techniques and databases. *PLoS Computational Biology, 3*, e42.

Thangakani, A. M., Nagarajan, R., Kumar, S., Sakthivel, R., Velmurugan, D., & Gromiha, M. M. (2016). CPAD, curated protein aggregation database: A repository of manually curated experimental data on protein and peptide aggregation. *PLoS One, 11*, e0152949.

Velecký, J., Hamsikova, M., Stourac, J., Musil, M., Damborsky, J., Bednar, D., & Mazurenko, S. (2022). SoluProtMut: A manually curated database of protein solubility changes upon mutations. *Computational Structural Biotechnology Journal, 20*, 6339–6347.

Wang, C. Y., Chang, P. M., Ary, M. L., Allen, B. D., Chica, R. A., Mayo, S. L., & Olafson, B. D. (2019). ProtaBank: A repository for protein design and engineering data. *Protein Science, 28*, 672.

Xu, X., Yu, H., Zhu, S., Li, P., Li, X., Gao, Y., Xiang, Y., Zhao, G., Simoncini, T., & Lin, H. (2023). Mutation of aspartic acid 262 on estrogen receptor abrogates estradiol signaling pathway. *Gynecological Endocrinology, 39*, 2250881.

Xu, Y., Verma, D., Sheridan, R. P., Liaw, A., Ma, J., Marshall, N. M., Mcintosh, J., Sherer, E. C., Svetnik, V., & Johnston, J. M. (2020). Deep dive into machine learning models for protein engineering. *Journal of Chemical Information and Modeling, 60*, 2773–2790.

Yang, K. K., Wu, Z., & Arnold, F. H. (2019). Machine-learning-guided directed evolution for protein engineering. *Nature Methods, 16*, 687–694.

Yun, C.-H., Boggon, T. J., Li, Y., Woo, M. S., Greulich, H., Meyerson, M., & Eck, M. J. (2007). Structures of lung cancer-derived EGFR mutants and inhibitor complexes: Mechanism of activation and insights into differential inhibitor sensitivity. *Cancer Cell, 11*, 217–227.

Zhou, M., Li, Q., & Wang, R. (2016). Current experimental methods for characterizing protein-protein interactions. *ChemMedChem, 11*, 738–756.

Part III

Disease causing mutations

https://doi.org/10.1142/9789811293269_0010

Chapter 10

Computational resources for understanding disease-causing mutations in proteins: applications to HIV

Sankaran Venkatachalam[1,#], Amit Phogat[1,#], and M. Michael Gromiha[1,*]

[1]*Department of Biotechnology, Bhupat and Jyoti Mehta School of Biosciences, Indian Institute of Technology Madras, Chennai, Tamil Nadu 600036, India*

Abstract

Amino acid substitutions in a protein alter its structure, which may affect the function and lead to diseases. Identifying disease-causing mutations is important for disease diagnosis and the development of personalized medicine. The advent of vast and diverse datasets and the availability of machine learning and deep learning algorithms has paved the way for the development of computational techniques to identify disease-causing

---

*Corresponding author
Tel: +91-44-2257-4138
Fax: +91-44-2257 4102
MMG: gromiha@iitm.ac.in
#Equal contribution

mutations. This chapter comprehends the list of available tools for identifying disease-causing mutations, with an emphasis on the role of mutations in conferring drug resistance in human immunodeficiency virus (HIV) infection.

10.1 Introduction

Mutations in a protein may disturb its structure and stability. Most of these mutations do not affect the function of a protein, which are known as neutral or benign mutations (Bromberg & Rost, 2007; Kulshreshtha *et al.*, 2016). On the other hand, mutations that affect protein function and lead to various diseases are called disease-causing mutations. For example, a novel Presenilin (PSEN)1 mutation, R278I was identified in familial Alzheimer's disease (AD), featuring language impairment and preserved memory. Schinzel–Giedion syndrome (SGS), a genetic disorder is associated with mutations in the Ski-like protein (SKI) gene, including missense changes such as G116E and G117R, disrupting key protein interactions in the SMAD2/3-binding and Dachshund-homology domains, and these alterations disrupt molecular interactions crucial for TGF-β signaling (Doyle *et al.*, 2012). Focal cortical dysplasia type II (FCDII), a major cause of intractable focal epilepsies, is associated with hyperactivation of the mTOR pathway in brain cells. The mutations in TSC1 (R22W, R204C) and TSC2 (V1547I) are identified in 12.5% of FCDII cases, leading to hyperactivation of the mTOR pathway by disrupting the TSC1-TSC2 complex (Koh & Lee, 2018).

Lethal neonatal encephalopathies linked to mitochondrial dysfunction are associated with biallelic large deletions in the contiguous ATAD3B and ATAD3A genes. A novel homozygous variant, L406R in ATAD3A, was identified in siblings with fatal neonatal cerebellar hypoplasia, seizures, and multi-organ involvement. It alters the protein structure, leading to reduced ATAD3A levels, and emphasizes the critical role of ATAD3A in mitochondrial biogenesis (Peralta *et al.*, 2019). Somatic mutations in hematopoietic stem and progenitor cells (HSPCs) leading to clonal hematopoiesis of indeterminate potential (CHIP) are strongly associated with coronary artery disease (CAD). The JAK2 V617F mutation induces accelerated atherosclerosis through cytokine-independent JAK-STAT pathway activation, emphasizing distinct pathways in CHIP-associated coronary disease (Heimlich & Bick, 2022). Further, major cancer-causing mutations are discussed in Chapter 11.

Several databases have been developed to accumulate information on the effect of mutations in proteins based on diseases. These data are effectively utilized for developing computational tools to predict disease-causing mutations. In this chapter, we discuss the databases, which contain information on disease-causing mutations along with tools developed for identifying such mutations. Further, the influence of mutations on drug resistance in human immunodeficiency virus (HIV) infection will be described.

10.2 Databases for disease-causing mutations

Information on disease-causing mutations has been rapidly increasing and the data are accumulated in several databases. **Table 10.1** lists a set of databases that contain details on disease-causing mutations.

Table 10.1 List of databases for disease-causing and neutral mutations

Name	Data type	Link	References
dbSNP	Nucleotide sequence variation including single nucleotide polymorphism, insertions/deletions, microsatellite repeats	http://www.ncbi.nlm.nih.gov/SNP.	Sherry *et al.* (2001)
OMIM	Catalog of human genes and genetic disorders	https://www.omim.org/	Hamosh *et al.* (2005)
SwissVar	Single amino acid polymorphism	https://www.uniprot.org/	Mottaz *et al.* (2010)
VariBench	Experimentally verified variation data was mapped to DNA, RNA, and protein sequences	http://structure.bmc.lu.se/VariBench	Sasidharan Nair and Vihinen (2013)
1000 Genomes	Genome and exome sequencing data	https://www.internationalgenome.org/home	Auton *et al.* (2015)

(Continued)

Table 10.1 (*Continued*)

Name	Data type	Link	References
ClinVar	Germline and somatic mutation data	https://www.ncbi.nlm.nih.gov/clinvar/	Landrum *et al.* (2016)
HuVarBase	Human variant data at protein and gene levels	https://www.iitm.ac.in/bioinfo/huvarbase/	Ganesan *et al.* (2019)
MaveDB	Multiplex assays of variant effects	https://www.mavedb.org/	Esposito *et al.* (2019)
HGMD	Germline mutations in nuclear genes	http://www.hgmd.org	Stenson *et al.* (2020)
gnomAD	Sequencing data from exomes and genomes	https://gnomad.broadinstitute.org/	Karczewski *et al.* (2020)

*Last accessed on November 20, 2023.

The dbSNP database (Sherry *et al.*, 2001) is a public repository for genetic variation, sequence information, and experimental conditions around polymorphisms for diverse research applications. The database encompasses various nucleotide sequence variations, including single nucleotide substitutions, insertions/deletions, microsatellite repeats, and named variants. It can be accessed at http://www.ncbi.nlm.nih.gov/SNP. Online Mendelian Inheritance in Man (OMIM) (Hamosh *et al.*, 2005) is a comprehensive knowledgebase of human genes and genetic disorders. The database has been integrated with the Entrez suite, providing timely information for human genetics research, education, and clinical genetics. Each OMIM entry includes a full-text summary of genetically determined phenotypes and links to other genetic databases. OMIM is available at https://www.omim.org/. SwissVar (Mottaz *et al.*, 2010) contains information on single amino acid polymorphisms and associated diseases within the UniProtKB/Swiss-Prot database. The portal allows querying for similar diseases, information about protein products, and the molecular details of each variant. The data previously hosted on SwissVar, including variant classification from literature reports, is now accessible through the UniProt Knowledgebase https://www.uniprot.org/.

VariBench (Sasidharan Nair & Vihinen, 2013) benchmark database suite contains experimentally verified, high-quality variation data sourced from literature and relevant databases. It provides mapping to different levels (sequences of protein, RNA, and DNA as well as protein structures) along with identifier mapping to relevant databases. The database is freely accessible at http://structure.bmc.lu.se/VariBench. The 1000 Genomes Project (Auton *et al.*, 2015) comprehensively describes common human genetic variation, utilizing whole-genome sequencing, deep exome sequencing, and microarray genotyping. The data from the project can be accessed at https://www.internationalgenome.org/home. ClinVar (Landrum *et al.*, 2016) is an open-access repository offering interpretations of the clinical significance of variants for reported conditions. It encompasses germline and somatic variants of diverse sizes, types, and genomic locations, with submissions from various sources, including clinical and research laboratories, OMIM®, GeneReviews™, UniProt, and expert panels. The database can be accessed at https://www.ncbi.nlm.nih.gov/clinvar/.

HuVarBase (HUmanVARiantdataBASE) (Ganesan *et al.*, 2019) is a comprehensive resource that integrates publicly available human variant data at both protein and gene levels. It combines protein-level information such as amino acid sequence, secondary structure, domain, function, and more with gene-level data, including gene name, chromosome number, DNA mutation, and disease class for disease-causing variants. HuVarBase offers search, display, visualization, and download options, enhancing the analysis of variants for disease prediction, prevention, and treatment. The database is accessible at https://www.iitm.ac.in/bioinfo/huvarbase/. MaveDB (Esposito *et al.*, 2019) is a repository for large-scale measurements of sequence variant impact, facilitating the discovery and distribution of multiplex assays of variant effect (MAVEs). MaveDB enables researchers to store and publish processed MAVE datasets, associated metadata, and linked raw data in a machine-readable, standardized, and searchable format. The user-friendly web interface enhances accessibility for the research community, facilitating applications in clinical studies, meta-analysis, and computational reanalysis. The database is accessible at https://www.mavedb.org/.

Human Gene Mutation Database (HGMD) (Stenson *et al.*, 2020) is a comprehensive database of published germline mutations in nuclear genes associated with human inherited diseases. The database serves as a valuable resource for accessing curated mutation data reported in the literature, aiding in both research and clinical applications. The database can be accessed at http://www.hgmd.org. Genome Aggregation Database (gnomAD) (Karczewski *et al.*, 2020) aggregates data from exomes and genomes, identifying high-confidence predicted loss-of-function variants after stringent filtering. This resource facilitates the analysis of genetic variants that inactivate protein-coding genes, offering insights into the phenotypic consequences of gene disruption. The database is available at https://gnomad.broadinstitute.org/.

10.3 Prediction of disease-causing mutations in proteins

Classifying benign and disease-causing mutations can help in understanding the role of mutations in diseases. Identifying disease-causing mutations experimentally is cumbersome and time-consuming. The increasing availability of large amounts of sequencing data helps for developing reliable computational tools (Kulshreshtha *et al.*, 2016). A list of currently available algorithms for predicting disease-causing mutations is presented in **Table 10.2**. The methods specifically designed for cancer-causing mutations are discussed in Chapter 11.

Table 10.2 List of currently available algorithms for predicting disease-causing mutations

Name	Features	Link	References
SNAP	Neural network-based method using biochemical properties, PSSM matrix, position-specific independent counts (PSIC), protein-family	http://www.rostlab.org/services/SNAP	Bromberg and Rost (2007)
PROVEAN	Alignment score utilizing homologous and distantly related sequences and clustering	http://provean.jcvi.org	Choi *et al.* (2012)
SIFT	Sequence alignment and evolutionary features	https://sift.bii.a-star.edu.sg/	Sim *et al.* (2012)

Table 10.2 (*Continued*)

Name	Features	Link	References
FATHMM	Multiple sequence alignment and hidden Markov models	http://fathmm.biocompute.org.uk	Shihab *et al.* (2013)
PolyPhen-2	Evolutionary and structural information	http://genetics.bwh.harvard.edu/pph2/	Adzhubei *et al.* (2013)
PredictSNP	Consensus of tools such as MAPP, nsSNPAnalyzer, PANTHER, PhD-SNP, PolyPhen-1, PolyPhen-2, SIFT, and SNAP	http://loschmidt.chemi.muni.cz/predictsnp	Bendl *et al.* (2014)
PON-P2	Random forest using amino acid features, evolutionary conservation, functional, and structural annotations	http://structure.bmc.lu.se/PON-P2/	Niroula *et al.* (2015)
AlphaMissense	Deep learning model using multiple sequence alignment	https://github.com/google-deepmind/alphamissense	Cheng *et al.* (2023)

Bromberg and Rost (2007) developed a neural network-based method called screening for non-acceptable polymorphisms (SNAP) for distinguishing disease-causing mutations from neutral mutations. The model uses various features such as biochemical properties (hydrophobicity, charge), encoded sequence information, transitioning frequencies, position-specific scoring matrix (PSSM) matrix, position-specific independent counts (PSIC), and family information. In a set of 80,000 mutants, SNAP correctly predicted 80% of disease-causing and 76% of neutral mutations. The tool is freely available at http://www.rostlab.org/services/SNAP.

Choi *et al.* (2012) developed a tool, Protein Variation Effect Analyzer (PROVEAN), for predicting the effects of point mutations as well as insertion, deletion, and multiple amino acid substitutions based on alignment scores. It utilizes homologous and distantly related sequences from the National Center for Biotechnology Information (NCBI) nonredundant protein database along with clustering using the CD-HIT program to generate an alignment-based score. In a set of 20,821 disease variants and 36,825 neutral mutations, PROVEAN was able to distinguish between neutral and pathogenic mutations with an area under the receiver operating characteristic curve (AUC) of 0.85. It is freely available at http://provean.jcvi.org.

Sim *et al.* (2012) developed a web server for sorting intolerant from tolerant (SIFT), which predicts the effect of single amino acid substitutions on protein function based on the assumption that amino acids crucial for protein function are conserved throughout the protein family (Ng & Henikoff, 2001). It performs the alignment for a query sequence and calculates the probability of the occurrence of each amino acid at any position. The algorithm provides an output score for the substitution, which is used to define a mutation as damaging or benign. The SIFT web server is freely available at https://sift.bii.a-star.edu.sg/.

Shihab *et al.* (2013) introduced the functional analysis through hidden Markov models (FATHMM) tool for predicting the effects of mutations on protein function. This algorithm utilizes JackHMMER for searching homologous sequences in the Uniref90 database and constructs hidden Markov models (probabilistic models), which capture information at each position from multiple sequence alignment. The model generates probability scores to predict the effect of mutation as pathogenic or neutral. The tool is freely available at http://fathmm.biocompute.org.uk.

Ramensky *et al.* (2002) developed Polymorphism Phenotyping (PolyPhen) for predicting the probability of damaging effects of missense mutations using evolutionary and structural information. Later (Adzhubei *et al.*, 2010) constructed a web server and software, PolyPhen-2, for predicting the effects of missense mutations on protein stability and function. The PolyPhen-2 differs from its previous version based on the alignment procedure, predictive features, and classification method. It utilizes evolutionary and structural features and a Naives Bayes classifier for predicting the probability of a mutation to be damaging or neutral. The tool is freely available at http://genetics.bwh.harvard.edu/pph2/.

Bendl *et al.* (2014) developed a consensus tool, PredictSNP, using prediction results obtained from other methods such as MAPP, nsSNPAnalyzer, PANTHER, PhD-SNP, PolyPhen-1, PolyPhen-2, SIFT, and SNAP. The consensus classifier combines the prediction scores of all tools and generates a PredictSNP score with +1 for deleterious and −1 for neutral predictions. The tool is freely available at http://loschmidt.chemi.muni.cz/predictsnp.

Niroula *et al.* (2015) presented a machine learning-based classifier PON-P2 for classifying disease-causing and neutral mutations. It is based on a random forest classifier using amino acid features, gene ontology, evolutionary conservation, and functional and structural annotations for classifying a mutation into pathogenic, neutral, and unknown classes. The tool is freely available at http://structure.bmc.lu.se/PON-P2/.

Cheng *et al.* (2023) introduced AlphaMissense, a deep learning model built on AlphaFold2 (protein structure prediction tool) for identifying disease-causing mutations. The model is trained on millions of sequences using multiple sequence alignment and it derives its predictions from structural aberrations that occur due to mutations. It is similar to AlphaFold, which is divided into two parts, such as (i) structure pre-training and (ii) fine-tuning the variants (benign and pathogenic). The prediction scores generated by the model are interpreted as the log-likelihood difference between wild and substituted amino acids. Further, a threshold is used to classify a mutation as "likely benign" or "likely pathogenic" or "ambiguous." The pre-annotated predictions and workflow of the AlphaMissense model are available at https://github.com/google-deepmind/alphamissense.

10.4 Influence of mutations for drug resistance in Human Immunodeficiency Virus infection

In this section, we will discuss the significance of mutations for drug resistance in HIV as well as available databases.

10.4.1 *Structure of HIV protease*

HIV protease (HIV PR) is one of the targets for mitigating HIV replication. **Figure 10.1** shows the 3D structure of HIV PR, which is a homodimer with 99 amino acids per monomer. Each of the monomers consists of functional components, namely the flap, hinge, fulcrum, and

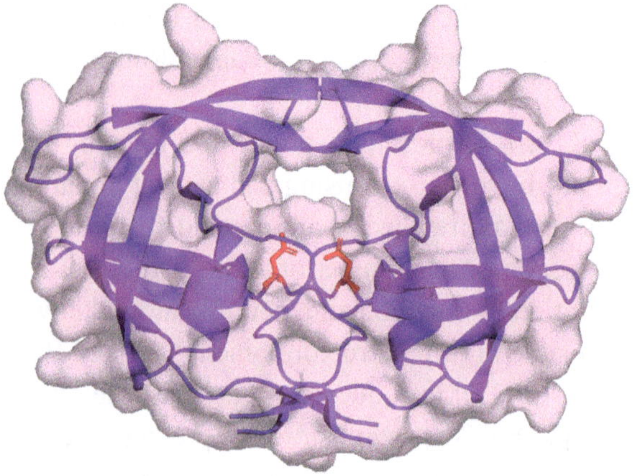

Figure 10.1 3D structure of HIV protease. The protein is shown in both cartoon and surface representations. The catalytic residues (Asp25/Asp25′) are shown in sticks

cantilever that work in synergy to achieve the function of binding the substrate. The binding site of the HIV PR is located at the dimer interface. The binding site is a conserved catalytic triad consisting of amino acids D25-T26-G27, which hydrolyzes the scissile peptide bond (Weber *et al.*, 2021; Dakshinamoorthy *et al.*, 2023).

10.4.2 *HIV protease inhibitors and effects of mutations*

The class of drugs that target HIV PR is called HIV protease inhibitors (PIs). These inhibitors are designed to mimic the binding of the substrate, and upon binding, they block the binding of the substrate to the protease, thereby mitigating the protease activity. PIs when bound to the protease, establish hydrogen bonds, hydrophobic, and van der Walls interactions with the binding site residues (Dakshinamoorthy *et al.*, 2023). However, the occurrence of mutations in the HIV PR has impaired the effect of these PIs, leading to drug resistance (Wensing *et al.*, 2010). These mutations occur due to the exposure of HIV PR to the drugs used for treatment. Moreover, they induce perturbations in the interaction network, leading to reduced drug susceptibility. Based on the site of occurrence, mutations are classified as (i) primary or major mutations and (ii) secondary or minor mutations. The sections below will describe the influence of these mutations on evading the inhibition mechanism and making the drug ineffective (Weber *et al.*, 2021).

10.4.2.1 *Role of primary mutations*

Primary mutations or major mutations occur in the active site of the HIV PR. In general, the primary mutations cause steric clashes, alter the hydrogen bond, hydrophobic, van der Waals interactions, and increase the electrostatic repulsion between the drug and the HIV PR, leading to the expansion of the binding site, thereby affecting the binding energy (Weber *et al.*, 2021; Dakshinamoorthy *et al.*, 2023). A list of primary or major mutations that are caused due to exposure to drugs is listed in **Table 10.3**.

Some of the active site mutations of HIV PR, when exposed to saquinavir (SQV), include G48VM, I54VTALM, V82AT, I84V, N88S, and L90M. In the G48V mutation, the conformational change between the wild-type and G48V HIV PR was observed on the protein flap. The change in the flap conformation occurred due to the steric clash between the drug and the protease, thereby exposing the hydrophobic residue F53 to the solvent (Wittayanarakul *et al.*, 2005). The mutation D30N

Table 10.3 List of major or primary mutations based on different drugs

Drug	Major mutations
Saquinavir (SQV*)	G48VM, I54VTALM, I84V, N88S, L90M, V82AT
Nelfinavir (NFV)	D30N, L33F, M46IL, G48VM, I47V, I54VTALM, V82ATSF, I84V N88DS, L90M
Fosamprenavir FPV*	L33F, V32I M46IL, I47VA, I50V, L76V, V82ATSF, I84V, L90M, I54VTALM
Indinavir (IDV*)	I54VTALM, L76V, I84V, N88S, L90M, V82ATFS, V32I, L33F
Tipranavir TPV*	I84V, I47VA, L33F, V32I, I84V, V82LT
Atazanavir (ATV*)	L90M, N88S, V82ATFS, I50L, I84V, I47V, V32I, L33F
Lopinavir (LPV*)	G48VM, I50V, L33F, V32I, I54VTALM, L76V, V82ATFS, M46IL, L90M
Darunavir (DRV*)	I50V, I54LM, I84V, L33F, I47VA, V32I, L76V

*Drug is given in combination with Ritonavir (RTV).

demolishes the hydrogen bond between the protease and the drug, nelfinavir (NFV), which affects the binding free energy and hence causes the flaps to open. This decreased the strength of the backbone hydrogen bonding carboxyl group in residue G48 with NFV (Santos & Soares, 2011). Similarly, in the cross-resistant mutant (mutation that occurs when exposed to more than one drug), L90M causes increased fluctuations of the flaps (Henes *et al.*, 2019). These primary mutations cause conformational changes in the active site (increased distance between the flap and the active site). In addition to decreasing drug susceptibility, these primary mutations also decrease the binding affinity for the substrate, leading to reduced viral activity.

10.4.2.2 *Role of secondary mutation*

While the primary or major mutations affect the protease activity by causing a decrease in the binding affinity of the natural substrates, the minor mutations also referred to as the secondary mutations, help the protease to restore the activity. The minor or secondary are referred to as non-active site mutations, that is, mutations not involving active site regions. A list of minor mutations corresponding to the drug molecules is listed in **Table 10.4**.

Table 10.4 List of minor or secondary mutations based on drugs

	Minor or secondary mutations
Saquinavir (SQV*)	V77I, I62V, A71VT, G73S, L24I, L10IRV, L38HL
Nelfinavir (NFV)	A71VT, V77I, M36I, L10FI
Fosamprenavir FPV*	L10FIRV, G73S
Indinavir (IDV*)	G73SA, M36I, L24I, K20MR, A71VT, V77I, L10IRV
Tipranavir TPV*	L10V, M36ILV, N83D, T74P, L89IVM, Q58E, K43T, H69KR
Atazanavir (ATV*)	I85V, I93LM, G73CSTA, F53LY, L24I, K20RMITV, L10IFVC, D60E, F53LY, M36ILV
Lopinavir (LPV*)	K20MR, L10FIRV, L63P, G73S, A71VT, F53L
Darunavir (DRV*)	V11I, T749, L89V

*Drug is given in combination with Ritonavir (RTV).

For instance, the residue Leu at the 10th position, located away from the active site, changed to L10IRV, L10FIRV, and L10FI, respectively, in the presence of saquinavir with indinavir, fosamprenavir with lopinavir, and Nlefnavir in combination with ritonavir. Further, the salt bridge between E35 and R57 in each of the monomers is essential for flap stability. The absence of an E35-R57 salt bridge in the PRs results in an apparent outward movement of the flaps, leading to the binding site expansion (Sherry *et al.*, 2021). Further, interaction of SQV with the HIV PR subtype C characterized by the insertion of two residues His and Leu at position 38, induces outward flap movement, causing decreased binding affinity (Venkatachalam *et al.*, 2023).

10.4.3 *Computational resources for mutations in HIV protease and drug resistance*

Predicting the drug resistance caused by HIV is of prime importance. Hence, several databases and computational tools have been developed for interpreting drug-resistant mutations.

Shafer (2006) developed a database and a system, HIVdb, for interpreting genotypic resistance information. HIVdb takes nucleotide sequences as input, performs sequence alignment, and computes a penalty score. It signifies the influence of individual mutations on drug susceptibility and the effect of combined mutation penalties on antiretroviral susceptibility on the most commonly occurring antiretroviral drug resistance mutation. Based on the penalty scores, the sequences are labeled as susceptible (score 0–9), potential low level of resistance (score 10–14), low-level resistance (score 15–30), intermediate resistance (score 30–59), and high-level resistance (score >60). Additionally, the HIVdb provides comments on (i) antiretroviral drug resistance that receives a mutation penalty score, (ii) mutations that are associated with decreased drug susceptibility, (iii) known drug resistance position with unusual mutations, and (iv) mutations associated with commonly used "genotypic susceptibility scores" developed for tipranavir, darunavir, and etravirine.

Geno2pheno (Döring *et al.*, 2018) is a web server that uses the mapped genotype to phenotype data of HIV protease and reverse transcriptase (RT) to predict resistance factors. It uses three different sets of sequences, such as (i) 652 genotypes, including 604 subtype B and 48 non-subtype B sequences, (ii) 184 GenBank sequences identified from patients not treated with any antiretroviral drug, and (iii) 178 sequences, including 124 subtype B and 54 non-subtype B trained on support vector machines (SVMs) to predict the drug-resistant factors.

For a submitted query, the web server returns a list of outputs such as (i) the alignment with the HXB2 reference strain, (ii) classification results, (iii) the predicted resistance factors, (iv) Z-scores relative to the treatment-naive, and (v) probability scores for all the drugs.

Riemenschneider *et al.* (2016) developed a tool, SHIVA for predicting HIV drug resistance that incorporates models for examining the resistance of PIs, nucleotide reverse transcriptase inhibitors (NRTIs), nonnucleotide reverse transcriptase inhibitors (NNRTIs), and integrase inhibitors (INIs). It takes the amino acid sequence or RNA/DNA sequence (covert to amino acid sequence) and computes hydrophobicity scores for prediction. The predictions are carried out using a random forest model that classifies the input sequences as resistant or susceptible. The output includes information about the fraction of resistance sequence for a given patient, which can be used to detect and treat minority variants. It is noteworthy that the models are trained for four classes of drugs, including all four types of inhibitors mentioned above that are used to treat HIV infection.

10.5 Conclusions

This chapter provides a summary of the available databases and tools to identify the disease-causing mutations in proteins. Further, different computational approaches, highlighting their advantages, are described. The second part of this chapter discusses the role of drug resistance associated with HIV infections. It includes details on the occurrence of mutations in the presence of PIs, along with computational tools to tackle the problem of drug resistance.

Acknowledgments

We thank the Department of Biotechnology and the Indian Institute of Technology Madras for computational facilities. The work is partially supported by the Department of Science and Technology, Government of India, under the BRICS project (DST/ICD/BRICS/Call-3/HIV protease/2019) to MMG, South African National Research Foundation (NRF) via the BRICS multilateral joint science and technology research collaboration (Grant number: 120452) to YS and the Conselho Nacional de Desenvolvimento Científico e Tecnológico (CNPq), Brazil (Grant number: 402817/2019-2) to VA.

References

Adzhubei, I., Jordan, D. M., & Sunyaev, S. R. (2013). Predicting functional effect of human missense mutations using PolyPhen-2. *Current Protocols in Human Genetics, 76*(1), 7–20.

Adzhubei, I. A., Schmidt, S., Peshkin, L., Ramensky, V. E., Gerasimova, A., Bork, P., Kondrashov, A. S., & Sunyaev, S. R. (2010). A method and server for predicting damaging missense mutations. *Nature Methods, 7*(4), 248–249.

Auton, A., Brooks, L. D., Durbin, R. M., Garrison, E. P., Kang, H. M., Korbel, J. O., Marchini, J. L., McCarthy, S., McVean, G. A., & Abecasis, G. R. (2015). A global reference for human genetic variation. *Nature, 526*, 68–74.

Bendl, J., Stourac, J., Salanda, O., Pavelka, A., Wieben, E. D., Zendulka, J., Brezovsky, J., & Damborsky, J. (2014). PredictSNP: Robust and accurate consensus classifier for prediction of disease-related mutations. *PLoS Computational. Biology, 10*(1), e1003440.

Bromberg, Y., & Rost, B. (2007). SNAP: Predict effect of non-synonymous polymorphisms on function. *Nucleic Acids Research, 35*, 3823–3835.

Cheng, J., Novati, G., Pan, J., Bycroft, C., Žemgulytė, A., Applebaum, T., Pritzel, A., Wong, L. H., Zielinski, M., Sargeant, T., Schneider, R. G., Senior, A. W., Jumper, J., Hassabis, D., Kohli, P., & Avsec, Ž. (2023). Accurate proteome-wide missense variant effect prediction with AlphaMissense. *Science, 381*(6664), eadg7492.

Choi, Y., Sims, G. E., Murphy, S., Miller, J. R., & Chan, A. P. (2012). Predicting the functional effect of amino acid substitutions and indels. *PLoS One, 7*, e46688.

Dakshinamoorthy, A., Asmita, A., & Senapati, S. (2023). Comprehending the structure, dynamics, and mechanism of action of drug-resistant HIV protease. *ACS Omega, 8*, 9748–9763.

Döring, M., Büch, J., Friedrich, G., Pironti, A., Kalaghatgi, P., Knops, E., Heger, E., Obermeier, M., Däumer, M., Thielen, A., Kaiser, R., Lengauer, T., & Pfeifer, N. (2018). geno2pheno[ngs-freq]: A genotypic interpretation system for identifying viral drug resistance using next-generation sequencing data. *Nucleic Acids Research, 46*, W271–W277.

Doyle, A. J., Doyle, J. J., Bessling, S. L., Maragh, S., Lindsay, M. E., Schepers, D., Gillis, E., Mortier, G., Homfray, T., Sauls, K., Norris, R. A., Huso, N. D., Leahy, D., Mohr, D. W., Caulfield, M. J., Scott, A. F., Destrée, A., Hennekam, R. C., Arn, P. H., Curry, C. J., Van Laer, L., McCallion, A. S., Loeys, B. L., & Dietz, H. C. (2012). Mutations in the TGF-β repressor SKI cause Shprintzen–Goldberg syndrome with aortic aneurysm. *Nature Genetics, 44*, 1249–1254.

Esposito, D., Weile, J., Shendure, J., Starita, L. M., Papenfuss, A. T., Roth, F. P., Fowler, D. M., & Rubin, A. F. (2019). MaveDB: An open-source platform to distribute and interpret data from multiplexed assays of variant effect. *Genome Biology, 20*, 1–11.

Ganesan, K., Kulandaisamy, A., Binny Priya, S., & Gromiha, M. M. (2019). HuVarBase: A human variant database with comprehensive information at gene and protein levels. *PLoS One, 14*, e0210475.

Hamosh, A., Scott, A. F., Amberger, J. S., Bocchini, C. A., & McKusick, V. A. (2005). Online Mendelian Inheritance in Man (OMIM), a knowledgebase of human genes and genetic disorders. *Nucleic Acids Research, 30*(1), 50–55.

Heimlich, J. B., & Bick, A. G. (2022). Somatic mutations in cardiovascular disease. *Circulation Research, 130*, 149–161.

Henes, M., Kosovrasti, K., Lockbaum, G. J., Leidner, F., Nachum, G. S., Nalivaika, E. A., Bolon, D. N. A., Kurt Yilmaz, N., Schiffer, C. A., & Whitfield, T. W. (2019). Molecular determinants of epistasis in HIV-1 protease: Elucidating the interdependence of L89V and L90M mutations in resistance. *Biochemistry, 58*, 3711–3726.

Karczewski, K. J., Francioli, L. C., Tiao, G., Cummings, B. B., Alföldi, J., Wang, Q., Collins, R. L., Laricchia, K. M., Ganna, A., Birnbaum, D. P., Gauthier, L. D., Brand, H., Solomonson, M., Watts, N. A., Rhodes, D., Singer-Berk, M., England, E. M., Seaby, E. G., Kosmicki, J. A., . . . MacArthur, D. G. (2020). The mutational constraint spectrum quantified from variation in 141,456 humans. *Nature, 581*, 434–443.

Koh, H. Y., & Lee, J. H. (2018). Brain somatic mutations in epileptic disorders. *Molecules Cells, 41*, 881–888.

Kulshreshtha, S., Chaudhary, V., Goswami, G. K., & Mathur, N. (2016). Computational approaches for predicting mutant protein stability. *Journal of Computer-Aided Molecular Design, 30*, 401–412.

Landrum, M. J., Lee, J. M., Benson, M., Brown, G., Chao, C., Chitipiralla, S., Gu, B., Hart, J., Hoffman, D., Hoover, J., Jang, W., Katz, K., Ovetsky, M., Riley,

G., Sethi, A., Tully, R., Villamarin-Salomon, R., Rubinstein, W., & Maglott, D. R. (2016). ClinVar: Public archive of interpretations of clinically relevant variants. *Nucleic Acids Research, 44*, D862–D868.

Mottaz, A., David, F. P. A., Veuthey, A.-L., & Yip, Y. L. (2010). Easy retrieval of single amino-acid polymorphisms and phenotype information using SwissVar. *Bioinformatics, 26*, 851–852.

Ng, P. C., & Henikoff, S. (2001). Predicting deleterious amino acid substitutions. *Genome Research, 11*, 863–874.

Niroula, A., Urolagin, S., & Vihinen, M. (2015). PON-P2: Prediction method for fast and reliable identification of harmful variants. *PLoS One, 10*(2), e0117380.

Peralta, S., González-Quintana, A., Ybarra, M., Delmiro, A., Pérez-Pérez, R., Docampo, J., Arenas, J., Blázquez, A., Ugalde, C., & Martín, M. A. (2019). Novel ATAD3A recessive mutation associated to fatal cerebellar hypoplasia with multiorgan involvement and mitochondrial structural abnormalities. *Molecular Genetics and Metabolism, 128*, 452–462.

Ramensky, V., Bork, P., & Sunyaev, S. (2002). Human non-synonymous SNPs: Server and survey. *Nucleic Acids Research, 30*, 3894–3900.

Riemenschneider, M., Hummel, T., & Heider, D. (2016). SHIVA — A web application for drug resistance and tropism testing in HIV. *BMC Bioinformatics, 17*(1), 1–6.

Santos, A. F. A., & Soares, M. A. (2011). The impact of the nelfinavir resistance-conferring mutation D30N on the susceptibility of HIV-1 subtype B to other protease inhibitors. *Memorias do Instituto Oswaldo Cruz, 106*, 177–181.

Sasidharan Nair, P., & Vihinen, M. (2013). VariBench: A benchmark database for variations. *Human Mutation, 34*, 42–49.

Shafer, R. W. (2006). Rationale and uses of a public HIV drug-resistance database. *Journal of Infectious Disorder. 194* Suppl, S51–S58.

Sherry, D., Worth, R., Ismail, Z. S., & Sayed, Y. (2021). Cantilever-centric mechanism of cooperative non-active site mutations in HIV protease: Implications for flap dynamics. *Journal of Molecular Graphics Modelling, 106*, 107931.

Sherry, S. T., Ward, M. H., Kholodov, M., Baker, J., Phan, L., Smigielski, E. M., & Sirotkin, K. (2001). dbSNP: The NCBI database of genetic variation. *Nucleic Acids Research, 29*, 308–311.

Shihab, H. A., Gough, J., Cooper, D. N., Stenson, P. D., Barker, G. L. A., Edwards, K. J., Day, I. N. M., & Gaunt, T. R. (2013). Predicting the functional, molecular, and phenotypic consequences of amino acid substitutions using hidden Markov models. *Human Mutation, 34*, 57–65.

Sim, N.-L., Kumar, P., Hu, J., Henikoff, S., Schneider, G., & Ng, P. C. (2012). SIFT web server: Predicting effects of amino acid substitutions on proteins. *Nucleic Acids Research, 40*, W452–W457.

Stenson, P. D., Mort, M., Ball, E. V, Chapman, M., Evans, K., Azevedo, L., Hayden, M., Heywood, S., Millar, D. S., Phillips, A. D., & Cooper, D. N. (2020). The Human Gene Mutation Database (HGMD(®)): Optimizing its use in a clinical diagnostic or research setting. *Human Genetics, 139*, 1197–1207.

Venkatachalam, S., Murlidharan, N., Krishnan, S. R., Ramakrishnan, C., Setshedi, M., Pandian, R., Barh, D., Tiwari, S., Azevedo, V., Sayed, Y., & Gromiha, M. M. (2023). Understanding drug resistance of wild-type and L38HL insertion mutant of HIV-1 C protease to saquinavir. *Genes, 14*(2), 533.

Weber, I. T., Wang, Y.-F., & Harrison, R. W. (2021). HIV protease: Historical perspective and current research. *Viruses, 13*(5), 839.

Wensing, A. M. J., van Maarseveen, N. M., & Nijhuis, M. (2010). Fifteen years of HIV protease inhibitors: Raising the barrier to resistance. *Antiviral Research, 85*, 59–74.

Wittayanarakul, K., Aruksakunwong, O., Saen-oon, S., Chantratita, W., Parasuk, V., Sompornpisut, P., & Hannongbua, S. (2005). Insights into saquinavir resistance in the G48V HIV-1 protease: Quantum calculations and molecular dynamic simulations. *Biophysical Journal, 88*, 867–879.

https://doi.org/10.1142/9789811293269_0011

Chapter 11

Databases and computational algorithms for identifying cancer hotspot residues and mutations in proteins

Medha Pandey[1,*], Suraj Kumar Shah[1], and M. Michael Gromiha[1,2,*]

[1]*Department of Biotechnology, Bhupat and Jyoti Mehta School of Biosciences, Indian Institute of Technology Madras, Chennai 600036, India*
[2]*International Research Frontiers Initiative, School of Computing, Tokyo Institute of Technology, Yokohama 226-8501, Japan*

Abstract

Advancements in the field of cancer genomics have yielded a wealth of information regarding genes and their associated mutations in the context of cancer. Researchers have conducted various experiments, including deep mutational scanning and polymerase chain reaction (PCR)-based methods, with the aim of pinpointing specific driver mutations. Some of

*Corresponding authors
MP: medhabioinfo@gmail.com
MMG: gromiha@iitm.ac.in

these driver mutations tend to congregate together, forming clusters known as "hotspots" due to their distinctive nature. Nevertheless, the task of precisely identifying the quantity and specific driver mutations and hotspots essential for each cancer type remains highly challenging. Additionally, the experimental methods employed for this purpose are both costly and time-consuming. To address these challenges, numerous meticulously organized databases have been established and regularly updated to catalog mutations and hotspots. Leveraging these databases, coupled with the application of machine learning and statistical techniques, various resources have been created to classify driver mutations and hotspots effectively. These methods draw upon significant sequence- and structure-based attributes of proteins and their evolutionary history to forecast their mutational impact, facilitating the computational prioritization of driver mutations and hotspots. This chapter serves as a comprehensive summary of cancer-specific mutation databases, encompassing critical experimental and computational methodologies as well as characteristic protein and mutation properties. These collective approaches represent significant strides toward advancing existing therapeutic strategies and enhancing the realm of personalized medicine.

Keywords: Databases; disease-causing mutations; machine-learning; cancer hotspots

11.1 Introduction

Hotspot residues and cancer-causing mutations constitute a captivating domain within the intricate realm of cancer research. Large-scale genome-wide studies have provided a huge scope for studies related to cancer mutations and hotspot residues. Hotspot residues are specific amino acid positions within critical genes or proteins that play a central role in the genesis and progression of cancer. These hotspots tend to occur in both gene-regulatory and protein-coding regions (Juul *et al.*, 2021). In protein-coding regions, a high frequency of single nucleotide polymorphisms (SNPs) has been observed. Based on the selective growth advantages to the cells, these missense mutations are known as drivers, whereas passenger mutations do not contribute to the selective growth. Mutations that activate oncogenes are highly specific and frequently reoccur at consistent positions among patients, as evidenced in oncogenes like KRAS and

BRAF (Burmer & Loeb, 1989; Davies *et al.*, 2002). These hotspots and mutations contribute to metastasis, malignancy, abnormalities in the cell signaling pathways, and cellular changes in healthy cells. They primarily impact genes and proteins crucial for fundamental cellular processes such as cell growth, division, and DNA repair, leading to affect their functions.

Identifying cancer-causing mutations and hotspot residues for different cancer types will help in prioritizing the potential therapeutic targets. It also provides unprecedented insights into the underlying mechanisms of tumorigenesis and advancements in cancer progression. This knowledge, in result, allows scientists and clinicians to develop targeted therapies tailored to the specific genetic alterations that underlie a patient's cancer. In this chapter, we focus on hotspot residues and cancer-causing mutations along with their effects on diverse cancer types, as well as the computational tools and databases used for the analysis. We also investigate the sequence, structural, and network-based properties that aid in identifying these genetic anomalies.

11.2 Cancer-causing mutations and hotspots

In the realm of cancer research, hotspot residues and driver mutations within key genes, also known as the Cancer Gene Census (CGC), offer critical insights into the molecular underpinnings of various malignancies. Among these genes, TP53, a renowned tumor suppressor, stands as a guardian of genomic stability, is not able to perform its protective role, and paves the way for cancer progression upon specific mutations. EGFR, a cell surface receptor, and KRAS, an oncogene, wield their influence over cell signaling. However, hotspot mutations in EGFR often dictate treatment responses in non-small cell lung cancer, and KRAS mutations drive relentless cell growth in pancreatic and colorectal cancers (Fitzgerald *et al.*, 2015; Lee *et al.*, 2016; Huang *et al.*, 2021). On the other hand, BRCA1 and BRCA2, pivotal in DNA repair, become susceptible to hotspot mutations, elevating the risk of hereditary breast and ovarian cancers (Yoshida & Miki, 2004; Saleem *et al.*, 2020). Along with BRCA1 and BRCA2, there are other genes such as CHEK2, ATM, and BRIP1 that are known to be associated with breast cancer risk (Castro *et al.*, 2013).

Further, BRAF, an oncogene in the MAPK pathway, undergoes hotspot mutations that propel melanoma and colorectal cancer development (Loo *et al.*, 2018; Ng *et al.*, 2019). Mutations at specific positions in these proteins, such as V600 in BRAF (melanoma and other cancer types) and G154, V157, R158, G245, R248, and R273 in P53 (lung cancer), are found to be hotspots and targeted for cancer therapies (Rivlin *et al.*, 2011; Masri *et al.*, 2022). These genes and their associated hotspot mutations not only reveal the intricate mechanisms of tumorigenesis but also offer avenues for targeted therapies in a new era of precision medicine in the fight against cancer.

11.3 Passenger hotspot mutations in cancer

Passenger hotspot mutations in cancer introduce an intriguing layer of complexity to our understanding of the disease. While driver mutations directly fuel tumorigenesis by altering critical genes and promoting uncontrolled cell growth, passenger mutations are those genetic alterations that occur coincidentally alongside driver mutations. The passenger mutations do not possess the same transformative power as the driver mutations. The passenger hotspots can influence tumor heterogeneity and evolution. Their presence underscores the intricate genetic landscape within cancer, demanding further exploration to decipher their role in shaping the tumor microenvironment and informing potential therapeutic strategies.

Passenger hotspot mutations are explored at the genomic level, considering a set of 295 CGC in 26 cancer types using statistical models based on the background frequency of mutations at specific positions. Hess *et al.* (2019) investigated these genes for the Long Normal Poisson (LNP) Model and checked for the conservation burden test (Pro[dN/dS≥1.2]) and found that 23 genes were having passenger hotspot mutations with high confidence. The mutations are observed in cancer census genes such as MB21D2, SLC27A5, BLCL2L12, PDE3A, and non-cancer census genes such as ERCC2, SOX17, and STK19. Yue *et al.* (2020) developed a database, dbCPM, for experimentally validated cancer passenger mutations.

11.4 Experimental methods to identify cancer-causing mutations and hotspots

Some of the key experimental methods employed to identify cancer-causing mutations and hotspots are sequencing techniques (whole genome, whole exome), site-directed mutagenesis, and functional assays (Meyerson *et al.*, 2010).

11.4.1 *Sequencing techniques*

Sequencing techniques are used to characterize the genetic mutations in tumors and to match the results with information obtained from other patient samples.

11.4.1.1 *Whole genome sequencing*

Whole genome sequencing (WGS) involves the sequencing of the entire genome of an individual (Nakagawa & Fujita, 2018), which provides information on both common and rare mutations in non-coding and coding regions. The comparison of cancer genomes with normal genomes reveals somatic mutations present only in cancer cells. The WGS method follows a systematic procedure such as DNA sample collection, library preparation and sequencing, identification of germline and somatic variants, validation of somatic variant calls, and generation of consensus sequences for the identification of mutations (Fujimoto *et al.*, 2021).

11.4.1.2 *Whole exome sequencing*

Whole exome sequencing (WES) concentrates on the exonic regions of the genome, which are the protein-coding areas where most cancer-driving mutations are typically located. It is more cost-effective than WGS and is suitable for identifying mutations in specific genes associated with cancer (Jelin & Vora, 2018).

11.4.1.3 *Targeted sequencing*

Targeted sequencing assays, such as amplicon sequencing (Ranjan *et al.*, 2016) or hybrid capture sequencing (Lu *et al.*, 2019), are used to selectively sequence a predefined set of genes or genomic regions. These methods are efficient for studying specific pathways or gene families linked to cancer.

11.4.1.4 *Single nucleotide polymorphism arrays*

SNP arrays detect genetic variations, including single nucleotide changes, insertions, deletions, and copy number variations (CNVs). They are helpful for identifying germline mutations associated with cancer susceptibility and somatic mutations in cancer genomes (Steventon-Jones *et al.*, 2022).

11.4.1.5 *Comparative genomic hybridization*

Comparative genomic hybridization (CGH) detects genomic imbalances by comparing the DNA of cancer cells to that of normal cells. It can identify copy number alterations (e.g., amplifications or deletions) in cancer genomes, highlighting potential driver mutations (Cheung & Bi, 2018).

11.4.1.6 *Fluorescent in situ hybridization*

Fluorescent in situ hybridization (FISH) is a cytogenetic method employing fluorescent probes to observe particular DNA sequences within the nucleus of a cell. It is valuable for detecting structural alterations, such as translocations, inversions, and gene fusions, which can drive cancer development (Ramos, 2022).

11.4.1.7 *Mutation-specific assays*

Mutation-specific assays are designed to detect specific mutations known to be involved in cancer, and it include PCR-based methods and functional assays (Ritterhouse & Barletta, 2015; Corless *et al.*, 2019).

11.4.2 *Polymerase chain reaction-based methods*

PCR can amplify and detect mutant alleles, like the BRAF V600E mutation found in melanoma (Ritterhouse & Barletta, 2015). Digital PCR provides ultra-sensitive detection of low-frequency mutations, whereas allele-specific PCR amplifies only the mutant allele while ignoring the wild-type allele (Corless *et al.*, 2019). Next-generation sequencing (NGS) panels focus on known cancer-associated genes and mutations. They are particularly useful for targeted screening and monitoring of specific mutations in cancer patients (Hu *et al.*, 2021).

11.4.3 *Functional assays*

These experiments involve introducing suspected mutations into cells or model organisms to assess their functional consequences. Functional assays can help determine whether a mutation is indeed oncogenic and how it contributes to cancer progression (Raraigh *et al.*, 2018).

11.5 Databases for cancer hotspots

A set of databases have been established to understand the influence of cancer hotspot residues in proteins and are listed in **Table 11.1**. Trevino (2020) developed a database, HotSpotAnnotations that includes information about a summary of mutation types, amino acid position and changes, cancer types, transcripts, the number of mutations, and information about false functional hotspots. It is available at https://bio.tools/hotspotannotations. The Cancer Hotspot database provides a dedicated catalog for genetic hotspots linked to cancer. It covers recurring mutations in various cancer-related genes, offering researchers and clinicians a comprehensive overview of these crucial genetic alterations (Chang *et al.*, 2016, 2018). This database can be accessed at https://www.cancerhotspots.org/. Diacofotaki *et al.* (2022) used cancer genome databases to study the mutations in the β-catenin gene across different types of tumors. The TP53 database is an online resource focusing on TP53 mutations (Leroy *et al.*, 2013), which is available at https://tp53.isb-cgc.org/. Gao *et al.* (2017) developed a 3D Hotspots database for mutational

Table 11.1 Description of the databases for mutation hotspots

Name	Characteristic features	Access link	Reference(s)
HotSpot Annotations	Database containing annotations for hotspot mutations in cancer.	https://bio.tools/hotspotannotations	Trevino (2020)
Cancer Hotspots	Database for genetic hotspots linked to cancer	https://www.cancerhotspots.org/	Chang *et al.* (2016, 2018)
The TP53 Database	Database for TP53 mutations	https://tp53.isb-cgc.org/	Leroy *et al.* (2013)
3D Hotspots	A resource for mutation hotspots in the 3D structures of proteins	https://www.3dhotspots.org/	Gao *et al.* (2017)

Last accessed on November 30, 2023.

clusters based on the 3D structures of proteins. It can be accessed at https://www.3dhotspots.org/.

11.6 Databases for cancer-causing mutations

Databases dedicated to cancer-causing mutations play a pivotal role in understanding the molecular basis of this complex disease and are listed in **Table 11.2**.

The Catalog of Somatic Mutations in Cancer (COSMIC) database offers details on mutations acquired somatically in human cancer. It compiles information from scientific literature and extensive experimental screenings conducted by the Cancer Genome Project at the Sanger Institute (Tate *et al.*, 2019). The cancer genome atlas (TCGA) program systematically documents cancer-associated genetic mutations through sophisticated genome sequencing and bioinformatics analysis (Wang *et al.*, 2016). Hudson *et al.* (2010) developed the International Cancer Genome Consortium (ICGC) database, which systematically analyzes cancer genomic, epigenomic, and transcriptomic data to reveal oncogenic

Table 11.2 List of the databases available for cancer-causing and neutral mutations

Name	Description	Link	References
COSMIC	Cancer-associated somatic mutations database in humans	https://cancer.sanger.ac.uk/cosmic	Tate *et al.* (2019)
TCGA	Database for genetic mutations responsible for cancer	https://www.cancer.gov/ccg/research/genome-sequencing/tcga	Tomczak *et al.* (2015)
ICGC	Database for identifying genetic abnormalities in cancer	https://dcc.icgc.org/	Hudson *et al.* (2010)
CBioPortal	Database for exploration and analysis of multidimensional cancer genomics data	http://cbioportal.org	Gao *et al.* (2013)
		https://genie.cbioportal.org	de Bruijn *et al.* (2023)
OncoKB	Database housing evidence-based information on individual somatic mutations and structural alterations found in patient tumors	http://oncokb.org/	Chakravarty *et al.* (2017)
dbCPM	Database for manually curated cancer passenger mutations	http://bioinfo.ahu.edu.cn:8080/dbCPM	Yue *et al.* (2020)

Last accessed on November 30, 2023.

mutations, mutagenic traces, and define clinically relevant cancer subtypes. The cBio Cancer Genomics Portal offers multidimensional cancer genomics datasets (Cerami *et al.*, 2012) with detailed guidelines for the analysis and visualization features of the portal in cancer genomics studies (Gao *et al.*, 2013). In a recent update, de Bruijn *et al.* (2023) incorporated additional real-world longitudinal data and broadened the capabilities of cBioPortal to facilitate the visualization and analysis of this extensive clinico-genomic dataset. OncoKB catalogs approximately 3,000 annotated unique mutations, fusions, and copy number alterations identified in 418 genes associated with cancer. The database also encapsulates comprehensive data regarding the biological and oncogenic impacts of these genetic alterations (Chakravarty *et al.*, 2017). Yue *et al.* (2020) constructed a comprehensive database, dbCPM, for experimentally validated passenger mutations reported in literature.

11.7 Role of sequence, structure, and network-based features

Prediction of driver mutations and hotspots in cancer relies on a multifaceted approach that integrates sequence, structure, and network-based features. Sequence-based analysis scrutinizes genomic data to identify mutations occurring at higher frequencies in cancer samples, hinting at their potential driving roles (Raphael *et al.*, 2014). Structural analysis dives deeper by exploring the three-dimensional conformation of proteins, unveiling how mutations may disrupt crucial interactions or protein functions (Engin *et al.*, 2016). Network-based features leverage the interconnectedness of biological systems, pinpointing mutations that perturb essential signaling pathways or protein–protein interactions (Alvarez *et al.*, 2016). Together, these approaches provide a comprehensive understanding of the genetic alterations underpinning cancer, enabling the discovery of key driver mutations and hotspots that guide targeted therapies and advance our knowledge of oncogenesis.

11.7.1 *Sequence-based features*

Sequence-based features have been employed in the development of numerous computational methods for predicting interactions among proteins, DNA, RNA, carbohydrates, and ligands (Ding & Kihara, 2018), as well as for protein secondary or tertiary structure prediction, thermodynamic stability, and predicting the functional impact of mutations in cancer and other diseases (Reva *et al.*, 2011). These features are shown to have an important role due to the availability of protein sequences. Some of the important sequence-based features are discussed as follows.

11.7.1.1 *Physicochemical properties*

Gromiha *et al.* (1999) assembled a collection of 49 properties related to amino acids, encompassing physical, chemical, conformational, and energetic attributes. These properties find extensive application in comprehending protein structure and function, along with forecasting the outcome of mutations (Gromiha, 2003; Charoenkwan *et al.*, 2021; Ahmed *et al.*, 2022). The AAindex comprises numerical indices associated with a diverse array of physicochemical and biochemical traits of amino acids (Kawashima & Kanehisa, 2000). In the context of mutations, the properties were obtained as the difference between the values of wild-type and mutant residues, as well as by including neighboring residues and then obtaining the difference (Gromiha *et al.*, 1999; Kulandaisamy *et al.*, 2021; Reddy *et al.*, 2023).

$$\Delta P_{\text{mutation}} = \Delta P_{\text{mutant}} - \Delta P_{\text{wild-type}} \quad (11.1)$$

$$P_{\text{ave}}(i) = \Sigma P_j / N \quad (11.2)$$

where, ΔP_{mutant} and $\Delta P_{\text{wild-type}}$ are the property values of the mutant and wild-type residue, respectively.

11.7.1.2 *Predicted secondary structure and solvent accessibility*

The detection of mutations causing diseases employs features such as the secondary structure and accessible surface area (ASA) of individual residues in a protein. The major secondary structure classifications include helix, strand, and coil. Based on ASA values, residues are categorized as either exposed (ASA > 25%) or buried (ASA < 25%). Commonly employed sequence-based methods for predicting secondary structure and solvent accessibility include JPred4 (Drozdetskiy *et al.*, 2015), SPIDER2 (Yang *et al.*, 2017), NetSurfP (Petersen *et al.*, 2009), and SARPred (Garg *et al.*, 2005).

11.7.1.3 *Motifs*

Motifs denote patterns within amino acid sequences and are generated by observing consecutive neighboring residues in dipeptides (denoted as XM, where X represents any residue and M denotes the residue of interest), tripeptides, and the introduction of gaps between them (e.g., X*M, where * signifies a gap). **Figure 11.1** contains the specific details of the motif extraction process.

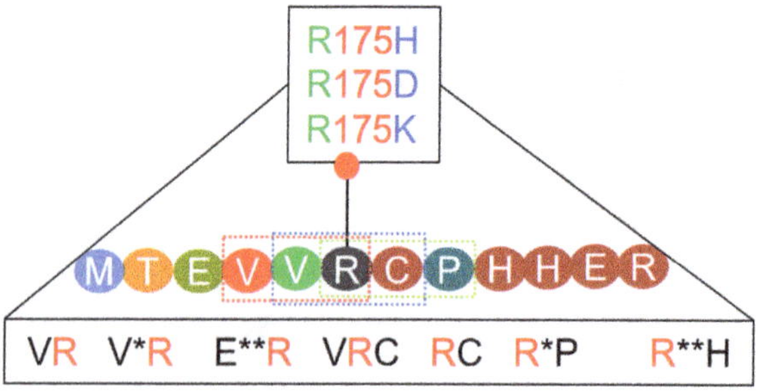

Figure 11.1 Extraction of various motifs from a protein sequence to identify disease-causing mutations and hotspot residues

The preferences for these motifs are quantified using an "odds score," defined as the ratio of a particular motif's occurrence in neutral and disease-prone sites (mutations). This ratio is given by the formula

$$\text{Odds ratio} = (N_{dP}/N_d)/(N_{nP}/N_n) \tag{11.3}$$

In this formula, N_{dP} represents the occurrence of a specific di- or tripeptide motif in disease-prone sites (mutations), while N_{nP} represents the occurrence of the same motif in neutral sites (or mutations). N_n and N_d are the total numbers of neutral and disease-prone sites (or mutations) in the dataset, respectively.

11.7.1.4 *Amino acid composition in the vicinity of hotspot residues and mutation sites*

The representation of amino acid composition involves tracking the occurrence of amino acid residues along the sequence, starting from the hotspot residue (or disease-causing mutation site) within specified window lengths. A window length of three residues, for instance, comprises the central residue and one residue on each side of the central residue (as illustrated by R175 in **Figure 11.1**, which includes VRC). Typically, window lengths ranging from 3 to 21 residues are considered for calculating the composition. The formula for this calculation is given as

$$\text{Comp}(i) = n(i)*100/N \tag{11.4}$$

In this equation, "i" stands for the 20 amino acid residues, "n(i)" indicates the number of residues for the amino acid, i, and "N" represents the overall number of residues.

11.7.1.5 *Domain location*

The positioning of a residue within a sequence is categorized into distinct domains: N-terminal (1%–30%), middle (31%–70%), or C-terminal (71%–100%). Singh *et al.* (2019) have documented that the highly conserved DNA-binding domain of TP53 harbors a cluster of somatic missense mutations, including R175, G245, R248, and R273. These mutations establish direct interactions with DNA and play an important role in maintaining the tertiary structure of a protein (Wong *et al.*, 1999). Likewise, the P152L mutation at the S3/S4 turn (opposite to the DNA-binding surface of the mutant p53) induces a conformational alteration within the DNA-binding domain.

In the case of EGFR (**Figure 11.2**), most of the mutations are located within the intracellular region (residues 669–1210), which encompasses

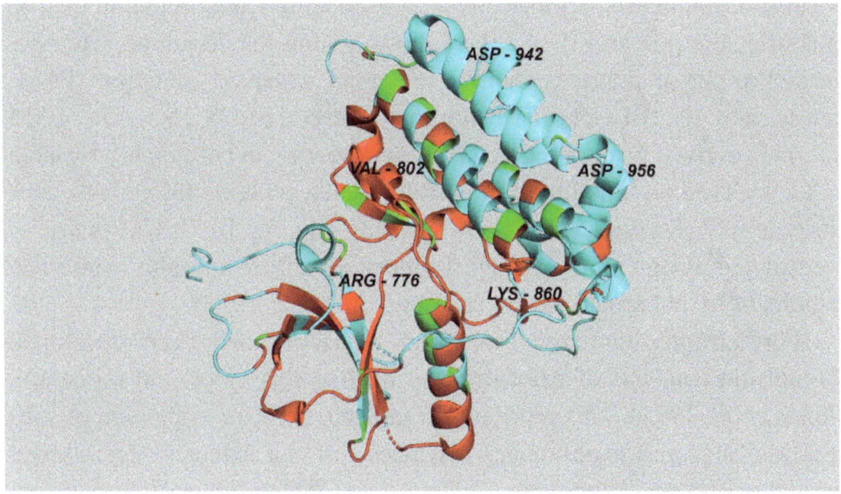

Figure 11.2 EGFR (PDB id: 2EB2) driver mutations (depicted in red) and passenger mutations (shown in green) within the intracellular domain, including the N- and C-termini, with specific residues labeled for reference (adapted from Anoosha *et al.*, 2015)

cytoplasmic and tyrosine kinase domains. These domains are renowned for their crucial involvement in initiating downstream signaling pathways and regulating EGFR function (Ferguson, 2008; Anoosha *et al.*, 2015). Comparable data have also been reported for other genes such as PIK3CA, PTEN, EGFR, and KRAS (Anoosha *et al.*, 2016).

11.7.1.6 *Position-specific scoring matrices profiles*

Position-specific scoring matrices (PSSM) profile represents a crucial component in the field of biology for deciphering the evolutionary insights embedded within a protein sequence. The PSSM relies on a multiple sequence alignment acquired through PSI-BLAST, giving substantial emphasis to conserved positions (Bhagwat & Aravind, 2007). The initial iteration produces a PSSM, which is then employed to query the database for fresh sequence matches in subsequent iterations. This iterative process is reiterated multiple times until the final matrix is achieved.

11.7.1.7 *Mutation matrices*

A substitution matrix, specifically a 20 × 20 amino acid mutation matrix, quantifies the probability of substitutions among the 20 amino acid residues. Prominent matrices include the point accepted mutation (PAM) matrix and the BLOcks SUbstitution Matrix (BLOSUM). The PAM matrix is derived from observed changes occurring in both highly variable and conserved regions found in closely related proteins throughout their entire length (Dayhoff *et al.*, 1978). Meanwhile, the BLOSUM matrix is constructed using amino acid substitutions identified in highly conserved regions of homologous sequences (Henikoff *et al.*, 1992).

Furthermore, numerous mutation matrices have been developed through the analysis of extensive and diverse sets of protein sequences (Jones *et al.*, 1992). This procedure entails the iterative evaluation of substitution rates and evolutionary distances through extensive pre-aligned pairwise sequence alignments (Müller *et al.*, 2002). Specialized matrices have also been crafted for specific genomic or protein families (Risler *et al.*, 1988), such as those tailored for AT-enriched genomes like *Plasmodium falciparum* and *Plasmodium Yoelii* (Brick & Pizzi, 2008; Paila *et al.*,

2008), integral membrane proteins (Jones *et al.*, 1994; Müller *et al.*, 2001), β-barrel transmembrane proteins (Ng *et al.*, 2000; Jimenez-Morales *et al.*, 2008), the rhodopsin family of G protein-coupled receptors (Rios *et al.*, 2015), hub proteins within protein–protein interaction networks (Renganayaki & Nair, 2017), and intrinsically disordered proteins (Radivojac *et al.*, 2002; Trivedi & Nagarajaram, 2020). The AAindex database houses a collection of such mutation matrices, which serve as vital features for the identification of disease-causing mutations.

11.7.1.8 *Conservation score*

The conservation score associated with a given residue indicates the likelihood of that residue appearing in the same position across various homologous sequences. Valdar (2002) introduced a web server called AACon, which is designed to compute a set of 17 distinct conservation scores. There are several other tools available for calculating conservation scores, such as AL2CO (Pei *et al.*, 2001) and ConSurf (Glaser *et al.*, 2003). These servers contribute significantly to comprehending the phylogenetic relationships among homologous sequences and identifying sites of functional or structural importance. **Table 11.3** lists the web servers available for predicting conservation scores.

Table 11.3 List of the web servers available for predicting conservation scores

Name	Description	Link	References
AL2CO	Calculating the conservation index at each position in a multiple sequence alignment of a protein	http://prodata.swmed.edu/al2co/al2co.php	Pei *et al.* (2001)
AACon	Predicting protein functional sites by calculating 17 different conservation scores	https://www.compbio.dundee.ac.uk/aacon/	Valdar (2002)
ConSurf	Identifying functional regions in proteins by estimating the degree of conservation of the amino-acid sites among their close sequence homologues	https://consurf.tau.ac.il/	Glaser *et al.* (2003)

Last accessed on November 30, 2023.

11.7.1.9 *Neighboring residue information based on amino acid groups*

The 20 amino acid residues are categorized into sets according to their physicochemical characteristics, including aliphatic, aromatic, polar, sulfur-containing, negatively charged, and positively charged residues. The inclination or bias toward these residue groups is assessed on either side of the mutant position within a defined window length (Gromiha, 2010; Anoosha *et al.*, 2015; Pandey & Gromiha, 2021).

11.7.2 *Structure-based features*

In structural analysis, three-dimensional structures are leveraged to calculate a multitude of attributes. The PDBparam web server (Nagarajan *et al.*, 2016) offers a range of structure-based parameters across four distinct categories: (i) physicochemical properties, (ii) inter-residue interactions, (iii) propensities related to secondary structures, and (iv) binding sites.

These specific features encompass information such as the secondary structures of amino acid residues using DSSP (Kabsch & Sander, 1983), identification of hydrogen bonds between atoms/residues based on HBPLUS (McDonald & Thornton, 1994), ASA for each atom/residue obtained from NACCESS (Hubbard & Thornton, 1993), contact order (Plaxco *et al.*, 1998), long-range order (Gromiha & Selvaraj, 2001), and surrounding hydrophobicity (Ponnuswamy, 1993). In addition, biological networks offer valuable insights for understanding the causes of mutations (Barabási *et al.*, 2011; Ozturk *et al.*, 2018). Ozturk and Carter (2022) used 16 network-based parameters to analyze missense mutations, including degree, closeness, betweenness, clustering coefficient, and eigenvector. Details of these parameters are discussed in our earlier review (Pandey *et al.*, 2024).

11.8 Computational approaches for identifying hotspots and mutations

Computational approaches for identifying hotspots and mutations have become indispensable tools in modern biology and medical research.

These methods help to understand the molecular basis of diseases as well as to prioritize mutations and hotspots. The computational tools reported in the literature for the identification of hotspots and mutations are discussed below.

11.8.1 *Identifying cancer hotspot residues*

Several methods have been developed for identifying cancer hotspot residues in proteins and are listed in **Table 11.4**.

Tokheim *et al.* (2016) developed Hotspot Missense mutation Areas in Protein Structure (HotMAPS), a computational tool for the high-throughput analysis of 3D hotspot regions in cancer missense mutations. Mutation3D is used to identify driver genes in cancer by detecting clusters

Table 11.4 List of computational tools for identifying cancer hotspot residues

Name	Description	Link	References
HotMAPS	High-throughput analysis of 3D hotspot regions in cancer missense mutations	https://github.com/KarchinLab/HotMAPS	Tokheim *et al.* (2016)
Mutation3D	Predicting driver genes in cancer by detecting clusters of amino acid substitutions in tertiary protein structures	http://mutation3d.org	Meyer *et al.* (2016)
QuartPAC	Identifying clusters of mutations in proteins using tertiary and quaternary structures	http://bioconductor.jp/packages/3.1/bioc/html/QuartPAC.html	Ryslik *et al.* (2016)
OncodriveCLUST	Predicting genes with mutation clustering ability in a protein sequence by assessing coding-silent mutations	http://bg.upf.edu/oncodriveclust	Tamborero *et al.* (2013)
HotSpot3d	Identifying spatial hotspots, which are local concentrations of mutations in 3D protein space, by integrating sequence mutations with 3D protein structures	https://github.com/ding-lab/hotspot3d	Niu *et al.* (2016)

(*Continued*)

Table 11.4 (*Continued*)

Name	Description	Link	References
CanProSite	Predict disease-prone sites using amino acid sequence features such as physicochemical properties, conservation scores, secondary structure, and peptide motifs	https://web.iitm.ac.in/bioinfo2/CanProSite/	Pandey and Gromiha (2021)
MutBLESS	Identifying cancer-prone sites using amino acid sequence-based features such as physicochemical properties, predicted secondary structure, conservation scores, and peptide motifs	https://web.iitm.ac.in/bioinfo2/MutBLESS/index.html	Pandey and Gromiha (2023)

Last accessed on November 30, 2023.

of amino acid substitutions within tertiary protein structures (Meyer *et al.*, 2016). The Quaternary Protein Amino acid Clustering (QuartPAC) implemented algorithms for identifying statistically significant mutation clusters in proteins using their tertiary and quaternary structures (Ryslik *et al.*, 2016). The OncodriveCLUST method is used to identify genes with mutation clustering ability within the protein sequence (Tamborero *et al.*, 2013). The HotSpot3d tool can identify spatial hotspots (clusters), which are local concentrations of mutations in 3D protein space (Niu *et al.*, 2016). Pandey and Gromiha (2021) developed CanProSite, employing deep neural networks to predict disease-prone sites using amino acid sequence features, including physicochemical properties, conservation scores, secondary structure, and di- and tri-peptide motifs. Later, this approach has been extended to 22 diverse cancer types to pinpoint disease-prone sites in cancer, and a web server, MutBLESS has been developed for identifying cancer-prone sites (Pandey & Gromiha , 2023).

11.8.2 *Predicting cancer-causing mutations*

A list of available tools for predicting cancer-causing mutations is presented in **Table 11.5.** The Cancer Driver Annotation (CanDrA) tool

Table 11.5 List of computational methods for cancer mutations

Name	Algorithm	Important features	References
CanDrA	SVM	UniProt annotations, structural features, predicted functional impact scores, and evolutionary features	Mao *et al.* (2013)
CScape	Machine learning	GC content, sequence spectra, dN/dS ratio, histone modifications	Rogers *et al.* (2017)
CHASMplus	Random forests	Interactions, charge ratio, substitution matrices, accessibility, UniProt annotations	Tokheim *et al.* (2019)
CADD	SVM	Conservation scores, epigenetic modification, functional prediction, genetic context of variants	Rentzsch *et al.* (2021)
MVP	Deep learning (ResNet)	GC content, conservation scores, mutation matrices, interactions, networks, predicted accessible surface area, and secondary structure	Qi *et al.* (2021)
GBM Driver	Deep learning	Motifs, conservation scores, and amino acid properties	Pandey *et al.* (2022)
OncodriveFM	Statistical method	Functional mutation bias	Mularoni *et al.* (2016)
IntOGen-mutations	Consensus	Functional mutation bias	Gonzalez-Perez *et al.* (2013)
MetaSVM	SVM	Gene expression	Kim *et al.* (2017)

Last accessed on November 30, 2023.

predicts missense driver mutations by leveraging structural and evolutionary features (Mao *et al.*, 2013). Rogers *et al.* (2017) developed the CScape tool for predicting oncogenic single-point mutations in the cancer genome using genomic and evolutionary features, including GC content, sequence spectra, and the ratio of non-synonymous to synonymous substitutions (dN/dS). Tokheim *et al.* (2019) developed

CHASMplus to predict driver missense mutations and putative rare driver mutations in a cancer-type-specific manner. Rentzsch *et al.* (2021) developed a method, combined annotation-dependent depletion (CADD) to predict the deleteriousness of variants in the cancer genome using a logistic regression model with conservation, epigenetic modification, functional prediction, and genetic context of variants. MVP is a deep learning-based approach to predict the pathogenicity of variants using GC content, mutation matrices, conservation scores, interactions in a protein structure, and predicted secondary structures (Qi *et al.*, 2021).

GBMDriver is a brain tumor-specific method to identify driver and passenger mutations using motif-based properties, sequence-based physicochemical properties, and conservation scores (Pandey *et al.*, 2022). OncodriveFM (Mularoni *et al.*, 2016) is a method to detect driver mutations in coding and non-coding genomic regions using a local functional mutation bias. IntOGen-mutations (Gonzalez-Perez *et al.*, 2013) predict cancer-causing mutations using functional mutation bias across the cohort of tumor samples. Kim *et al.* (2017) developed MetaSVM to identify consensus genes associated with different diseases using microarray datasets and support vector machines.

11.9 Designing mutation-specific inhibitors

Several investigations have been carried out to design mutation-specific inhibitors for the targets involved in cancer. Anoosha *et al.* (2017) developed quantitative structure activity relationship (QSAR) models for identifying inhibitors specific to mutations A289V, G598V, G719S, P753S, R832C, S768I, T751I, and L858R/T790M in EGFR. Panicker *et al.* (2017) developed structure-based 3D pharmacophore models for identifying next-generation inhibitors against clinically relevant mutations, L858R, T790M, L858R, and delE746-750, using chemical features, hydrogen bond donors, hydrogen bond acceptors, and aromatic rings. Zhong *et al.* (2023) developed 2D-SAR and 3D-QSAR models to identify inhibitors specific to key amino acids K428, D756, Y761, and R765 in a channel protein, anoctamin 1.

11.10 Conclusions

In this chapter, we provided a list of cancer mutation databases, which are important for understanding the molecular mechanism and developing computational algorithms. The influence of cancer-causing mutations is studied using various sequence- and structure-based parameters of proteins, which include patterns, conservation, amino acid properties, neighboring residues along the sequence, and surrounding residues in protein structures. The information gained in sequence and structure-based analysis will be utilized for identifying cancer hotspots, cancer-causing mutations, and designing mutation-specific inhibitors for targets, and we have extensively discussed the available computational resources in the literature. The information provided in this chapter will be helpful for developing strategies for targeted therapies.

Acknowledgments

The authors wish to acknowledge the Indian Institute of Technology Madras for computational facilities. MP and SKS thank the Ministry of Education for providing fellowships.

References

Ahmed, Z., Hasan-Zulfiqar, H., Tang, L., & Lin, H. (2022) A Statistical analysis of the sequence and structure of thermophilic and non-thermophilic proteins. *International Journal of Molecular Sciences*, *23*, 10116.

Alvarez, M. J., Shen, Y., Giorgi, F. M., Lachmann, A., Ding, B. B., Ye, B. H., & Califano, A. (2016). Functional characterization of somatic mutations in cancer using network-based inference of protein activity. *Nature Genetics, 48*(8), 838–847.

Anoosha, P., Huang, L. T., Sakthivel, R., Karunagaran, D., & Gromiha, M. M. (2015). Discrimination of driver and passenger mutations in epidermal growth factor receptors in cancer. *Mutation Research, 780*, 24–34.

Anoosha, P., Sakthivel, R., & Michael Gromiha, M. (2016). Exploring preferred amino acid mutations in cancer genes: Applications to identify potential drug targets. *Biochimica Biophysica Acta, 1862*(2), 155–165.

Anoosha, P., Sakthivel, R., & Gromiha, M. M. (2017). Investigating mutation-specific biological activities of small molecules using quantitative structure-activity relationship for epidermal growth factor receptor in cancer. *Mutation Research, 806*, 19–26.

Barabási, A. L., Gulbahce, N., & Loscalzo, J. (2011). Network medicine: A network-based approach to human disease. *Nature Reviews Genetics*, *12*(1), 56–68.

Bhagwat, M., & Aravind, L. (2007). PSI-BLAST tutorial. *Methods in Molecular Biology, 395*, 177–186.

Brick, K., & Pizzi, E. (2008). A novel series of compositionally biased substitution matrices for comparing Plasmodium proteins. *BMC Bioinformatics, 9*, 236.

Burmer, G. C., & Loeb, L. A. (1989). Mutations in the KRAS2 oncogene during progressive stages of human colon carcinoma. *Proceedings of the National Academy of Science, 86*, 2403–2407.

Castro, E., Goh, C., Olmos, D., *et al.* (2013). Germline BRCA mutations are associated with higher risk of nodal involvement, distant metastasis, and poor survival outcomes in prostate cancer. *Journal of Clinical Oncology*, *31*(14), 1748–1757.

Cerami, E., Gao, J., Dogrusoz, U., Gross, B. E., Sumer, S. O., Aksoy, B. A., Jacobsen, A., Byrne, C. J., Heuer, M. L., Larsson, E., Antipin, Y., Reva, B., Goldberg, A. P., Sander, C., & Schultz, N. (2012). The cBio cancer genomics portal: An open platform for exploring multidimensional cancer genomics data. *Cancer Discovery*, *2*(5), 401–404.

Chakravarty, D., Gao, J., Phillips, S. M., *et al.* (2017). OncoKB: A precision oncology knowledge base. *JCO Precision Oncology, 2017*, PO.17.00011.

Chang, M. T., Asthana, S., Gao, S. P., Lee, B. H., Chapman, J. S., Kandoth, C., Gao, J., Socci, N. D., Solit, D. B., Olshen, A. B., Schultz, N., & Taylor, B. S. (2016). Identifying recurrent mutations in cancer reveals widespread lineage diversity and mutational specificity. *Nature Biotechnology*, *34*(2), 155–163.

Chang, M. T., Bhattarai, T. S., Schram, A. M., Bielski, C. M., Donoghue, M. T. A., Jonsson, P., Chakravarty, D., Phillips, S., Kandoth, C., Penson, A., Gorelick, A., Shamu, T., Patel, S., Harris, C., Gao, J., Sumer, S. O., Kundra, R., Razavi, P., Li, B. T., Reales, D. N., … Taylor, B. S (2018). Accelerating discovery of functional mutant alleles in cancer. *Cancer Discovery, 8*(2), 174–183.

Charoenkwan, P., Chotpatiwetchkul, W., Lee, V. S., Nantasenamat, C., Shoombuatong, W. (2021). A novel sequence-based predictor for identifying and characterizing thermophilic proteins using estimated propensity scores of dipeptides. *Scientific Report, 11*, 23782.

Cheung, S. W., & Bi, W. (2018). Novel applications of array comparative genomic hybridization in molecular diagnostics. *Expert Review of Molecular Diagnostics, 18*(6), 531–542.

Corless, B. C., Chang, G. A., Cooper, S., Syeda, M. M., Shao, Y., Osman, I., Karlin-Neumann, G., & Polsky, D. (2019). Development of novel mutation-specific droplet digital PCR assays detecting TERT promoter mutations in tumor and plasma samples. *Journal of Molecular Diagnostics, 21*(2), 274–285.

Davies, H., Bignell, G. R., Cox, C., Stephens, P., Edkins, S., Clegg, S., Teague, J., Woffendin, H., Garnett, M. J., Bottomley, W., Davis, N., Dicks, E., Ewing, R., Floyd, Y., Gray, K., Hall, S., Hawes, R., Hughes, J., Kosmidou, V., Menzies, A., … Futreal, P. A (2002). Mutations of the BRAF gene in human cancer. *Nature, 417*(6892), 949–954.

Dayhoff, M. O., Schwartz, R. M., Orcutt, B. C. (1978). A model of evolutionary change in proteins. *National Biomedical Research, 5*(3), 345–352.

de Bruijn, I., Kundra, R., Mastrogiacomo, B., Tran, T. N., Sikina, L., Mazor, T., Li, X., Ochoa, A., Zhao, G., Lai, B., Abeshouse, A., Baiceanu, D., Ciftci, E., Dogrusoz, U., Dufilie, A., Erkoc, Z., Garcia Lara, E., Fu, Z., Gross, B., Haynes, C., … Schultz, N. (2023). Analysis and visualization of longitudinal genomic and clinical data from the AACR project GENIE biopharma collaborative in cBioPortal. *Cancer Res.*

Diacofotaki, A., Loriot, A., & De Smet, C. (2022). Identification of tissue-specific gene clusters induced by DNA demethylation in lung adenocarcinoma: More than germline genes. *Cancers, 14*(4), 1007.

Ding, Z., & Kihara, D. (2018). Computational methods for predicting protein–protein interactions using various protein features. *Current Protocols Protein Science, 93*(1), e62.

Drozdetskiy, A., Cole, C., Procter, J., & Barton, G. J. (2015). JPred4: A protein secondary structure prediction server. *Nucleic Acids Research, 43*, W389–W394.

Engin, H. B., Kreisberg, J. F., & Carter, H. (2016). Structure-based analysis reveals cancer missense mutations target protein interaction interfaces. *PloS One*, *11*(4), e0152929.

Ferguson, K. M. (2008). Structure-based view of epidermal growth factor receptor regulation. *Annual Review of Biophysics, 37*, 353–373.

Fitzgerald, T. L., Lertpiriyapong, K., Cocco, L., Martelli, A. M., Libra, M., Candido, S., Montalto, G., Cervello, M., Steelman, L., Abrams, S. L., &

McCubrey, J. A. (2015). Roles of EGFR and KRAS and their downstream signaling pathways in pancreatic cancer and pancreatic cancer stem cells. *Advances in Biological Regulation*, *59*, 65–81.

Fujimoto, A., Wong, J. H., Yoshii, Y., Akiyama, S., Tanaka, A., Yagi, H., Shigemizu, D., Nakagawa, H., Mizokami, M., & Shimada, M. (2021). Whole-genome sequencing with long reads reveals complex structure and origin of structural variation in human genetic variations and somatic mutations in cancer. *Genome Medicine, 13*(1), 65.

Gao, J., Aksoy, B. A., Dogrusoz, U., Dresdner, G., Gross, B., Sumer, S. O., Sun, Y., Jacobsen, A., Sinha, R., Larsson, E., Cerami, E., Sander, C., & Schultz, N. (2013). Integrative analysis of complex cancer genomics and clinical profiles using the cBioPortal. *Science Signaling, 6*(269), pl1.

Gao, J., Chang, M. T., & Johnsen, H. C., Gao, S. P., Sylvester, B. E., Sumer, S. O., Zhang, H., Solit, D. B., Taylor, B. S., Schultz, N., & Sander, C. (2017). 3D clusters of somatic mutations in cancer reveal numerous rare mutations as functional targets. *Genome Medicine*, 9(1), 4.

Garg, A., Kaur, H., & Raghava, G. P. S. (2005). Real value prediction of solvent accessibility in proteins using multiple sequence alignment and secondary structure. *Proteins, 61*, 318–324.

Glaser, F., Pupko, T., Paz, I., Bell, R. E., Bechor-Shental, D., Martz, E., & Ben-Tal, N. (2003). ConSurf: Identification of functional regions in proteins by surface-mapping of phylogenetic information. *Bioinformatics, 19*, 163–164.

Gonzalez-Perez, A., Perez-Llamas, C., Deu-Pons, J., Tamborero, D., Schroeder, M. P., Jene-Sanz, A., Santos, A., & Lopez-Bigas, N. (2013). IntOGen-mutations identifies cancer drivers across tumor types. *Nature Methods, 10*(11), 1081–1082.

Gromiha, M. M. (2003) Importance of native-state topology for determining the folding rate of two-state proteins. *Journal of Chemical Information Computer. Scientists. 43*, 1481–1485.

Gromiha, M. M. (2010). Protein bioinformatics In –Gromiha, M. M (Ed.), *Protein Structure analysis* (Ch. 3, pp. 63–105), Academic Press, ISBN 9788131222973.

Gromiha, M. M., & Nagarajan, R. (2013). Computational approaches for predicting the binding sites and understanding the recognition mechanism of protein-DNA complexes. *Advances In Protein Chemistry Structural Biology*, *91*, 65–99.

Gromiha, M. M., Oobatake, M., Kono, H., Uedaira, H., & Sarai, A. (1999). Role of structural and sequence information in the prediction of protein stability changes: Comparison between buried and partially buried mutations. *Protein Engineering, 12*(7), 549–555.

Gromiha, M. M., Oobatake, M., *et al.* (1999). Important amino acid properties for enhanced thermostability from mesophilic to thermophilic proteins. *Biophysical Chemistry, 82*, 51–67.

Gromiha, M. M., Ridha, F., *et al.* (2024) "Protein structural bioinformatics: An overview," *Encyclopedia of Bioinformatics and Computational Biology*, 2nd Edition (in press)

Gromiha, M. M. & Selvaraj, S. (2001). Comparison between long-range interactions and contact order in determining the folding rate of two-state proteins: Application of long-range order to folding rate prediction. *Journal of Molecular Biology, 310*, 27–32.

Gromiha, M. M., & Selvaraj, S. (2004). Inter-residue interactions in protein folding and stability. *Progress Biophysics Molecular Biology, 86*(2), 235–277.

Henikoff, S., & Henikoff, J. G. (1992). Amino acid substitution matrices from protein blocks. *Proceedings of the National Academy Science of the USA, 89*, 10915–10919.

Hess, J. M., Bernards, A., Kim, J., Miller, M., Taylor-Weiner, A., Haradhvala, N. J., Lawrence, M. S., & Getz, G. (2019). Passenger hotspot mutations in cancer. *Cancer Cell, 36*(3), 288–301.e14.

Hu, T., Chitnis, N., Monos, D., & Dinh, A. (2021). Next-generation sequencing technologies: An overview. *Human Immunology, 82*(11), 801–811.

Huang, L., Guo, Z., Wang, F., & Fu, L. (2021). KRAS mutation: From undruggable to druggable in cancer. *Signal Transduction Targeted Therapy, 6*(1), 386.

Hubbard, S. J., & Thornton, J. (1993). *NACCESS, Computer Program.* The University of Manchester.

Hudson, T. J., Anderson, W., Artez, A., *et al.* (2010). International network of cancer genome projects. *Nature, 464*(7291), 993–998.

Jelin, A. C., & Vora, N. (2018). Whole exome sequencing: Applications in prenatal genetics. *Obstetrics Gynecology Clinics North America, 45*(1), 69–81.

Jimenez-Morales, D., Adamian, L., & Liang, J. (2008). Detecting remote homologues using scoring matrices calculated from the estimation of amino acid substitution rates of beta-barrel membrane proteins. *Annual International Conference of IEEE Engineering in Medicine and Biology Society, 2008*, 1347–1350.

Jones, D. T., Taylor, W. R., & Thornton, J. M. (1992). The rapid generation of mutation data matrices from protein sequences. *Computer Applications in the Biosciences, 8*, 275–282.

Jones, D. T., Taylor, W. R., & Thornton, J. M. (1994). A mutation data matrix for transmembrane proteins. *FEBS Letters, 339*, 269–275.

Juul, R. I., Nielsen, M. M., Juul, M., Feuerbach, L., & Pedersen, J. S. (2021). The landscape and driver potential of site-specific hotspots across cancer genomes. *NPJ Genomic Medicine, 6*(1), 33.

Kabsch, W., & Sander, C. (1983). Dictionary of protein secondary structure: Pattern recognition of hydrogen-bonded and geometrical features. *Biopolymers, 22*, 2577–2637.

Kawashima, S., & Kanehisa, M. (2000). AAindex: Amino acid index database. *Nucleic Acids Research, 28*, 374.

Kim, S., Jhong, J. H., Lee, J., & Koo, J. Y. (2017). Meta-analytic support vector machine for integrating multiple omics data. *BioData Mining, 10*, 2.

Kulandaisamy, A., Zaucha, J., Frishman, D., & Gromiha, M. M. (2021). MPTherm-pred: Analysis and prediction of thermal stability changes upon mutations in transmembrane proteins. *Journal of Molecular Biology, 433*(11), 166646.

Lee, S., Heinrich, E. L., Lu, J., Lee, W., Choi, A. H., Luu, C., Chung, V., Fakih, M., & Kim, J. (2016). Epidermal growth factor receptor signaling to the mitogen activated protein kinase pathway bypasses ras in pancreatic cancer Cells. *Pancreas, 45*(2), 286–292.

Leroy, B., Fournier, J. L., Ishioka, C., Monti, P., Inga, A., Fronza, G., & Soussi, T. (2013). The TP53 website: An integrative resource centre for the TP53 mutation database and TP53 mutant analysis. *Nucleic Acids Research, 41*(Database issue), D962–D969.

Loo, E., Khalili, P., Beuhler, K., Siddiqi, I., & Vasef, M. A. (2018). BRAF V600E mutation across multiple tumor types: Correlation between DNA-based sequencing and mutation-specific immunohistochemistry. *Applied Immunohistochemistry Molecular Morphology, 26*(10), 709–713.

Lu, W., Zhu, M., Chen, Y., & Bai, Y. (2019). A novel approach to improving hybrid capture sequencing targeting efficiency. *Molecular and Cellular Probes, 46*, 101424.

Mao, Y., Chen, H., Liang, H., Meric-Bernstam, F., Mills, G. B., & Chen, K. (2013). CanDrA: Cancer-specific driver missense mutation annotation with optimized features. *PloS One, 8*(10), e77945.

Masri, R., Al Housseiny, A., Aftimos, G., & Bitar, N. (2022). Incidence of BRAF V600E gene mutation among Lebanese population in melanoma and colorectal cancer: A retrospective study between 2010 and 2019. *Cureus, 14*(9), e29315.

McDonald, I. K., & Thornton, J. M. (1994). Satisfying hydrogen bonding potential in proteins. *Journal of Molecular Biology, 238*, 777–793.

Meyer, M. J., Lapcevic, R., Romero, A. E., Yoon, M., Das, J., Beltrán, J. F., Mort, M., Stenson, P. D., Cooper, D. N., Paccanaro, A., & Yu, H. (2016). Mutation3D: Cancer gene prediction through atomic clustering of coding variants in the structural proteome. *Human Mutation, 37*(5), 447–456.

Meyerson, M., Gabriel, S., & Getz, G. (2010). Advances in understanding cancer genomes through second-generation sequencing. *Nature Reviews Genetics, 11*(10), 685–696.

Mularoni, L., Sabarinathan, R., Deu-Pons, J., Gonzalez-Perez, A., & López-Bigas, N. (2016). OncodriveFML: A general framework to identify coding and non-coding regions with cancer driver mutations. *Genome Biology, 17*(1), 128.

Müller, T., Rahmann, S., & Rehmsmeier, M. (2001). Non-symmetric score matrices and the detection of homologous transmembrane proteins. *Bioinformatics, 17*, S182–S189.

Müller, T., Spang, R., & Vingron, M. (2002). Estimating amino acid substitution models: A comparison of Dayhoff's estimator, the resolvent approach and a maximum likelihood method. *Molecular Biology and Evolution, 19*, 8–13.

Nagarajan, R., Archana, A., Thangakani, A. M., Jemimah, S., Velmurugan, D., & Gromiha, M. M. (2016). PDBparam: Online resource for computing structural parameters of proteins. *Bioinformatics and Biology Insights, 10*, 73–80.

Nakagawa, H., & Fujita, M. (2018). Whole genome sequencing analysis for cancer genomics and precision medicine. *Cancer Science, 109*(3), 513–522.

Ng, J. Y., Lu, C. T., & Lam, A. K. (2019). BRAF mutation: Current and future clinical pathological applications in colorectal carcinoma. *Histology Histopathology, 34*(5), 469–477.

Ng, P. C., Henikoff, J. G., Henikoff, S. (2000). PHAT: A transmembrane-specific substitution matrix. Predicted hydrophobic and transmembrane. *Bioinformatics, 16*, 760–766.

Niu, B., Scott, A. D., Sengupta, S., Bailey, M. H., Batra, P., Ning, J., Wyczalkowski, M. A., Liang, W. W., Zhang, Q., McLellan, M. D., Sun, S. Q., Tripathi, P., Lou, C., Ye, K., Mashl, R. J., Wallis, J., Wendl, M. C., Chen, F., & Ding, L. (2016). Protein-structure-guided discovery of functional mutations across 19 cancer types. *Nature Genetics, 48*(8), 827–837.

Ozturk, K., & Carter, H. (2022). Predicting functional consequences of mutations using molecular interaction network features. *Human Genetics, 141*(6), 1195–1210.

Ozturk, K., Dow, M., Carlin, D. E., Bejar, R., & Carter, H. (2018). The emerging potential for network analysis to inform precision cancer medicine. *Journal of Molecular Biology*, *430*(18 Pt A), 2875–2899.

Paila, U., Kondam, R., & Ranjan, A. (2008). Genome bias influences amino acid choices: Analysis of amino acid substitution and recompilation of substitution matrices exclusive to an AT-biased genome. *Nucleic Acids Research*, *36*, 6664–6675.

Pandey, M., Anoosha, P., Yesudhas, D., & Gromiha, M. M. (2022). Identification of potential driver mutations in glioblastoma using machine learning. *Briefings in Bioinformation, 23*(6), bbac451.

Pandey, M., & Gromiha, M. M. (2021). Predicting potential residues associated with lung cancer using deep neural networks. *Mutation Research*, *822*, 111737.

Pandey, M., & Gromiha, M. M. (2023). MutBLESS: A tool to identify disease-prone sites in cancer using deep learning. *Biochimica et Biophysica Acta Molecular Basis Disease*, *1869*(6), 166721.

Pandey, M., Shah, S. K., & Gromiha, M. M. (2024) Computational approaches for identifying disease-causing mutations in proteins. *Advance in Protein Chemistry Structural Biology.* (in press)

Panicker, P. S., Melge, A. R., Biswas, L., Keechilat, P., & Mohan, C. G. (2017). Epidermal growth factor receptor (EGFR) structure-based bioactive pharmacophore models for identifying next-generation inhibitors against clinically relevant EGFR mutations. *Chemical Biology Drug Design, 90*(4), 629–636.

Pei, J., & Grishin, N. V. (2001). AL2CO: Calculation of positional conservation in a protein sequence alignment. *Bioinformatics, 17*, 700–712.

Petersen, B., Petersen, T. N., Andersen, P., Nielsen, M., & Lundegaard, C. (2009). A generic method for assignment of reliability scores applied to solvent accessibility predictions. *BMC Structural Biology, 9*, 51.

Plaxco, K. W., Simons, K. T., & Baker, D. (1998). Contact order, transition state placement and the refolding rates of single domain proteins. *Journal of Molecular Biology, 277*(4), 985–994.

Ponnuswamy, P. K. (1993). Hydrophobic characteristics of folded proteins. *Progress in Biophysics and Molecular Biology, 59*, 57–103.

Qi, H., Zhang, H., Zhao, Y., Chen, C., Long, J. J., Chung, W. K., Guan, Y., & Shen, Y. (2021). MVP predicts the pathogenicity of missense variants by deep learning. *Nature Communications, 12*(1), 510.

Radivojac, P., Obradovic, Z., Brown, C. J., & Dunker, A. K. (2002). Improving sequence alignments for intrinsically disordered proteins. *Pacific Symposium Biocomputing, 2002*, 589–600.

Ramos, J. M. (2022). Fluorescent In Situ Hybridization (FISH). *Methods Molecular Biology, 2422*, 179–189.

Ranjan, R., Rani, A., Metwally, A., McGee, H. S., & Perkins, D. L. (2016). Analysis of the microbiome: Advantages of whole genome shotgun versus 16S amplicon sequencing. *Biochemical Biophysical Research Communications, 469*(4), 967–977.

Raphael, B. J., Dobson, J. R., Oesper, L., & Vandin, F. (2014). Identifying driver mutations in sequenced cancer genomes: Computational approaches to enable precision medicine. *Genome Medicine*, *6*(1), 5.

Raraigh, K. S., Han, S. T., Davis, E., Evans, T. A., Pellicore, M. J., McCague, A. F., Joynt, A. T., Lu, Z., Atalar, M., Sharma, N., Sheridan, M. B., Sosnay, P. R., & Cutting, G. R. (2018). Functional assays are essential for interpretation of missense variants associated with variable expressivity. *American Journal of Human Genetics, 102*(6), 1062–1077.

Reddy, P. R., Kulandaisamy, A., & Michael Gromiha, M. (2023). TMH Stab-pred: Predicting the stability of α-helical membrane proteins using sequence and structural features. *Methods, 218*, 118–124.

Renganayaki, G., & Nair, A. (2017). Hubsm: A novel amino acid substitution matrix for comparing hub proteins. *International Journal of Advanced Research in Computer Science Software Engineering, 7*, 211–218.

Rentzsch, P., Schubach, M., Shendure, J., & Kircher, M. (2021). CADD-Splice-improving genome-wide variant effect prediction using deep learning-derived splice scores. *Genome Medicine, 13*(1), 31.

Reva, B., Antipin, Y., & Sander, C. (2011). Predicting the functional impact of protein mutations: Application to cancer genomics. *Nucleic Acids Research, 39*(17), e118.

Rios, S., Fernandez, M. F., Caltabiano, G., Campillo, M., Pardo, L., & Gonzalez, A. (2015). GPCRtm: An amino acid substitution matrix for the transmembrane region of class A G Protein-Coupled Receptors. *BMC Bioinformatics, 16*, 206.

Risler, J. L., Delorme, M. O., Delacroix, H., & Henaut, A. (1988). Amino acid substitutions in structurally related proteins. A pattern recognition approach. Determination of a new and efficient scoring matrix. *Journal of Molecular Biology, 204*, 1019–1029.

Ritterhouse, L. L., & Barletta, J. A. (2015). BRAF V600E mutation-specific antibody: A review. *Seminar in Diagnostic Pathology, 32*(5), 400–408.

Rivlin, N., Brosh, R., Oren, M., & Rotter, V. (2011). Mutations in the p53 tumor suppressor gene: Important milestones at the various steps of tumorigenesis. *Genes Cancer*, 2(4), 466–474.

Rogers, M. F., Shihab, H. A., Gaunt, T. R., & Campbell, C. (2017). CScape: A tool for predicting oncogenic single-point mutations in the cancer genome. *Scientific Reports, 7*(1), 11597.

Ryslik, G. A., Cheng, Y., Modis, Y., & Zhao, H. (2016). Leveraging protein quaternary structure to identify oncogenic driver mutations. *BMC Bioinformatics, 17*, 137.

Saleem, M., Ghazali, M. B., Wahab, M. A. M. A., Yusoff, N. M., Mahsin, H., Seng, C. E., Khalid, I. A., Rahman, M. N. G., & Yahaya, B. H. (2020). The BRCA1 and BRCA2 Genes in Early-Onset Breast Cancer Patients. *Adv Exp Med Biol., 1292*, 1–12.

Singh, S., Kumar, M., Kumar, S., Sen, S., Upadhyay, P., Bhattacharjee, S., M, N., Tomar, V. S., Roy, S., Dutt, A., & Kundu, T. K. (2019). The cancer-associated, gain-of-function TP53 variant P152Lp53 activates multiple signaling pathways implicated in tumorigenesis. *Journal of Biological Chemistry, 294*, 14081–14095.

Steventon-Jones, V., Stavish, D., Halliwell, J. A., Baker, D., & Barbaric, I. (2022). Single nucleotide polymorphism (SNP) arrays and their sensitivity for detection of genetic changes in human pluripotent stem cell cultures. *Current Protocols, 2*(11), e606.

Tamborero, D., Gonzalez-Perez, A., & Lopez-Bigas, N. (2013). OncodriveCLUST: Exploiting the positional clustering of somatic mutations to identify cancer genes. *Bioinformatics*, *29*(18), 2238–2244.

Tate, J. G., Bamford, S., Jubb, H. C., *et al.* (2019). COSMIC: The catalogue of somatic mutations in cancer. *Nucleic Acids Research, 47*(D1), D941–D947.

Tokheim, C., Bhattacharya, R., Niknafs, N., Gygax, D. M., Kim, R., Ryan, M., Masica, D. L., & Karchin, R. (2016). Exome-scale discovery of hotspot mutation regions in human cancer using 3D protein structure. *Cancer Research*, *76*(13), 3719–3731.

Tokheim, C., & Karchin, R. (2019). CHASMplus reveals the scope of somatic missense mutations driving human cancers. *Cell Systems, 9*(1), 9–23.e8.

Tomczak, K., Czerwińska, P., & Wiznerowicz, M. (2015). The cancer genome atlas (TCGA): An immeasurable source of knowledge. *Contemporary Oncology, 19*(1A), A68–A77.

Trevino, V. (2020). HotSpotAnnotations — A database for hotspot mutations and annotations in cancer. *Database*, *2020*, baaa025.

Trivedi, R., & Nagarajaram, H. A. (2020). Substitution scoring matrices for proteins — an overview. *Protein Science, 29*, 2150–2163.

Valdar, W. S. (2002). Scoring residue conservation. *Proteins, 48*(2), 227–241.

Wang, Z., Jensen, M. A., & Zenklusen, J. C. (2016). A practical guide to the cancer genome atlas (TCGA). *Methods in Molecular Biology, 1418*, 111–141.

Wong, K. B., DeDecker, B. S., Freund, S. M., Proctor, M. R., Bycroft, M., & Fersht, A. R. (1999). Hot-spot mutants of p53 core domain evince characteristic local structural changes. *Proceedings of the National Academy of Science USA, 96*, 8438–8442.

Yang, Y., Heffernan, R., Paliwal, K., Lyons, J., Dehzangi, A., Sharma, A., Wang, J., Sattar, A., & Zhou, Y. (2017). SPIDER2: A package to predict secondary structure, accessible surface area, and mainchain torsional angles by deep neural networks. *Methods Molecular Biology, 1484*, 55–63.

Yoshida, K., & Miki, Y. (2004). Role of BRCA1 and BRCA2 as regulators of DNA repair, transcription, and cell cycle in response to DNA damage. *Cancer Science, 95*(11), 866–871.

Yue, Z., Zhao, L., & Xia, J. (2020). dbCPM: A manually curated database for exploring the cancer passenger mutations. *Brief Bioinformatics, 21*(1), 309–317.

Zhong, J., Xuan, W., Lu, S., Cui, S., Zhou, Y., Tang, M., Qu, X., Lu, W., Huo, H., Zhang, C., Zhang, N., & Niu, B. (2023) Discovery of ANO1 Inhibitors based on Machine learning and molecule docking simulation approaches. *European Journal of Pharmaceutical Sciences, 184*, 106408.

https://doi.org/10.1142/9789811293269_0012

Chapter 12

Experimental and computational approaches for deciphering disease-causing mutations in membrane proteins

A. Kulandaisamy[1,*], P Ramakrishna Reddy[1], Dmitrij Frishman[2], and M. Michael Gromiha[1,*]

[1]*Department of Biotechnology, Bhupat and Jyoti Mehta School of Biosciences, Indian Institute of Technology Madras, Chennai, Tamil Nadu 600 036, India*
[2]*Department of Bioinformatics, Technische Universität München, Wissenschaftszentrum Weihenstephan, Freising, Germany*

Abstract

Transmembrane proteins (TMPs) are integral components of cellular and organelle membranes, playing pivotal roles as transporters, enzymes, receptors, and mediators in essential cellular functions. Mutations occurring in TMPs can disrupt their functions, and leading to various diseases such as cystic fibrosis, cancer, and many others. Experimental information on disease-associated and neutral mutations in TMPs has been

*Corresponding authors
AK: kulandai28@gmail.com
MMG: gromiha@iitm.ac.in

curated and cataloged in databases such as MutHTP and TMSNP. These repositories offer comprehensive features, encompassing the sequence, structure, topology, and disease associations of TMPs. Relating these features with disease-causing mutations provides intricate relationships between sequence/structural parameters and diseases and the development of computational tools. Machine learning-based computational tools utilize diverse features, including evolutionary data, physiochemical properties, atomic contacts, contact potentials, and energetic contributions for identifying disease-causing mutations. These specialized tools serve to characterize the impact of novel variants across the entire human membrane proteome. In this chapter, we discuss about the identification of genetic variants using experimental methods, recent progress on computational databases and tools for disease-causing and neutral mutations in TMPs. These information provide additional insights for designing mutation-specific strategies for different diseases.

12.1 Introduction

Genomes typically encompass approximately 20% to 30% of genes responsible for encoding transmembrane proteins (TMPs) (Almeida *et al.*, 2017). These TMPs traverse cellular or organelle membranes via the phospholipid bilayer, relying on their structural motifs, including α-helical or β-barrel configurations (Bowie, 2004). Their primary function lies in the regulation of vital processes such as the translocation of ions and molecules across the cellular membrane, the initiation of signal transduction cascades, the facilitation of intercellular communication, and the catalysis of pivotal biochemical reactions. The functional attributes of TMPs hinge upon both their protein sequence and the integrity of their structural conformation (Ponnuswamy & Gromiha, 1993; Almén *et al.*, 2009; Gromiha & Ou, 2014).

During cellular development, genetic sequences can undergo alterations or mutations, which may include insertions, deletions, and single nucleotide polymorphisms (SNPs), with the latter being the most common type. While several of these genetic variations confer advantages to cellular survival, some exert deleterious effects on normal cellular and physiological functions (Reich *et al.*, 2003; Loewe & Hill, 2010). Variants recurrently encountered within disease samples, as well as those influencing the roles of TMPs, are designated as "disease-causing," whereas others assume a "neutral" status. It is noteworthy that nearly 60% of TMPs operate as potential therapeutic targets across a spectrum of diseases, encompassing diverse malignancies, retinal dystrophies, and Parkinson's disease, among others (Hamosh *et al.*, 2005; Ng *et al.*, 2012; Gao *et al.*, 2015; Yin & Flynn, 2016; Dobson *et al.*, 2018; Zaucha *et al.*, 2021). As an illustration, consider the fibroblast growth factor receptor 3 (FGFR3) protein carrying the G380R missense mutation, which results in an increase in charge within the membrane-spanning regions. This specific functional alteration is correlated with the development of achondroplasia, which is the prevalent form of human dwarfism (He *et al.*, 2011). The identification and comprehensive characterization of these disease-causing variants within specific specimens is imperative. Furthermore, elucidating the underlying molecular mechanisms of these pathogenic mutations, through a synergy of experimental and computational methodologies, illuminates prospective avenues for the design of mutation-specific therapeutic strategies against a myriad of genetic diseases.

In recognition of the paramount importance of comprehending disease-causing mutations within primary drug targets, specifically TMPs, this chapter focuses on four key aspects: (i) experimental approaches for identifying the genetic variants, (ii) the available computational databases containing the experimental information of disease-causing and neutral mutations, (iii) large-scale analytical studies performed to investigate disease-causing and neutral mutations in membrane proteins, and (iv) the specific computational tools for characterizing the phenotypic effects of mutations in human health.

12.2 Experimental approaches for identifying disease-causing and neutral variants

The earlier Sanger sequencing method and the newer next-generation sequencing (NGS) technologies are employed for detecting genetic variants in both human disease and normal samples (Buermans & Den Dunnen, 2014; Goodwin *et al.*, 2016; Huss *et al.*, 2018; Koboldt, 2020). The primary distinction between these approaches lies in their sequencing capacity. Unlike the Sanger method, which sequences one

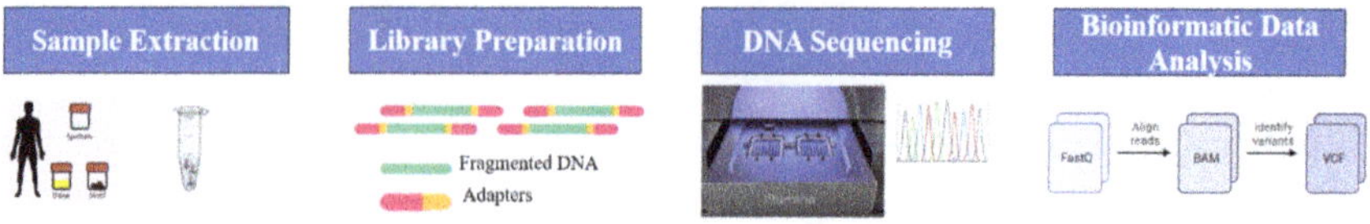

Figure 12.1 Whole-exome sequencing pipeline for identifying the genetic variants

DNA fragment at a time, NGS operates on a massive parallel scale, allowing the simultaneous sequencing of millions of fragments during a single run. These sequencing methods have applications in various contexts, including the identification of variants associated with specific genes (such as those encoding TMPs), whole exome sequencing, and complete genome sequencing. **Figure 12.1** illustrates the whole-exome sequencing technology, which involves four primary stages.

(i) **Sample extraction:** This initial step entails isolating and purifying the DNA from the chosen healthy or disease sample of interest.
(ii) **Library preparation:** In this phase, polymerase chain reaction (PCR) amplification is employed to generate a collection of DNA fragments of specific sizes, known as a library. Subsequently, adaptor molecules are ligated to these fragments.
(iii) **Sequencing:** The prepared library sequences are loaded into a sequencer machine (e.g., Illumina) for the actual sequencing process.
(iv) **Bioinformatic data analysis:** The raw sequenced reads undergo a series of computational analyses. This includes quality assessment to ensure data reliability, alignment of the reads against the human reference genome, variant calling to identify genetic variations, and subsequent analysis and interpretation of these variants. Furthermore, the identified variants are subject to further scrutiny through various functional and biochemical studies to characterize their effects and significance.

12.3 Computational databases for protein-coding disease-causing mutations

Advancements in sequencing technology, coupled with statistical and functional studies, have enabled the identification of genetic variants that play roles in causing diseases or remain neutral. This curated information is readily accessible through various databases, such as Humsavar, SwissVar, ClinVar, 1000 Genomes, and COSMIC (Mottaz *et al.*, 2010; Landrum *et al.*, 2016; Tate *et al.*, 2019; UniProt Consortium, 2019). These databases serve as invaluable resources for investigating genetic variations and their links to diseases, offering crucial insights into how specific variants impact human health. Researchers use these databases to study genetic variant prevalence, assess functional consequences, and determine clinical significance across populations. Within these databases, the categorization of a mutation as disease-causing or neutral hinges on two primary factors: the presence of these mutations in clinical samples and the existence of experimental evidence demonstrating their functional implications.

Public computational databases play a crucial role in advancing our understanding of membrane proteins (TMPs) due to their vital functions and connections to various diseases. These dedicated databases explore aspects like structure (Kozma *et al.*, 2012; Lomize *et al.*, 2012; Bittrich *et al.*, 2022), function (Gromiha *et al.*, 2009; Marsico *et al.*, 2010; Saier *et al.*, 2016; Isberg *et al.*, 2016), stability (Kulandaisamy *et al.*, 2021), and topology (Tusnady *et al.*, 2007; Kozma *et al.*, 2012; Lomize *et al.*, 2012; Dobson *et al.*, 2015; Bittrich *et al.*, 2022; Sun *et al.*, 2023) of TMPs. Specific databases like MutHTP (Kulandaisamy *et al.*, 2018; 2022) and TMSNP (Garcia-Recio *et al.*, 2021) focus on disease-causing and neutral mutations within TMPs. These repositories provide curated data about these mutations, including their associated diseases, alongside a wealth of protein sequence and structure details. In this section, we provide an overview of MutHTP (Kulandaisamy *et al.*, 2018) and TMSNP (Garcia-Recio *et al.*, 2021), outlining their contents, features, and potential applications.

12.3.1 *MutHTP*

12.3.1.1 *Data collection and curation*

Membrane proteins were sourced from the UniProt database using a keyword search for "Transmembrane" in the subcellular location field (UniProt Consortium, 2019). The mutations, including insertions, deletions, and missense that are associated with membrane proteins, were collected from multiple databases: Humsavar (http://www.uniprot.org/docs/humsavar), SwissVar (Mottaz *et al.*, 2010), 1000 Genomes (UniProt Consortium, 2019), COSMIC (Tate *et al.*, 2019), and ClinVar (Landrum *et al.*, 2016).

In the case of Humsavar and SwissVar, the disease-causing or neutral mutations were mainly defined based on manual annotations derived from literature reports. From the UniProt human variants file, 1000 Genome data was extracted, providing information about mutation origins, disease associations, and UniProt accession numbers. COSMIC, a cancer-specific repository, was used to obtain meticulously curated data, specifically focusing on disease-causing mutations based on their occurrence. Mutations occurring more than once in a single cancer gene were considered as disease mutations (Gnad *et al.*, 2013). ClinVar evaluates mutations in patient samples, providing clinical significance assessments and contributor details. In both the 1000 Genome and ClinVar databases, "pathogenic" or "likely pathogenic" mutations are deemed disease-causing, while those marked "benign" or "likely benign" are considered neutral. The overall pipeline of MutHTP construction, features, and applications is shown in **Figure 12.2**.

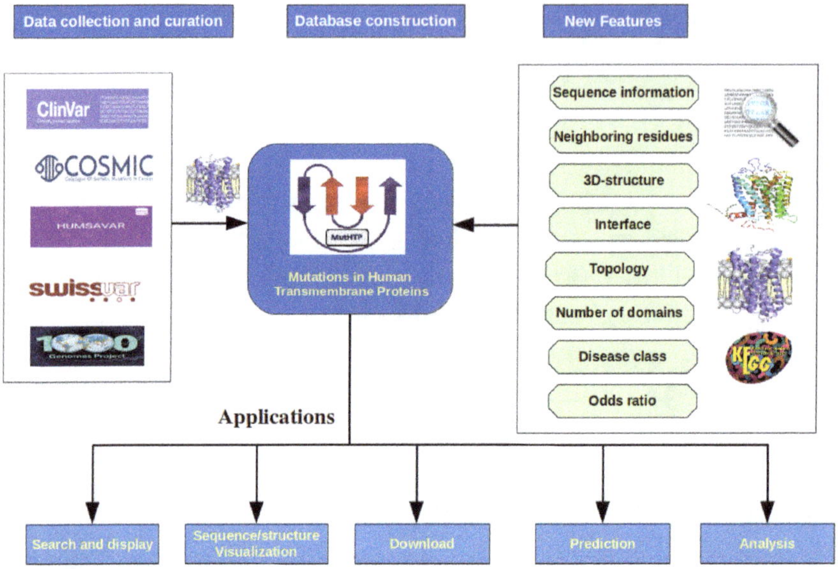

Figure 12.2 A detailed workflow for the MutHTP database. The figure was taken from Kulandaisamy *et al.* (2018)

12.3.1.2 *Contents of MutHTP*

MutHTP has 1,84,500 disease-causing mutations that are associated with 2,700 diseases and 17,827 neutral mutations in membrane proteins. Most of the disease-causing mutations are involved in different types of cancer (89%), followed by nervous system diseases (4%), and congenital disorders (3%). This database is publicly available at http://www.iitm.ac.in/bioinfo/MutHTP/. Each entry in the MutHTP database provides the following comprehensive features and characteristics at both gene and protein levels:

- Gene name, chromosome number and genomic location, mutation details, and origin (somatic or germline) of the mutation.
- Details of protein name, sequence and structure, UniProt, and PDB accession numbers, functional importance (conservation score) of a mutation site, neighboring amino acid residues around the mutation site, and annotation of the mutation site on protein–protein interactions.
- Membrane topology-specific features, such as the number of transmembrane (TM) segments in the protein, the location of the mutation site with respect to the membrane bilayer, and the type of pass (single or multi-pass).
- Name of the disease(s) associated with a mutation, classification of diseases into 14 classes based on the KEGG database, and source of the database.
- Visualization of mutation sites at protein sequence and structure information using the JSmol applet.
- Furthermore, MutHTP stands out as the first membrane protein-specific mutational database, offering diverse search and display options for efficient data retrieval and download capabilities.

12.3.2 *TMSNPdb*

The human membrane proteins were collected from the UniProt database. For each of these membrane proteins, disease-causing mutations associated with Mendelian disorders are retrieved from the ClinVar and SwissVar databases. To ensure specificity for membrane proteins, only mutations occurring in the TM helices are considered, as these regions exhibit distinct characteristics compared to globular proteins. The ranges of TM segments were obtained from the UniProt database. Additionally, nonpathogenic missense mutations and their population allele frequency are retrieved from the GnomAD (Gudmundsson *et al.*, 2022) and ClinVar (Landrum *et al.*, 2016) databases.

The culmination of this data collection process resulted in the creation of a comprehensive database containing 196,705 nonpathogenic, 2,624 pathogenic, and 437 likely pathogenic mutations within the TM region of membrane proteins. This database is accessible at http://lmc.uab.es/tmsnp/tmsnpdb.

12.4 Analysis of disease-causing and neutral mutations in membrane proteins

In the past two decades, there are significant number of computational analytical large-scale studies have been conducted to investigate the potential impact of membrane protein mutations in disease and normal samples. A pioneer study by Partridge *et al.* (2004) assembled a dataset of 240 protein mutations that are associated with 80 membrane proteins from Human Gene Mutation Database (HGMD) for comparing the disease-causing and neutral mutations between membrane and soluble proteins. This study revealed that mutations in Ser and Arg residues had a more deleterious effect on TMPs than soluble ones (Partridge *et al.*, 2004).

Utilizing the Humsavar database, Molnár *et al.* (2016) constructed a specific dataset of 8,561 disease-causing and 10,952 neutral mutations in human membrane proteins. It reported a higher rate of disease-causing mutations in membrane proteins compared to non-membrane proteins. In addition to this, membrane proteins with 2, 4, 6, and >8 TM segments containing a large number of disease-causing mutations than neutral ones. Based on the membrane bilayer, most of the disease-causing mutations are clustered in the middle of the membrane than neutral mutations (Molnár *et al.*, 2016).

Dobson *et al.* (2018) compiled a dataset of 20,608 disease-causing germline mutations and 28,499 neutral mutations linked to TM, ordered, and disordered protein regions. Of these, 8,276 disease-causing and 7,055 neutral mutations pertained to TM regions. Their findings revealed that disease-causing mutations are predominantly enriched in membrane-proximal areas of the cytoplasmic phase of TMPs.

In a targeted investigation, Nastou *et al.* (2019) conducted a specific analysis of disease-causing and neutral mutations within human potassium, sodium, and calcium channels. These mutations were categorized based on topological domain annotations, revealing a higher prevalence of disease-causing mutations in the pore-forming loop and TM segments. Notably, particular amino acid substitutions, such as R→C and A→V mutations in the pore loop, as well as E→K, R→H, and R→Q mutations in the voltage sensor domain of voltage-gated ion channels (VGICs), were linked to disease-causing mutations. Furthermore, there was a greater occurrence of charged to polar and nonpolar to nonpolar amino acid changes in disease-causing mutations within calcium, potassium, and sodium channels (Nastou *et al.*, 2019).

Our recent study examined 172,218 disease-causing and 17,512 neutral missense mutations sourced from the MutHTP database. We classified these mutations based on protein structure (alpha-helical and beta-barrel), topology (extra-cellular, membrane-spanning, intracellular), function (GPCR, transporters), and diseases (KEGG-based categories). This classification is further used for analyzing the frequency of amino acid residues and their changes (mutations) in disease-causing and neutral cases (Kulandaisamy *et al.*, 2019).

12.4.1 *Frequency of occurrence of amino acids as wild-type and mutants*

We calculated the frequency of the occurrence of amino acid residues in both wild-type and mutated positions. Among all human membrane proteins, Arg was the most commonly mutated residue in disease cases, followed by Gly, Ser, and Ala. In neutral mutations, Arg dominated, followed by Val and Ala. The high frequency of Arg mutations could stem from the CpG dinucleotide mutation and its stability-enhancing hydrogen bonds. Arg also exhibits strong protein–protein binding. Leu, Ser, and Lys are favored in disease mutations, while Val, Ser, and Thr are favored in neutral mutations. Cys, Phe, and Lys mutants are disease-associated, and Ala, Gly, and Val mutants are common in neutral ones. **Table 12.1**

Table 12.1 Preferred amino acid residues in wild-type and mutant positions of disease-causing and neutral mutations

Features/types of proteins	Type of classification	Disease-causing mutations		Neutral mutations	
		Wild-type	Mutant	Wild-type	Mutant
Whole human membrane proteome	—	Arg, Gly, and Ser	Leu, Ser, and Lys	Arg, Val, and Ala	Val, Ser, and Thr
Secondary structure	α-helical	Arg, Gly, and Ser	Leu, Ser, and Lys	Arg, Val, and Ala	Val, Ser, and Thr
	β-barrel	Glu, Arg, and Met	Lys, Cys, and His	Asp, Ala, and Ile	Val, Tyr, and Trp
Number of times protein cross the cell membrane	Single pass	Arg, Gly, and Pro	Lys, Cys, and Phe	Val, Thr, and Arg	Arg, Ala, and Val
	Multi-pass	Arg, Gly, and Glu	Lys, Cys, and Phe	Ala, Val, and Ile	Ala, Gly, and Val
Topology	Inside or cytoplasm	Arg, Glu, and Gly	Lys, His, and Trp	Ala, Pro, and Arg	Ser, Val, and Thr
	Membrane	Ala, Leu, and Gly	Arg, Asp, and Glu	Ile, Val, and Phe	Val, Met, and Thr
	Outside or extracellular	Gly, Pro, and Glu	Lys, Cys, and Leu	Val, Thr, and Ala	Ser, Ile, and His
Function	GPCR family proteins	Arg, Gly, and Ser	Ile, Phe, and Cys	Ala, Val, and Arg	Ser, Met, and Thr
	Transporters	Arg, Gly, and Glu	Cys, Lys, and Phe	Ala, Val, and Arg	Ser, Met, and Thr

Data are obtained from Kulandaisamy *et al.* (2022).

lists frequently occurring wild-type and mutant residues for various protein types.

12.4.2 *Preferred disease-causing and neutral mutations in human membrane proteins*

Kulandaisamy *et al.* (2019) assessed the preference of mutations in membrane proteins, utilizing the frequency of amino acid substitutions,

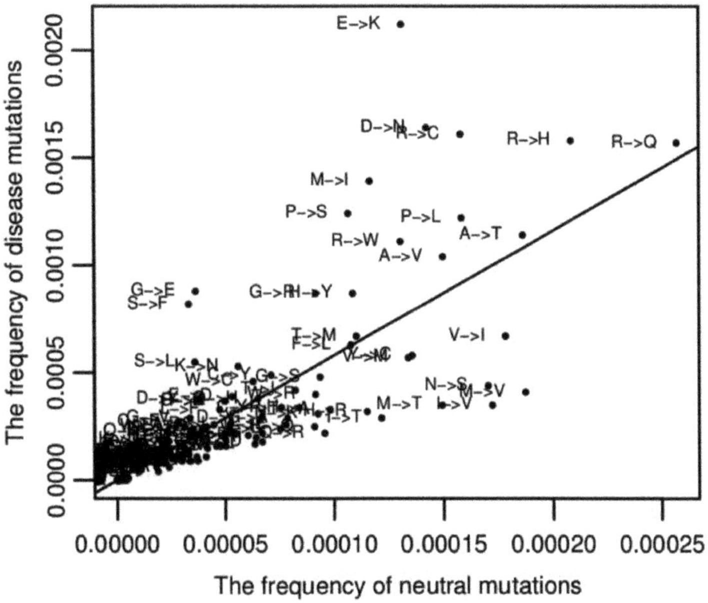

Figure 12.3 Preference of disease-causing and neutral missense mutations in human membrane proteins. Note: Mutations are located above the diagonal line that are more frequent in disease, whereas mutations below the line are more frequent in neutral cases. The figure was taken from Kulandaisamy *et al.* (2019)

as depicted in **Figure 12.3**. Notably, E→K ranked highest in disease-associated mutations, followed by D→N and R→C. Conversely, in neutral mutations, M→V was most frequent, trailed by V→I and I→V. This suggests that alterations involving charged residues (E, D, and R) more frequently lead to disease (Kulandaisamy *et al.*, 2019). Such changes in charge can impact protein structure/function, affecting selectivity, interactions, and enzymatic activity, culminating in diseases like cancer or neurodegeneration (Tang *et al.*, 1993; Colegio *et al.*, 2002; Perocchi *et al.*, 2010; Chen *et al.*, 2010; Nishi *et al.*, 2013). For instance, an E64K mutation switches claudin-15 from transporting Na^+ to Cl^- ions (Colegio *et al.*, 2002). The observed mutation preference pattern remains consistent across different protein types, including α-helical, single-pass, and multi-pass proteins, as well as within the GPCR and transporter families.

12.4.3 *Preference of mutations in different disease classes*

The analysis was broadened to encompass 14 distinct disease categories, revealing notable trends. Specifically, the mutations E→K and D→N were notably enriched in cancer cases, whereas cysteine mutations (C→Y and C→R) were prevalent in congenital disorders of metabolism (CDM) and immune system diseases (ISD). Notably, Nishi *et al.* (2013) highlighted that cancer-linked missense mutations disrupt electrostatic interactions in protein–protein and protein–nucleic acid complexes. Additionally, mutations like R→Q, R→H, R→C, and R→W prominently featured across more than 10 disease classes.

12.4.4 *Comparison of mutations between topologically similar regions of globular and membrane proteins*

A comparative analysis was conducted to examine the mutation patterns within analogous regions of both globular and membrane proteins. This analysis encompassed distinct scenarios: buried versus membrane-spanning, exposed versus internal, and exposed versus external regions. Notably, disease-causing mutations exhibited a preference for transitions from polar (Ser, Thr) to nonpolar residues, specifically within membrane regions. The significance of polar residues in membrane protein stability, folding, cation coordination, and inter-helical interactions has been documented (Nicoll *et al.*, 1996; Senes *et al.*, 2000; Murtazina *et al.*, 2001; North *et al.*, 2006). Conversely, disease-causing mutations often featured conversions from nonpolar to charged residues (e.g., G→E, G→R), corroborating prior findings (Molnár *et al.*, 2016). In buried regions, the Cys residue, pivotal for disulfide bond formation and protein stabilization, was frequently substituted by nonpolar residues (**Figure 12.4**). Conversely, alterations involving nonpolar residues were prominent in inside/cytoplasmic regions, whereas exposed regions demonstrated a propensity for interchanges between charged (Asp, Glu) and nonpolar (Pro, Gly) residues. Additionally, changes encompassing Cys and positively charged residues (Lys, Arg) were preferred in extracellular and exposed regions, respectively. Among neutral mutations, membrane regions exhibited a predilection for polar to nonpolar changes and vice versa, while buried

	A	C	D	E	F	G	H	I	K	L	M	N	P	Q	R	S	T	V	W	Y	Total	Mut
A		0 1	162 231	91 120	0 1	88 137	0 0	0 0	0 1	1 0	0 0	1 0	102 188	0 0	0 0	237 354	1178 1569	1085 1497	0 0	0 0	2945 4099	0.053 0.024
C	0 1		0 0	0 0	52 280	24 134	0 1	0 0	0 0	0 2	0 0	0 0	0 0	0 0	101 359	46 207	0 1	0 1	18 130	109 559	350 1675	0.020 0.019
D	5 24	0 0		22 68	0 1	39 87	21 85	0 0	0 0	0 0	0 0	157 297	0 0	0 0	0 0	0 0	0 0	11 43	0 0	37 92	292 697	0.063 0.027
E	11 18	0 0	36 76		0 0	27 37	0 0	0 0	221 355	0 0	0 0	0 0	0 0	35 53	0 0	0 0	0 0	8 26	0 0	0 0	338 565	0.067 0.024
F	0 0	74 179	0 0	0 0		0 1	0 0	89 104	0 1	573 680	0 0	0 0	0 0	0 0	0 0	164 229	1 0	143 165	0 0	63 75	1107 1434	0.020 0.011
G	86 115	100 122	264 420	525 504	0 1		0 1	1 0	6 10	1 2	0 0	1 3	0 0	0 0	613 654	343 418	0 0	220 277	57 59	0 0	2217 2586	0.056 0.028
H	0 0	0 0	12 32	0 0	0 0	0 0		0 0	0 0	8 26	0 0	17 47	19 40	31 60	44 120	0 0	0 0	0 0	0 0	101 193	232 518	0.043 0.023
I	0 0	0 0	0 1	0 0	125 191	0 0	0 0		20 34	73 137	164 376	81 215	1 0	0 0	10 24	53 139	257 511	342 627	0 0	0 1	1126 2256	0.019 0.012
K	0 0	0 0	0 0	10 25	0 0	0 0	0 0	4 4		0 1	6 4	42 50	0 0	5 11	6 23	0 0	11 15	0 0	0 1	0 0	84 134	0.023 0.014
L	0 0	0 1	0 0	0 0	723 970	0 1	58 114	277 335	0 0		211 268	0 0	427 791	81 154	166 380	66 139	0 0	286 449	18 33	0 0	2313 3635	0.020 0.011
M	0 0	0 0	0 0	0 0	0 0	0 0	0 0	469 470	59 51	63 86		0 1	0 0	0 0	39 64	0 0	115 164	166 239	0 0	0 0	911 1075	0.045 0.020
N	0 0	0 0	43 65	0 0	0 1	0 0	16 34	24 38	60 76	0 0	0 0		0 0	0 0	1 0	89 163	16 26	0 0	0 0	22 34	271 437	0.026 0.016
P	17 52	0 0	0 0	0 0	5 4	0 0	60 67	0 1	0 0	337 458	1 0	0 0		17 32	39 89	389 408	45 82	0 0	0 0	0 0	910 1193	0.067 0.031
Q	0 0	0 0	0 0	12 30	0 0	0 0	47 77	0 0	32 37	14 19	0 0	0 0	15 37		30 69	0 0	0 0	0 0	0 0	0 0	150 269	0.023 0.014
R	0 0	221 280	0 0	1 0	0 0	25 70	191 265	8 20	17 47	37 60	5 7	0 1	13 41	152 284		29 44	6 22	0 0	104 199	0 0	809 1340	0.174 0.077
S	31 30	102 85	0 0	0 0	755 442	46 60	0 0	53 54	0 1	393 350	0 0	96 113	100 128	0 0	92 105		47 62	0 0	22 18	126 89	1863 1537	0.051 0.022
T	151 157	0 0	0 0	0 0	0 0	0 0	0 0	266 249	39 47	1 0	371 358	66 66	74 81	0 0	55 49	76 79		0 0	0 0	0 0	1099 1086	0.035 0.017
V	294 550	0 0	63 99	52 111	134 266	107 287	1 0	903 1038	1 0	266 460	670 993	0 0	0 0	0 1	0 0	0 1	0 0		0 2	0 0	2491 3808	0.037 0.018
W	0 0	101 178	0 0	0 0	0 0	15 45	0 0	0 0	0 0	37 86	0 1	0 0	0 0	0 0	73 161	25 30	0 0	0 0		0 0	251 501	0.022 0.015
Y	0 0	238 370	34 50	0 0	42 57	0 0	132 193	0 0	0 1	0 0	0 0	38 62	0 0	0 1	0 0	26 61	0 0	0 0	0 0		510 795	0.023 0.014
Mem **Buried**	**595** **947**	**836** **1216**	**614** **974**	**713** **858**	**1836** **2214**	**371** **859**	**526** **837**	**2094** **2313**	**455** **661**	**1804** **2367**	**1428** **2007**	**499** **855**	**751** **1306**	**321** **596**	**1269** **2097**	**1543** **2272**	**1676** **2452**	**2261** **3324**	**219** **442**	**458** **1043**	**20269** **29640**	

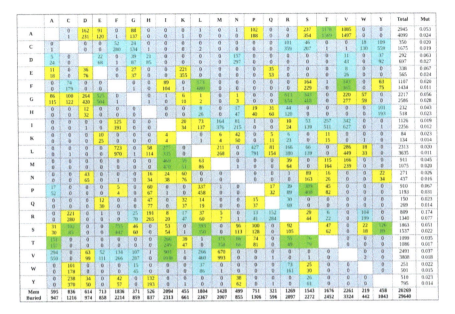

Figure 12.4 Comparison of disease-causing mutations in the membrane region of TMPs and the buried region of globular proteins. Note: Mem: Membrane; Mut: Relative mutability. The top and bottom rows of each cell refer to the number of mutations in the membrane and buried regions, respectively. Green blocks represent statistically significant mutations that are enriched in the membrane region of TMPs. Cyan blocks indicate statistically significant mutations enriched in the buried region of globular proteins. Yellow blocks show nonsignificant mutations. Light steel blue blocks represent the mutations that are not present either in buried or membrane regions. The figure was taken from Kulandaisamy *et al.* (2019)

regions witnessed conversions from hydrophobic to smaller hydrophobic residues. Exposed regions, in turn, displayed a prevalence of alterations between charged and polar residues.

12.5 Features for discriminating disease-causing and neutral mutations

Several sequence- and structure-based features have been developed for predicting the effects of missense variants. These features include physicochemical properties, contact potentials, amino acid substitution matrices, solvent accessibility, secondary structure, conservation scores, and energetic parameters (Ng *et al.*, 2003; Capriotti *et al.*, 2006; Bromberg *et al.*, 2008; Adzhubei *et al.*, 2013; De Beer *et al.*, 2013; Shihab *et al.*, 2013; Choi & Chan, 2015).

12.5.1 *Physicochemical properties*

Kulandaisamy *et al.* (2020) compiled a set of 685 features encompassing physical, chemical, energetic, and conformational attributes of amino acid residues from AAindex database and 49 parameters derived from relevant literature (Kawashima & Kanehisa, 2000; Gromiha, 2005). Additionally, a set of 70 parameters is included in ProtScale and related references (Gasteiger *et al.*, 1999; Morita *et al.*, 2011; Xiao & Shen, 2015; Simm *et al.*, 2016). Features that are not relevant to TMPs and redundant properties with a correlation of >0.85 are removed. This meticulous refinement yielded a well-defined set of 253 properties, which are normalized within a range of 0 to 1.

The change in property values ($\Delta P_{\text{mutation}}$) due to the mutation is computed using Equation 12.1:

$$\Delta P_{\text{mutation}} = P_{\text{mutant}} - P_{\text{wild-type}} \qquad \text{(Eq. 12.1)}$$

Here, $P_{\text{wild-type}}$ and P_{mutant} denote the property values associated with the wild-type and mutant residues, respectively.

12.5.2 *Substitution matrices and contact potentials*

A set of 94 substitution matrices and 47 pair-wise contact potential matrices for amino acids are obtained from the AAindex database (Kawashima & Kanehisa, 2000). The substitution matrices are specifically allocated to each mutation type. As for contact potentials, the alteration in an amino acid's contact potential resulting from a mutation is calculated by subtracting the contact potential value of the N/C-terminal neighbors of the mutated position from the wild-type residue from the corresponding values associated with the N/C-terminal neighbors of the mutant residue.

12.5.3 *Evolutionary information*

Evolutionary information serves as a crucial indicator of the functional significance of a residue in a protein sequence. We computed the 18 distinct types of normalized conservation scores using the standalone AACon tool (Valdar, 2002; Manning *et al.*, 2008). In addition to this, we generated the position-specific scoring matrix (PSSM) profiles of each protein through PSI-BLAST (Altschul *et al.*, 1997) using the UniRef90 database sequences (Suzek *et al.*, 2014), with three rounds of iteration and an E-value cutoff of 0.001. The properties encompassed the value of the wild-type residue at the mutation site ($PSSM_{wild\text{-}type}$), the probability of the mutant residue being positioned at the mutation site ($PSSM_{mutant}$), and the quantified difference between these two values ($PSSM_{mutant} - PSSM_{wild\text{-}type}$) were considered.

12.5.4 *Neighboring residue attributes*

The neighboring residues surrounding the mutant residue were incorporated by the following equation:

$$\Delta P_{local} = P_i(\text{mutant}) - \frac{\sum_{n=i-j}^{i+j} P_n(\text{wild-type})}{2j+1} \qquad \text{(Eq. 12.2)}$$

Here, j takes values of 1, 2, and 3, corresponding to window lengths of 3, 5, and 7, respectively. The total window length is $2j + 1$, and "I" represents the position of the mutation.

In addition, the amino acids were categorized into six groups based on their physicochemical properties: aliphatic (G, A, L, I, and V), aromatic (F, Y, and W), sulfur-containing (M and C), polar (N, Q, S, T, and P), negatively charged (D and E), and positively charged (R, H, and K) residues. Furthermore, we utilized the distribution of amino acid categories surrounding the mutation site within window lengths of 3 to 21 as supplementary features. Furthermore, we decoded the wild-type and mutant residues, the position of the mutation, and the atomic frequencies in each residue as features.

12.5.5 *Features specific to membrane proteins*

The membrane protein-specific features, including TM segment count, mutation site topology, and residue fractions for signal-peptide, cytoplasmic, membrane, and extracellular regions were considered. Topology data was sourced from the CCTOP (Bernsel *et al.*, 2009) and TOPCONS (Dobson *et al.*, 2015) servers. Additionally, we utilized substitution matrices for human membrane proteins, along with region-specific matrices from our prior study (Kulandaisamy *et al.*, 2019).

12.5.6 *Feature selection and classification*

For feature selection and classification, we employed the Waikato Environment for Knowledge Analysis (WEKA) platform (Hall *et al.*, 2009). Within WEKA, feature selection involved two methods: (i) CfsSubsetEval evaluator with genetic and BestFirst search, and (ii) Consistency subset evaluator with genetic and BestFirst search. Subsequently, classification was performed using various available methods in WEKA. Based on cross-validation performance, we chose the voting algorithm to classify mutations as disease-causing or neutral. Furthermore, the model's performance was validated using an independent test dataset comprising 20% of the original data.

12.6 Model evaluation and validation

The assessment of classification models was conducted using a 10-fold group-wise cross-validation strategy. Mutations were partitioned based on protein sequence identity clusters, forming 10 distinct groups. Each cluster was treated as a single group, ensuring evolutionary independence between training and validation proteins across the cross-validation folds. Performance evaluation metrics encompassed sensitivity, specificity, accuracy, balanced accuracy, and Matthew's correlation coefficient (MCC) — critical gauges of the classification model's robustness:

$$\text{Sensitivity (SN)} = \frac{\text{TP}}{\text{TP} + \text{FN}} \tag{Eq. 12.3}$$

$$\text{Specificity (SP)} = \frac{\text{TN}}{\text{TN} + \text{FP}} \quad \text{(Eq. 12.4)}$$

$$\text{Accuracy (ACC)} = \frac{\text{TP} + \text{TN}}{\text{TP} + \text{TN} + \text{FP} + \text{FN}} \quad \text{(Eq. 12.5)}$$

$$\text{Balanced accuracy (BAC)} = \frac{\text{SN} + \text{SP}}{2} \quad \text{(Eq. 12.6)}$$

$$\text{MCC} = \frac{(\text{TP} \times \text{TN}) - (\text{FP} \times \text{FN})}{\sqrt{(\text{TP} + \text{FP})(\text{TP} + \text{FN})(\text{TN} + \text{FP})(\text{TN} + \text{FN})}} \quad \text{(Eq. 12.7)}$$

Here, TP, TN, FP, and FN denote true positives, true negatives, false positives, and false negatives, respectively. Disease-causing mutations are considered the positive class, while neutral mutations are treated as the negative class. Additionally, we constructed receiver operating characteristic (ROC) curves by plotting the true positive rate against the false positive rate.

12.7 Computational algorithms/tools for discriminating disease-causing and neutral mutations

Several *in silico* tools, incorporating sequence and structural analysis, have been developed for predicting the effects of missense variants. However, accurately predicting the impact of variants in membrane proteins remains a challenge. This might be due to the oversight of generic methods designed for globular proteins, which fail to consider the specific physiochemical requirements for membrane protein integration into the lipid bilayer. To address this limitation, specialized machine learning tools have been developed for discriminating the disease-causing and neutral mutations within membrane proteins.

12.7.1 *Pred-MutHTP*

Pred-MutHTP is a primary sequence-based classification method for discriminating against disease-causing and neutral mutations (Kulandaisamy *et al.*, 2020), trained using missense mutations sourced from the MutHTP database (Kulandaisamy *et al.*, 2018). A reliable dataset was constructed by considering the mutations that are present in a minimum of two databases among ClinVar (Landrum *et al.*, 2016), COSMIC (Tate *et al.*, 2019), Humsavar, SwissVar (Mottaz *et al.*, 2010), and 1000 Genomes. Further, a nonredundant dataset was then established through considerations of protein sequence similarity, mutation position conservation, and amino acid substitution type. This dataset encompasses 11,846 disease-causing and 9,533 neutral mutations spanning 1,014 and 2,958 proteins, respectively. Additionally, the dataset was grouped by the location of mutations with respect to the membrane topology (such as extracellular/outside, TM, and intracellular/inside) and used them for model development (**Table 12.2**).

The machine-learning model with 19 different sequence-based features has been effectively distinguishing disease-causing and neutral mutations within human membrane proteins. This model achieved a sensitivity of 76.3%, a specificity of 72.4%, an accuracy of 74.6%, and an AUC of 0.82 in a 10-fold group-wise cross-validation (**Table 12.2**). Furthermore, the model's performance is consistent and robust across

Table 12.2 Performance of classification models on different types of datasets

Data set	Number of features	Data type/ validation	SN	SP	ACC	BAC	MCC	AUC
Whole data	20	10-fold-group-wise	76.32	72.46	74.6	74.39	0.480	0.82
		Test	78.1	78.6	78.4	78.3	0.566	0.856
Inside	15	10-fold	72.3	73.54	72.93	72.92	0.454	0.796
		Test	75.4	76.1	75.6	75.75	0.476	0.813
Membrane	15	10-fold-group-wise	81.38	74.81	79.3	78.095	0.543	0.842
		Test	86.6	83.8	85.4	85.2	0.70	0.913
Outside	19	10-fold-group-wise	73.35	74.84	74.5	74.095	0.437	0.809
		Test	78.7	74.9	77.2	76.8	0.529	0.835

Note: SN: sensitivity; SP: specificity; ACC: Accuracy; BAC: Balanced accuracy; MCC: Matthew's correlation coefficient; and AUC: Area under the curve. Data are taken from Kulandaisamy *et al.* (2020).

different divisions of the dataset based on TMPs function, the number of associated mutations per protein, the count of TM segments, and distinct disease classes. The topology-specific models demonstrated an accuracy of 87% for membrane regions, 75% for intracellular regions, and 79% for extracellular regions in a blind test set. When compared to 11 existing methods in the literature, Pred-MutHTP showcased a notable accuracy improvement of 4% to 11% (Kulandaisamy *et al.*, 2020).

Pred-MutHTP is freely accessible at https://www.iitm.ac.in/bioinfo/PredMutHTP/. Users can predict the effects of mutations in membrane proteins by inputting the protein's UniProt ID along with mutation details (wild-type residue position and alternative/mutant amino acids). Pred-MutHTP outputs results in table format, presenting the query sequence, topology details, mutation classification (disease-associated or neutral), and confidence score for each prediction. Additionally, predictions for all potential amino acid variants in membrane protein sequences are precomputed and available at https://www.iitm.ac.in/bioinfo/PredMutHTP/pred_db_search.php, enabling users to search and download prediction outcomes (Kulandaisamy *et al.*, 2020).

12.7.2 *TMSNP*

The TMSNP predictor, available at http://lmc.uab.es/tmsnp/, is a sequence-based tool developed using a nonredundant dataset of 2,704 disease-causing and 19,292 neutral mutations sourced from the TMPSNP database (Garcia-Recio *et al.*, 2021). TMSNP employs an array of features, including residue conservation, physicochemical properties, substitution matrices, and UniProt Pfam accession codes. Among different machine-learning methods, the TMSNP predictor achieved an impressive accuracy of 88% and a MCC of 0.76 when employing the random forest (RF) algorithm.

12.7.3 *BorodaTM*

BorodaTM (Popov *et al.*, 2019) integrates the sequence, structure, and energetic features of TMPs for distinguishing between disease-causing and neutral missense mutations within TMPs. The method was trained with human membrane protein missense mutations obtained from the

Humsavar database (https://www.uniprot.org/docs/humsavar) and the final dataset encompasses 392 disease-causing and 154 neutral mutations spread across 64 proteins.

BorodaTM uses physicochemical properties, four substitution matrices, atomic contacts, solvent-accessible surface area, secondary structure, packing density, and distinct non-covalent bond energy contributions. This method is constructed by a gradient-boosting decision tree algorithm from the XGBoost software. Impressively, BorodaTM attains an accuracy of 94% and a MCC of 0.88 for the entire dataset, surpassing existing methods with an approximately 10% accuracy improvement in TMP missense mutation prediction. The BorodaTM's precomputed prediction results for roughly 1.8 million missense mutations across 379 human membrane protein structures are accessible at https://www.iitm.ac.in/bioinfo/MutHTP/boroda.php.

12.7.4 *mCSM-membrane*

mCSM-membrane stands as a structure-based pathogenicity predictor, based on graph-based signatures of the wild-type residue environment, perturbations in intramolecular interactions due to mutations, amino acid composition, substitution matrices, and other pertinent physicochemical properties (Pires *et al.*, 2020). This method uses the same dataset that the BorodaTM method constructed. and mCSM-membrane has achieved a remarkable accuracy of 95% alongside a MCC of 0.73 when assessed on the test set. The mCSM-membrane method is available as a web server at http://biosig.unimelb.edu.au/mcsm_membrane/. Users can provide input in the form of protein structures in ".pdb" format or a PDB accession number, along with the protein sequence and list of mutation(s). The generated output encompasses mutation predictions (disease-causing or neutral), molecular visualization, and membrane topology information.

12.7.5 *MutTMPredictor*

MutTMPredictor developed by Ge *et al.* (2021) utilizing approximately 60 sequence- and structure-based features to differentiate between disease-causing and neutral mutations (Ge *et al.*, 2021). This method is distinct for harnessing a dataset acquired from the BorodaTM, Pred-MutHTP, and TMSNP methods. In addition, it incorporates an extensive feature set that amalgamates those employed in the BorodaTM approach, PSSM profiles, as well as prediction results from SIFT, PROVEAN, PolyPhen-2, and FATHMM methods (Ng *et al.*, 2003; Adzhubei *et al.*, 2013; De Beer *et al.*, 2013; Shihab *et al.*, 2013; Choi & Chan, 2015). This comprehensive approach empowers MutTMPredictor to achieve an exceptional accuracy of 96% and a MCC of 0.89 for 546 mutations, outperforming other existing methods. This method is accessible at http://csbio.njust.edu.cn/bioinf/muttmpredictor/.

12.8 Potential applications of computational databases and tools

The above-discussed computational databases and tools discussed in this chapter, coupled with the insights derived from extensive analyses, hold significant potential for various applications. They can serve as foundational resources for developing algorithms aimed at predicting the impact of mutations (both disease-causing and neutral) in membrane proteins. By investigating a diverse range of disease-causing mutations across different medical conditions, these resources contribute to a deeper comprehension of mutational effects and the intricate structure–function relationships within membrane proteins. Additionally, the protein–protein interface information provided by databases like MutHTP enables the exploration of how interface residues are influenced by disease-causing mutations, shedding light on their impact on crucial protein interactions. The computational tools available within these resources play a crucial role in annotating the phenotypic consequences of genetic variants in the context of human health and diseases, facilitating the interpretation of how mutations contribute to specific conditions. In summary, these databases and tools offer insights into the role of mutations in diseases and have the potential to inform the design of mutation-specific treatment strategies, thereby spanning applications ranging from algorithm development to disease understanding, protein interaction analysis, variant annotation, and targeted therapeutic approaches.

12.9 Conclusions

TMPs distinguish themselves from water-soluble proteins through distinctive structural arrangements in their subcellular locations, unique physiochemical properties, and specialized functions. Notably, TMPs exhibit significant associations with various diseases. Here, we explored the disease-causing mutation databases MutHTP and TMSNP, about their scope, contents, features, and practical applications. Furthermore, we delve into the developmental processes, performance, validation, and usability of computational tools, including Pred-MutHTP and BorodaTM, specifically designed for TMPs. Moreover, a large-scale analysis of disease-causing and neutral mutations, providing insights into the occurrence and roles of these mutations in disease pathogenesis has been presented. Collectively, these informations offer a comprehensive understanding of the role of mutations in sequence, structure, functions, and diseases in TMPs. These key findings shed light on the design of mutation-specific strategies for combating specific diseases.

Acknowledgments

We express our gratitude to the members of the Protein Bioinformatics Lab for their invaluable input and extend our acknowledgment to the Indian Institute of Technology Madras for providing the computational resources. We also acknowledge the support of the Department of Science and Technology, Government of India (INT/RUS/RSF/P-09), and the Russian Science Foundation for a research grant (16-44-02002).

References

Adzhubei, I., Jordan, D. M., & Sunyaev, S. R. (2013). Predicting functional effect of human missense mutations using PolyPhen-2. *Current Protocols in Human, 76*(1), 7–20.

Almeida, J. G., Preto, A. J., Koukos, P. I., Bonvin, A. M., & Moreira, I. S. (2017). Membrane proteins structures: A review on computational modeling tools. *Biochimica Biophysica Acta Biomembranes, 1859*(10), 2021–2039.

Almén, M. S., Nordström, K. J., Fredriksson, R., & Schiöth, H. B. (2009). Mapping the human membrane proteome: A majority of the human membrane proteins can be classified according to function and evolutionary origin. *BMC Biology, 7*(1), 50.

Altschul, S. F., Madden, T. L., Schäffer, A. A., Zhang, J., Zhang, Z., Miller, W., & Lipman, D. J. (1997). Gapped BLAST and PSI-BLAST: A new generation of protein database search programs. *Nucleic Acids Research, 25*(17), 3389–3402.

Bernsel, A., Viklund, H., Hennerdal, A., & Elofsson, A. (2009). TOPCONS: Consensus prediction of membrane protein topology. *Nucleic Acids Research, 37*(S2), W465–W468.

Bittrich, S., Rose, Y., Segura, J., Lowe, R., Westbrook, J. D., Duarte, J. M., & Burley, S. K. (2022). RCSB Protein Data Bank: Improved annotation, search and visualization of membrane protein structures archived in the PDB. *Bioinformatics, 38*(5), 1452–1454.

Bowie, J. U. (2004). Membrane proteins: A new method enters the fold. *Proceedings of the National Academy Sciences USA, 101*(12), 3995–3996.

Bromberg, Y., Yachdav, G., & Rost, B. (2008). SNAP predicts effect of mutations on protein function. *Bioinformatics, 24*(20), 2397–2398.

Buermans, H. P. J., & Den Dunnen, J. T. (2014). Next generation sequencing technology: Advances and applications. *Biochimica Biophysica Acta, 1842*, 1932–1941.

Capriotti, E., Calabrese, R., & Casadio, R. (2006). Predicting the insurgence of human genetic diseases associated to single point protein mutations with support vector machines and evolutionary information. *Bioinformatics, 22*(22), 2729–2734.

Chen, Y., Salem, R. M., Rao, F., Fung, M. M., Bhatnagar, V., Pandey, B., Mahata, M., Waalen, J., Nievergelt, C. M., Lipkowitz, M. S., & Hamilton, B. A. (2010). Common charge-shift mutation Glu65Lys in K+ channel β1-Subunit KCNMB1: Pleiotropic consequences for glomerular filtration rate and progressive renal disease. *American Journal of Nephrology, 32*(5), 414–424.

Choi, Y., & Chan, A. P. (2015). PROVEAN web server: A tool to predict the functional effect of amino acid substitutions and indels. *Bioinformatics, 31*(16), 2745–2747.

Colegio, O. R., Van Itallie, C. M., McCrea, H. J., Rahner, C., & Anderson, J. M. (2002). Claudins create charge-selective channels in the paracellular pathway between epithelial cells. *American Journal of Physiology Cell Physiology, 283*, C142–C147.

De Beer, T. A., Laskowski, R. A., Parks, S. L., Sipos, B., Goldman, N., & Thornton, J. M. (2013). Amino acid changes in disease-associated variants differ radically from variants observed in the 1000 genomes project dataset. *PLoS Computational Biology, 9*, e1003382.

Dobson, L., Mészáros, B., & Tusnády, G. E. (2018). Structural principles governing disease-causing germline mutations. *Journal of Molecular Biology, 430*(24), 4955–4970.

Dobson, L., Reményi, I., & Tusnády, G. E. (2015). CCTOP: A consensus constrained TOPology prediction web server. *Nucleic Acids Research, 43*(W1), W408–W412.

Dobson, L., Reményi, I., & Tusnády, G. E. (2015). The human transmembrane proteome. *Biology Direct, 10*(1), 1–18.

Gao, M., Zhou, H., & Skolnick, J. (2015). Insights into disease-associated mutations in the human proteome through protein structural analysis. *Structure, 23*(7), 1362–1369.

Garcia-Recio, A., Gómez-Tamayo, J. C., Reina, I., Campillo, M., Cordomí, A., & Olivella, M. (2021). TMSNP: A web server to predict pathogenesis of missense mutations in the transmembrane region of membrane proteins. *NAR Genomics and Bioinformatics, 3*(1), lqab008.

Gasteiger, E., Hoogland, C., Gattiker, A., Duvaud, S. E., Wilkins, M. R., Appel, R. D., & Bairoch, A. (1999). Protein identification and analysis tools on the ExPASy server. *Methods in Molecular Biology, 112*:531–52.

Ge, F., Zhu, Y. H., Xu, J., Muhammad, A., Song, J., & Yu, D. J. (2021). MutTMPredictor: Robust and accurate cascade XGBoost classifier for prediction of mutations in transmembrane proteins. *Computational and Structural Biotechnology Journal, 19*, 6400–6416.

Gnad, F., Baucom, A., Mukhyala, K., Manning, G., & Zhang, Z. (2013). Assessment of computational methods for predicting the effects of missense mutations in human cancers. *BMC Genomics, 14*(3), 1–13.

Goodwin, S., McPherson, J. D., & McCombie, W. R. (2016). Coming of age: Ten years of next- generation sequencing technologies. *Nature Reviews Genetics, 17*(6), 333–351.

Gromiha, M. M. (2005). A statistical model for predicting protein folding rates from amino acid sequence with structural class information. *Journal of Chemical Information and Model, 45*(2), 494–501.

Gromiha, M. M., & Ou, Y. Y. (2014). Bioinformatics approaches for functional annotation of membrane proteins. *Brief Bioinformatics, 15*, 155–168.

Gromiha, M. M., Yabuki, Y., Suresh, M. X., Thangakani, A. M., Suwa, M., & Fukui, K. (2009). TMFunction: Database for functional residues in membrane proteins. *Nucleic Acids Research, 37*, D201–D204.

Gudmundsson, S., Singer-Berk, M., Watts, N. A., Phu, W., Goodrich, J. K., Solomonson, M., Genome Aggregation Database Consortium, Rehm, H. L., MacArthur, D. G., & O'Donnell-Luria, A. (2022). Variant interpretation using population databases: Lessons from gnomAD. *Human Mutation, 43*(8), 1012–1030.

Hall, M., Frank, E., Holmes, G., Pfahringer, B., Reutemann, P., & Witten, I. H. (2009). The WEKA data mining software: An update. *ACM SIGKDD Explorations Newsletter, 11*(1), 10–18.

Hamosh, A., Scott, A. F., Amberger, J. S., Bocchini, C. A., & McKusick, V. A. (2005). Online Mendelian Inheritance in Man (OMIM), a knowledgebase of human genes and genetic disorders. *Nucleic Acids Research, 33*, D514–D517.

He, L., Shobnam, N., Wimley, W. C., & Hristova, K. (2011). FGFR3 heterodimerization in achondroplasia, the most common form of human dwarfism. *Journal of Biological Chemistry, 286*(15), 13272–13281.

Huss, W. J., Hu, Q., Glenn, S. T., Gangavarapu, K. J., Wang, J., Luce, J. D., Quinn, P. K., Brese, E. A., Zhan, F., Conroy, J. M., & Paragh, G. (2018). Comparison of sureselect and nextera exome capture performance in single-cell sequencing. *Hum Heredity, 83*(3), 153–162.

Isberg, V., Mordalski, S., Munk, C., Rataj, K., Harpsøe, K., Hauser, A. S., Vroling, B., Bojarski, A. J., Vriend, G., & Gloriam, D. E. (2016). GPCRdb: An information system for G protein-coupled receptors. *Nucleic Acids Research, 44*(D1), D356–D364.

Kawashima, S., & Kanehisa, M. (2000). AAindex: Amino acid index database. *Nucleic Acids Research, 28*(1), 374–374.

Koboldt, D. C. (2020). Best practices for variant calling in clinical sequencing. *Genome Medicine, 12*(1), 1–13.

Kozma, D., Simon, I., & Tusnady, G. E. (2012). PDBTM: Protein Data Bank of transmembrane proteins after 8 years. *Nucleic Acids Research, 41*(D1), D524–D529.

Kulandaisamy, A., Binny Priya, S., Sakthivel, R., Tarnovskaya, S., Bizin, I., Hönigschmid, P., Frishman, D., & Gromiha, M. M. (2018). MutHTP: Mutations in human transmembrane proteins. *Bioinformatics, 34*(13), 2325–2326.

Kulandaisamy, A., Priya, S. B., Sakthivel, R., Frishman, D., & Gromiha, M. M. (2019). Statistical analysis of disease-causing and neutral mutations in human membrane proteins. *Proteins, 87*(6), 452–466.

Kulandaisamy, A., Ridha, F., Frishman, D., & Gromiha, M. M. (2022). Computational approaches for investigating disease-causing mutations in membrane proteins: Database development, analysis and prediction. *Current Topics in Medicinal Chemistry, 22*(21), 1766–1775.

Kulandaisamy, A., Sakthivel, R., & Gromiha, M. M. (2021). MPTherm: Database for membrane protein thermodynamics for understanding folding and stability. *Brief Bioinformatics, 22*(2), 2119–2125.

Kulandaisamy, A., Zaucha, J., Sakthivel, R., Frishman, D., & Gromiha, M. M. (2020). Pred-MutHTP: Prediction of disease-causing and neutral mutations in human transmembrane proteins. *Human Mutation, 41*(3), 581–590.

Landrum, M. J., Lee, J. M., Benson, M., Brown, G., Chao, C., Chitipiralla, S., Gu, B., Hart, J., Hoffman, D., Hoover, J., & Jang, W. (2016). ClinVar: Public archive of interpretations of clinically relevant variants. *Nucleic Acids Research, 44*(D1), D862–D868.

Loewe, L., & Hill, W. G. (2010). The population genetics of mutations: Good, bad and indifferent. *Philosophical Transactions of the Royal Society London B Biological Sciences, 365*(1544), 1153–1167.

Lomize, M. A., Pogozheva, I. D., Joo, H., Mosberg, H. I., & Lomize, A. L. (2012). OPM database and PPM web server: Resources for positioning of proteins in membranes. *Nucleic Acids Research, 40*(D1), D370–D376.

Manning, J. R., Jefferson, E. R., & Barton, G. J. (2008). The contrasting properties of conservation and correlated phylogeny in protein functional residue prediction. *BMC Bioinformatics, 9*(1), 51.

Marsico, A., Scheubert, K., Tuukkanen, A., Henschel, A., Winter, C., Winnenburg, R., & Schroeder, M. (2010). MeMotif: A database of linear motifs in α-helical transmembrane proteins. *Nucleic Acids Research, 38*, D181–D189.

Molnár, J., Szakács, G., & Tusnády, G. E. (2016). Characterization of disease-associated mutations in human transmembrane proteins. *PLoS One, 11*(3), e0151760.

Morita, M., Katta, A. M., Ahmad, S., Mori, T., Sugita, Y., & Mizuguchi, K. (2011). Lipid recognition propensities of amino acids in membrane proteins from atomic resolution data. *BMC Biophysics, 4*(1), 21.

Mottaz, A., David, F. P., Veuthey, A. L., & Yip, Y. L. (2010). Easy retrieval of single amino- acid polymorphisms and phenotype information using SwissVar. *Bioinformatics, 26*(6), 851–852.

Murtazina, R., Booth, B. J., Bullis, B. L., Singh, D. N., & Fliegel, L. (2001). Functional analysis of polar amino-acid residues in membrane associated regions of the NHE1 isoform of the mammalian Na+/H+ exchanger. *European Journal of Biochemistry, 268*, 4674–4685.

Nastou, K. C., Batskinis, M. A., Litou, Z. I., Hamodrakas, S. J., & Iconomidou, V. A. (2019). Analysis of single-nucleotide polymorphisms in human voltage-gated ion channels. *Journal of Proteome Research, 18*(5), 2310–2320.

Ng, D. P., Poulsen, B. E., & Deber, C. M. (2012). Membrane protein misassembly in disease. *Biochimica Biophysica Acta, 1818*, 1115–1122.

Ng, P. C., & Henikoff, S. (2003). SIFT: Predicting amino acid changes that affect protein function. *Nucleic Acids Research, 31*(13), 3812–3814.

Nicoll, D. A., Hryshko, L. V., Matsuoka, S., Frank, J. S., & Philipson, K. D. (1996). Mutation of amino acid residues in the putative transmembrane segments of the cardiac sarcolemmal Na+- Ca2+ exchanger. *Journal of Biological Chemistry, 271*, 13385–13391.

Nishi, H., Tyagi, M., Teng, S., Shoemaker, B. A., Hashimoto, K., Alexov, E., Wuchty, S., & Panchenko, A. R. (2013). Cancer missense mutations alter binding properties of proteins and their interaction networks. *PLoS One, 8*, e66273.

North, B., Cristian, L., Stowell, X. F., Lear, J. D., Saven, J. G., & DeGrado, W. F. (2006). Characterization of a membrane protein folding motif, the Ser zipper, using designed peptides. *Journal of Molecular Biology, 359*, 930–939.

Partridge, A. W., Therien, A. G., & Deber, C. M. (2004). Missense mutations in transmembrane domains of proteins: Phenotypic propensity of polar residues for human disease. *Proteins, 54*(4), 648–656.

Perocchi, F., Gohil, V. M., Girgis, H. S., Bao, X. R., McCombs, J. E., Palmer, A. E., & Mootha, V. K. (2010). MICU1 encodes a mitochondrial EF hand protein required for Ca2+ uptake. *Nature, 467*, 291–296.

Pires, D. E., Rodrigues, C. H., & Ascher, D. B. (2020). mCSM-membrane: Predicting the effects of mutations on transmembrane proteins. *Nucleic Acids Research, 48*(W1), W147–W153.

Ponnuswamy, P. K., & Gromiha, M. M. (1993). Prediction of transmembrane helices from hydrophobic characteristics of proteins. *International Journal of Peptide and Protein Research, 42*(4), 326–341.

Popov, P., Bizin, I., Gromiha, M. M., & Frishman, D. (2019). Prediction of disease-associated mutations in the transmembrane regions of proteins with known 3D structure. *PloS One, 14*(7), e0219452.

Reich, D. E., Gabriel, S. B., & Altshuler, D. (2003). Quality and completeness of SNP databases. *Nature Genetics, 33*(4), 457–458.

Saier, M. H., Reddy, V. S., Tsu, B. V., Ahmed, M. S., Li, C., & Moreno-Hagelsieb, G. (2016). The transporter classification database (TCDB): Recent advances. *Nucleic Acids Research, 44*, D372–D379.

Senes, A., Gerstein, M., & Engelman, D. M. (2000). Statistical analysis of amino acid patterns in transmembrane helices: The GxxxG motif occurs frequently and in association with β-branched residues at neighboring positions. *Journal of Molecular Biology, 296*, 921–936.

Shihab, H. A., Gough, J., Cooper, D. N., Stenson, P. D., Barker, G. L., Edwards, K. J., Day, I. N., & Gaunt, T. R. (2013). Predicting the functional, molecular, and phenotypic consequences of amino acid substitutions using hidden Markov models. *Human Mutation, 34*(1), 57–65.

Simm, S., Einloft, J., Mirus, O., & Schleiff, E. (2016). 50 years of amino acid hydrophobicity scales: Revisiting the capacity for peptide classification. *Biological Research, 49*(1), 31.

Sun, J., Kulandaisamy, A., Liu, J., Hu, K., Gromiha, M. M., & Zhang, Y. (2023). Machine learning in computational modelling of membrane protein sequences and structures: From methodologies to applications. *Computational Structural Biotechnology Journal, 21*, 1205–1226.

Suzek, B. E., Wang, Y., Huang, H., McGarvey, P. B., Wu, C. H., & UniProt Consortium. (2014). UniRef clusters: A comprehensive and scalable alternative for improving sequence similarity searches. *Bioinformatics, 31*(6), 926–932.

Tang, S. H. A. O. Q. I. N. G., Mikala, G., Bahinski, A., Yatani, A., Varadi, G., & Schwartz, A. (1993). Molecular localization of ion selectivity sites within the pore of a human L-type cardiac calcium channel. *Journal of Biological Chemistry, 268*, 13026–13029.

Tate, J. G., Bamford, S., Jubb, H. C., Sondka, Z., Beare, D. M., Bindal, N., Boutselakis, H., Cole, C. G., Creatore, C., Dawson, E., & Fish, P. (2019). COSMIC: The catalogue of somatic mutations in cancer. *Nucleic Acids Research, 47*(D1), D941–D947.

Tusnady, G. E., Kalmar, L., & Simon, I. (2007). TOPDB: Topology data bank of transmembrane proteins. *Nucleic Acids Research, 36*, D234–D239.

UniProt Consortium. (2019). UniProt: A worldwide hub of protein knowledge. *Nucleic Acids Research, 47*(D1), D506–D515.

Valdar, W. S. (2002). Scoring residue conservation. *Proteins: Structure, Function, and Bioinformatics, 48*(2), 227–241.

Xiao, F., & Shen, H. B. (2015). Prediction enhancement of residue real-value relative accessible surface area in transmembrane helical proteins by solving the output preference problem of machine learning-based predictors. *Journal of Chemical Information and Modeling, 55*(11), 2464–2474.

Yin, H., & Flynn, A. D. (2016). Drugging membrane protein interactions. *Annual Review of Biomedical Engineering, 18*, 51–76.

Zaucha, J., Heinzinger, M., Kulandaisamy, A., Kataka, E., Salvádor, Ó. L., Popov, P., Rost, B., Gromiha, M. M., Zhorov, B. S., & Frishman, D. (2021). Mutations in transmembrane proteins: Diseases, evolutionary insights, prediction and comparison with globular proteins. *Brief Bioinformatics, 22*(3), bbaa132.

https://doi.org/10.1142/9789811293269_0013

Chapter 13

Decoding the evolution of COVID-19 through mutational studies on SARS-CoV-2

Divya Sharma[1], Puneet Rawat[2], and M. Michael Gromiha[1,3]

[1]*Protein Bioinformatics Lab, Department of Biotechnology, Bhupat and Jyoti Mehta School of Biosciences, Indian Institute of Technology Madras, Chennai, Tamil Nadu 600036, India*

[2]*University of Oslo and Oslo University Hospital, Oslo, Norway*

[3]*International Research Frontiers Initiative, School of Computing, Tokyo Institute of Technology, Yokohama, 226-8501, Japan*

Abstract

The COVID-19 pandemic had a major global impact and caused significant loss of lives. SARS-CoV-2, the causative virus for COVID-19, harbors spike protein that interacts with the human ACE2 receptor to infect the host cell. Mutations in the spike protein led to the emergence of several SARS-CoV-2 variants, such as Alpha, Beta, Gamma, Delta, and Omicron, which led to increased transmission rates, virulence, and evasion from the immune response. In this chapter, we reviewed the progress in different directions: (i) phylogeny of SARS-CoV-2, (ii) mutational effects at the interface of spike protein-ACE2 and spike protein-antibodies, (iii) computational tools and techniques to enhance our understanding of SARS-CoV-2 mutations, (iv) genomic studies that offer insights into the evolving patterns of mutations over time, and (iv) mutations that led to reduced neutralization efficacy of vaccines. This comprehensive review reveals the effect of mutations on virus–host interaction, which is pivotal for improved treatments and understanding the evolutionary path of the virus and immune evasion, and highlights the need to enhance our knowledge about viral mutations to be better prepared for future health crises.

13.1 Background

Coronaviruses have been around for a long time and impacted human lives, some being mild led to the common cold, such as HCoV-OC43, HCoV-HKU1, HCoV-NL63, and HCoV-229E, while others being more pathogenic and leading to epidemics and pandemics, such as SARS-CoV, MERS-CoV, and SARS-CoV-2 (Rawat *et al.*, 2021a; Yesudhas *et al.*, 2021). The first known coronavirus outbreak originated in China during 2002 to 2003 due to SARS-CoV and caused severe respiratory illness with a high fatality rate. It was eventually contained through public health measures and case identification. The MERS-CoV outbreak in 2012 originated in the Middle East and caused severe respiratory illness, but its human-to-human transmission was limited, preventing large-scale pandemics (Zhu *et al.*, 2020). The most recent coronavirus outbreak due to the SARS-CoV-2 virus was reported in Wuhan, China, in late 2019. The virus quickly spread globally, leading to the declaration of the COVID-19 pandemic by the World Health Organization (WHO) in March 2020 (Hu *et al.*, 2021). The pandemic had a profound impact on public health, economies, and societies worldwide and led to 770 million cases and 6.9 million deaths worldwide (https://covid19.who.int/). SARS-CoV, MERS-CoV, and SARS-CoV-2 all target the respiratory tract as their primary site of infection, where they bind to specific receptors on the surface of human cells. This interaction is the initial step in the process of viral entry and infection. The SARS-CoV-2 virus attaches to the angiotensin-converting enzyme 2 (ACE2) receptor, which allows the virus to enter the host cells and replicate, leading to the characteristic symptoms of COVID-19, ranging from mild respiratory symptoms to severe pneumonia and acute respiratory distress syndrome (ARDS) (Mehta *et al.*, 2020; Varghese *et al.*, 2020).

SARS-CoV-2 is a member of the genus Betacoronavirus, belonging to the family Coronaviridae. The betacoronavirus genus encompasses various coronaviruses such as OC43, HKU1, SARS-CoV, MERS-CoV, and SARS-CoV-2, which have been known to infect mammals, including humans. It shares approximately 79% of its genome sequence with SARS-CoV, the virus responsible for the severe acute respiratory syndrome outbreak in 2002 to 2003 (Hu *et al.*, 2021). Additionally, SARS-CoV-2 shows approximately 50% genome sequence identity with MERS-CoV, the virus that caused the middle east respiratory syndrome outbreak in 2012 (Hu *et al.*, 2021). Despite these genetic differences, SARS-CoV-2 shares a common genome organization with other betacoronaviruses. The genome organization of SARS-CoV-2 follows the characteristic pattern of betacoronaviruses, which includes genes encoding structural proteins like the spike (S), envelope (E), membrane (M), and nucleocapsid (N) proteins, along with other non-structural and accessory proteins (Singh & Yi, 2021). The SARS-CoV-2 genome is a positive-sense single-stranded ribonucleic acid (RNA) that is ~30,000 base pair long encoding for approximately 9700 amino acids. It has four structural proteins (S, E, M, and N) for virus assembly and 16 nonstructural proteins required for replication or transcription of the virus (Raj, 2021; Low *et al.*, 2022). The spike protein contains the receptor-binding domain (RBD), which binds to the ACE2 receptor in the human cell and mediates entry into the host. The genomic characterization of SARS-CoV-2 is shown in **Figure 13.1**. The entry into the host cell is guided by a protein called the spike (S) glycoprotein (Tortorici & Veesler, 2019). The S protein is found on the virus's surface, and it is made up of two parts, S1 and S2. Initially, the S protein is in a trimeric pre-fusion form, but later it is cleaved into S1 and S2 subunits by a host enzyme called furin (Bosch *et al.*, 2003). The S1 part contains the RBD, which attaches to ACE2 on the target cell membrane with the help of another host enzyme called TMPRSS2. Further, the S2 subunit of the S protein goes through a structural change that allows the virus to fuse with the host cell's membrane (Mariano *et al.*, 2020). This fusion enables the virus to enter the host cell and mediates the infection.

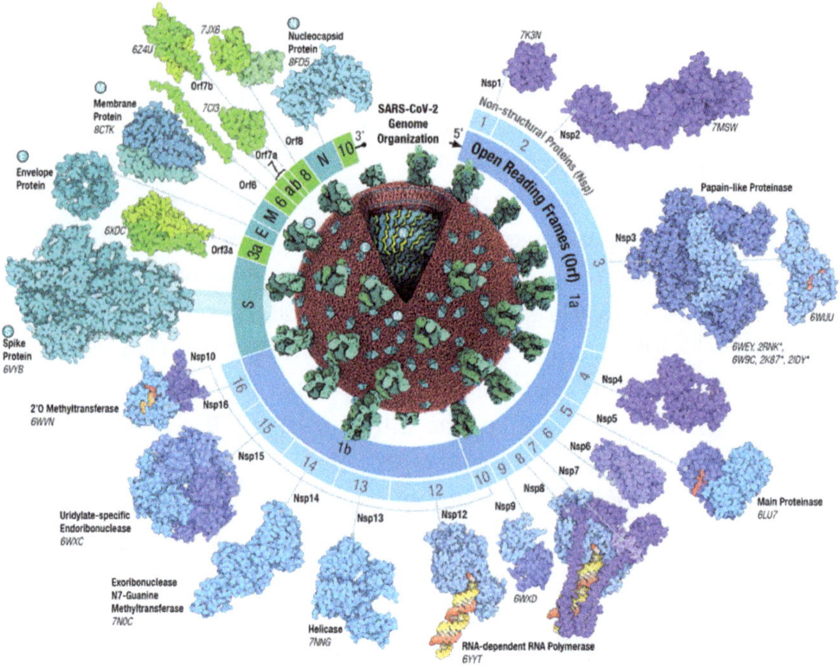

Figure 13.1 Genomic organization of SARS-Cov-2. The figure was taken from RCSB.org/covid19 PDB structures (https://www.rcsb.org/news/feature/5e74d55d2d410731e9944f52)

To control the spread of the virus, governments implemented various measures worldwide, including lockdowns, travel restrictions, and social distancing (Prabakaran *et al.*, 2021). While these measures were effective in slowing the transmission, they also had significant economic and social consequences, leading to job losses, mental health issues, and disruptions in education (Onyeaka *et al.*, 2021). The urgency of the COVID-19 pandemic prompted a rapid global response to develop therapeutics and vaccines. Researchers employed antibody and drug repurposing, which identified antibodies and drugs originally intended for different diseases but found to be potentially effective for COVID-19, such as Remdesivir, hydroxychloroquine, Ivermectin, and so on (Guy *et al.*, 2020; Muralidharan *et al.*, 2021; Rawat *et al.*, 2021b; Shanmugam *et al.*, 2022). Some of these drugs were approved or given emergency use authorization (EUA) by the

Table 13.1 FDA-approved therapeutics for COVID-19*

Approved drugs	EUA-drugs	EUA-monoclonal antibodies
Actemra (tocilizumab)	Lagevrio (molnupiravir)	REGEN-COV (casirivimab and imdevimab)
Veklury (remdesivir)	Kineret (anakinra)	Sotrovimab
Olumiant (baricitinib)	Olumiant (baricitinib)	Bamlanivimab and etesevimab
Paxlovid (nirmatrelvir and ritonavir)	Actemra (tocilizumab)	Bebtelovimab
	Gohibic (vilobelimab)	Evusheld (tixagevimab co-packaged with cilgavimab)

*Data are obtained from the FDA website, https://www.fda.gov/drugs/emergency-preparedness-drugs/coronavirus-covid-19-drugs (accessed on September, 14, 2023).

Food and Drug Administration (FDA) (Toussi *et al.*, 2023). **Table 13.1** shows the list of approved and EUA therapeutics for COVID-19 as of September 2023. Innovative approaches such as mRNA technology and codon deoptimization were used for vaccine development, enabling the creation of effective vaccines in record time (Dolgin, 2021; Wang *et al.*, 2021; Sharma *et al.*, 2023a). Several vaccines, including those developed by Pfizer-BioNTech, Moderna, AstraZeneca, Johnson & Johnson, Covaxin, and others, received EUAs and were deployed worldwide (Quinn *et al.*, 2021; Hadj Hassine, 2022). Vaccination campaigns played a crucial role in reducing severe cases, hospitalizations, and deaths (Rahmani *et al.*, 2022).

During the pandemic, although vaccination efforts were in full swing, several SARS-CoV-2 variants emerged through mutations in the viral genome. Some variants, such as the Alpha (B.1.1.7), Beta (B.1.351), Gamma (P.1), and Delta (B.1.617.2) variants, were more transmissible and, in some cases, partially evaded immunity from previous infections or vaccines (known as immune escape) (Jacobs *et al.*, 2023). Vaccine manufacturers responded by modifying their vaccines to target specific variants, and booster doses were recommended to enhance protection against these variants (Mbaeyi *et al.*, 2021). The ability of the SARS-CoV-2 virus to mutate introduced new variants, and their immune escape posed a persistent challenge to managing the pandemic. Understanding how mutations impact viral binding is crucial for developing better treatments and for gaining insights into how the virus evolves and escapes the immune system. This underscores the need to advance our knowledge about viral mutations to be better prepared for future health challenges.

13.2 Phylogeny of SARS-CoV-2

Mutations in the SARS-CoV-2 spike protein have led to several lineages and sub-lineages from time to time which affected the infectivity, pathogenicity, and antigenicity of the virus (Li *et al.*, 2020). The WHO classified these lineages into variants of concern (VOCs) and variants of interest (VOIs) based on their disease transmission capability, severity of the disease, and evasion of the immune response elicited by vaccines or convalescent plasma (www.who.int). The phylogenetic tree of SARS-CoV-2 is shown in **Figure 13.2**. Fortunately, all VOCs have been de-escalated based on the decreased severity and infectivity of the disease, and only a few VOIs remain, which are Omicron sublineages (as of September, 14, 2023). The previous VOCs are discussed in detail below (**Table 13.2**).

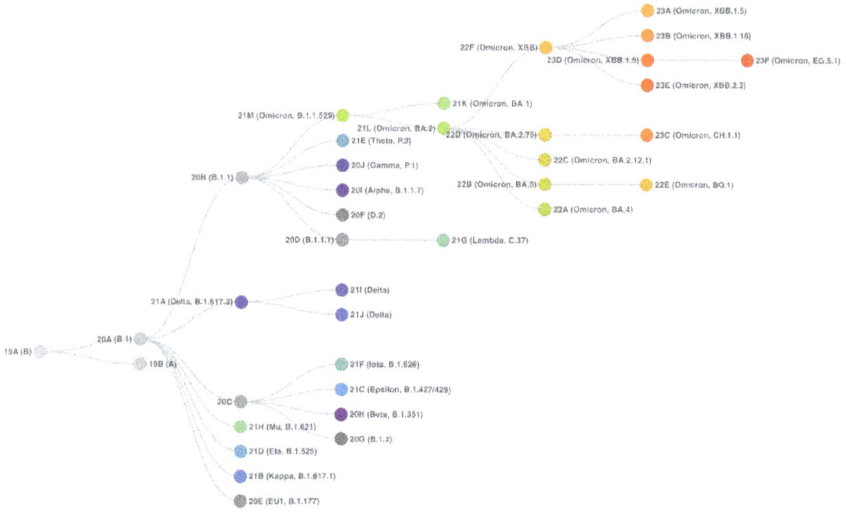

Figure 13.2 Phylogenetic relationships of SARS-CoV-2 clades. This figure is adopted from CoVariants (https://covariants.org/) which is a free and open-source project (https://covariants.org/)

Table 13.2 Details on reported variants of concern (VOCs)

WHO label	PANGO lineage*	Spike mutations of interest	First detected	Date de-escalated	Impact on transmissibility	Impact on immunity	Impact on severity
Alpha	B.1.1.7	N501Y, D614G, P681H	September 2020, United Kingdom	March 9, 2022	Increased	Similar	Increased
Beta	B.1.351	K417N, E484K, N501Y, D614G, A701V	September 2020, South Africa	March 9, 2022	Increased	Increased	Increased
Gamma	P.1	K417T, E484K, N501Y, D614G, H655Y	December 2020, Brazil	March 9, 2022	Increased	Increased	Increased
Delta	B.1.617.2	L452R, T478K, D614G, P681R	December 2020, India	June, 7, 2022	Increased	Increased	Increased
Omicron	B.1.1.529	(a)	November 2021, South Africa	March, 14, 2023	Increased	Increased	Reduced

(a): A67V, Δ69-70, T95I, G142D, Δ143-145, N211I, Δ212, ins215EPE, G339D, S371L, S373P, S375F, K417N, N440K, G446S, S477N, T478K, E484A, Q493R, G496S, Q498R, N501Y, Y505H, T547K, D614G, H655Y, N679K, P681H, N764K, D796Y, N856K, Q954H, N969K, L981F.

*PANGO lineage is a classification system used to categorize and track the genetic lineages of SARS-CoV-2 variants.

The data are obtained from https://www.ecdc.europa.eu/en/covid-19/variants-concern and https://www.who.int/

Alpha: Alpha variant (B.1.1.7 lineage) was first detected in the United Kingdom during September. It is characterized by nine spike protein mutations compared to the original Wuhan-Hu1 virus from China. The mutations N501Y, D614G, and P681H in the alpha variant led to increased viral transmissibility, higher hospitalization rates, and case fatality (Flores-Vega *et al.*, 2022).

Beta: This variant (B.1.351) was first identified in September 2020 in South Africa with K417N, E484K, N501Y, D614G, and A701V as the main mutations. The beta variant had nine mutations as compared to the original SARS-CoV-2 strain. These mutations led to increased transmission and severity of the disease with increased escape from host immunity (Mistry *et al.*, 2021).

Gamma: The P.1 variant of SARS-CoV-2 was identified in Japan and Brazil in November 2020, emerged from lineage B.1.1.28, and carried 12 mutations in its spike protein. Notable mutations in this variant were K417T, E484K, N501Y, D614G, and H655Y. These mutations had significant implications for the virus's transmissibility and the risk of reinfection. Moreover, they had been shown to reduce the efficacy of monoclonal antibody (mAb) therapies, leading to challenges in treatment. Convalescent patient plasma and sera from vaccinated individuals also exhibited reduced neutralizing activity against this variant (Scovino *et al.*, 2022).

Delta: The sub-lineage B.1.617.2, known as the Delta variant, was first identified in India in October 2020. This highly transmissible variant had eleven mutations in the spike protein and belonged to the B.1.617 sub-lineage. The Delta variant lacks the mutations at positions 501 and 484 in the RBD of the spike protein, which were associated with the evasion of neutralizing antibodies. Despite this, the Delta variant had spread globally, infecting both fully vaccinated and unvaccinated individuals, and became dominant by June 2021. Mutations in the RBD domain have been shown to reduce the binding of some mAbs, impacting the effectiveness of certain treatments (Cox *et al.*, 2023). Convalescent individual sera exhibited fourfold less potency against the Delta variant compared to the Alpha variant (Planas *et al.*, 2021). *In vitro* experiments with the sera of vaccinated individuals showed reduced binding and neutralizing antibody titers against the Delta variant, particularly with additional mutations like K417N in the B.1.617.2 sub-lineage. Mutations L452R and T478K localized in the antigenic site of RBD, and the P681R mutation enhanced spike protein cleavage, contributed to the Delta variant's reduced vaccine efficacy by hampering neutralizing antibody binding and enhancing viral infection through the ACE2 receptor (Planas *et al.*, 2021; Singanayagam *et al.*, 2022).

Omicron: The B.1.1.529 lineage, known as the Omicron variant, was first identified in Botswana, Africa, on November, 11, 2021. This variant stands out due to its more than 30 mutations in the spike protein (B.1.1.529/BA.1). Omicron has given rise to at least three genetically related sub-lineages (BA.1, BA.2, and BA.3) that diverged from the B.1.1.529 lineage. Many of these mutations are also found in the Delta and Alpha variants, which have been associated with increased infectivity and immune evasion against infection-blocking antibodies and vaccine-induced antibodies. Despite the significant number of spike protein mutations, Omicron has been linked to a high rate of transmissibility, immune resistance, and an increased risk of reinfection, along with lower lung infectivity and reduced pathogenicity compared to the Delta variant (Fan *et al.*, 2022). Omicron shares some mutations with other VOCs, such as N501Y, K417N, and E484K, associated with enhanced transmissibility and immune evasion (Pondé, 2022). Studies have shown that Omicron replicates at a lower rate in lung cells compared to the B.1.617.2 lineage, but it is considered more infectious than the Delta variant, evading the humoral immune response in fully vaccinated individuals, even with booster doses. However, a third dose of the Pfizer or Moderna vaccine and combining the Johnson and Johnson vaccine with a Pfizer booster have shown promise in increasing neutralizing antibody efficacy against Omicron (Garcia-Beltran *et al.*, 2022; McCallum *et al.*, 2022; Viana *et al.*, 2022).

13.3 Impact of COVID variants on pathogenicity and infectivity

As the SARS-CoV-2 virus spreads and replicates, it accumulates genetic mutations, resulting in the emergence of new variants with different characteristics. Several COVID-19 variants surfaced over the course of the ongoing COVID-19 pandemic, exhibiting varying degrees of transmissibility, virulence, and immune evasion and resulting in challenges for public health responses and vaccination efforts (Zhou *et al.*, 2022). Notably, variants like the Alpha (B.1.1.7) and Delta (B.1.617.2) have displayed increased transmissibility, resulting in higher infection rates and more rapid outbreaks (Chen *et al.*, 2020; Teng *et al.*, 2021; Volz *et al.*, 2021). Furthermore, the Beta (B.1.351) and Gamma (P.1) variants have exhibited enhanced virulence, leading to more severe illness and elevated rates of hospitalization and mortality in certain populations (Ong *et al.*, 2022; Carabelli *et al.*, 2023). Perhaps one of the most concerning aspects is their ability to partially evade immunity induced by prior infections or vaccinations, leading to breakthrough infections in vaccinated individuals and potentially reducing vaccine efficacy against specific variants (Thorne *et al.*, 2022; Andeweg *et al.*, 2023; He *et al.*, 2023). Consequently, efforts to control the pandemic have necessitated variant-specific vaccines and booster doses to enhance protection against evolving strains, requiring continuous monitoring of variant prevalence. These variants have also posed diagnostic challenges, as some have exhibited alterations in the genes targeted by diagnostic tests, potentially leading to false-negative results (Sharma *et al.*, 2023b). This prompted extensive research, surveillance efforts, and collaborative initiatives among public health agencies and researchers. Their collective actions and high vaccination coverage played a pivotal role in the implementation of effective measures to mitigate the impact of emerging variants on global health.

13.4 Experimental studies for mutational analysis of SARS-CoV-2

Understanding the effects of mutations in the SARS-CoV-2 genome is of great interest to the scientific community as it provides insights into virus transmissibility, infectivity, immune response evasion, and potential impact on diagnostics, treatments, and vaccines. To study these mutations, several experimental approaches were employed, such as *in vitro* cell culture, pseudovirus assays, animal models, antibody binding assays, immune escape studies, structural studies, and so on. The mutational studies could be divided based on the SARS-CoV-2 spike protein interface with either ACE2 or antibodies.

13.4.1 *Mutational studies at the binding interface of spike protein and ACE2 receptor*

The interaction between SARS-CoV-2 spike protein and ACE2 (**Figure 13.3**) is a key step in viral infection, where RBD binds with ACE2, allowing entry into the host cell. Researchers have studied the spike protein–ACE2 interface to understand the transmissibility, pathogenicity, and severity of COVID-19. In vitro cell culture studies have been employed to cultivate SARS-CoV-2 in controlled cell cultures and introduce specific mutations

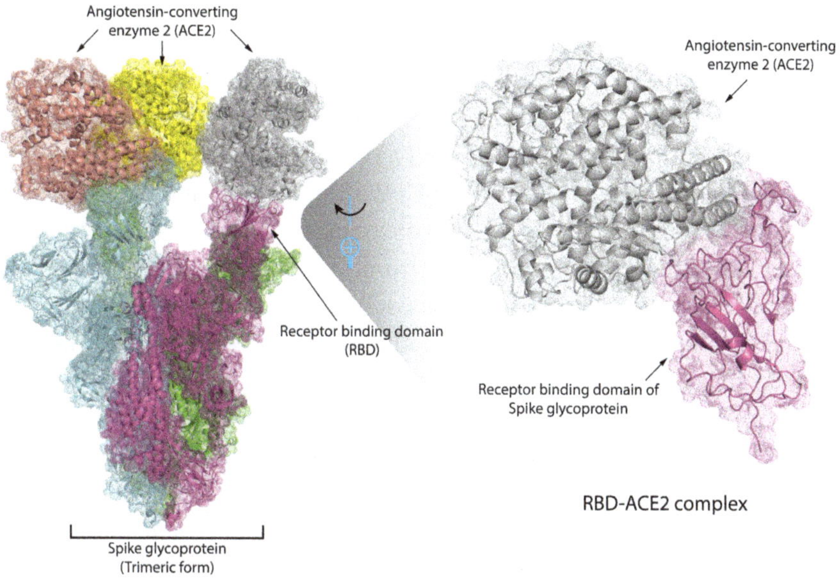

Figure 13.3 Binding of the SARS-CoV-2 spike protein to the human ACE2 receptor. SARS-CoV-2 spike protein is a trimeric protein (left figure) and contains the receptor binding domain (RBD), responsible for interaction with the ACE2 receptor on the host cell. The figure on the right shows the close-up interaction between ACE2 and the RBD region of spike protein. The RBD–ACE2 interaction area is around 1861 $Å^2$ and involves 17 residues on the RBD surface interacting with ACE2 at a 4 Å distance cutoff

to observe changes in viral behavior. The specific amino acids within the spike protein were deliberately mutated to observe how these modifications influence the strength and nature of the interaction between the spike protein and the ACE2 receptor, this is known as site-directed mutagenesis (Li *et al.*, 2020). Pseudovirus or live virus entry assays have been utilized to assess how mutations in the spike protein impact the capability of the virus to enter host cells effectively by interacting with the ACE2 receptor (Wang *et al.*, 2023). Moreover, binding assays have been used to measure the binding affinity between different spike protein variants and the ACE2 receptor. To gauge the binding affinity, researchers have employed experimental techniques such as surface plasmon resonance (SPR) (Raghu *et al.*, 2022) and isothermal titration calorimetry (ITC) (Saponaro, 2018). Further, techniques like X-ray crystallography and cryo-electron microscopy provide detailed visualizations of the three-dimensional structure of the spike protein–ACE2 receptor complex, elucidating how mutations alter the conformation of the binding site (Bate *et al.*, 2022). Gobeil *et al.* delved into the impact of naturally occurring mutations in SARS-CoV-2 on the structure, conformation, and antigenicity of the spike protein (Gobeil *et al.*, 2021). They conducted a comprehensive analysis, evaluating the interaction with both the ACE2 receptor and antibodies for 19 SARS-CoV-2 spike RBD constructs encompassing amino acid modifications identified in variants ranging from alpha to epsilon. Notably, all examined variants exhibited enhanced binding to the ACE2 receptor. Likewise, Kim *et al.* employed pseudoviruses carrying distinct SARS-CoV-2 spike protein substitutions and utilized cell lines expressing ACE2 from nine diverse animal species (Kim *et al.*, 2022). Their approach aimed to elucidate the repercussions of mutations seen in VOCs, including Omicron, on viral entry and the ability to be neutralized by antibodies. Their findings revealed a consistent pattern across all variants, indicating enhanced binding to SARS-CoV-2.

13.4.2 *Mutational studies at the spike protein–antibody interface*

Understanding the immune evasion of SARS-CoV-2 due to mutations in its genome is important to curb the virus. Numerous investigations have collectively enriched our current comprehension of how mutations within the SARS-CoV-2 spike protein impact its neutralization (Weisblum *et al.*, 2020; Andreano *et al.*, 2021). These inquiries encompass various approaches, including classical investigations into escape mutations that pinpoint mutations arising in virus populations after exposure to mAbs or convalescent plasma rich in polyclonal antibodies (McCarthy *et al.*, 2021; Thomson *et al.*, 2021). Additionally, focused analyses have delved into specific mutations, while broader studies have spanned extensive evaluations of numerous circulating variants or the comprehensive spectrum of potential amino acid substitutions within the RBD (Li *et al.*, 2020; Greaney *et al.*, 2021b). Antibody-binding assays such as enzyme-linked immunosorbent assays (ELISAs) and binding experiments assess how mutations affect the interaction between the spike protein and antibodies, informing the potential for immune evasion (Datta *et al.*, 2021). These assays with deep mutational scanning and yeast surface display revealed the mutations that escape antibody binding or neutralization (Starr *et al.*, 2021).

For assessing the impact of genetic variations on the structure and function of a protein, deep mutational scanning is an essential experimental technique (Fowler & Fields, 2014). This approach involves systematically introducing a diverse array of mutations across a protein sequence and assessing how each alteration influences its behavior. In the context of SARS-CoV-2, deep mutational scanning provides a comprehensive understanding of the effects of mutations within the viral spike protein, shedding light on critical aspects such as viral entry, immune recognition, and potential escape mechanisms. Starr *et al.* used deep mutational scanning to empirically gauge the impact of every conceivable amino acid mutation within the SARS-CoV-2 RBD on both ACE2-binding strength and protein expression, serving as a marker for protein folding stability (Starr *et al.*, 2020). The outcome of their work, represented as sequence-phenotype maps, offered insights into the evolutionary dynamics of SARS-CoV-2. Notably, these maps quantified the constraints exerted on antibody epitopes and suggested that most of the changes observed in the RBD of human SARS-CoV-2 isolates were driven by a natural selection process that favored genetic variations that did not harm the survival of the virus or fitness. Similarly, Greaney *et al.* assessed the impact of RBD mutations on antibody escape and generated interactive antibody escape maps to better understand the regions in SARS-CoV-2 RBD that lead to immune escape (Greaney *et al.*, 2021a). They found that the mutations exerting the most profound impact on antibody binding tend to localize within a restricted subset of regions in the receptor-binding motif (RBM) of the RBD. Among them, E484 was the most important, as at this site, neutralization by convalescent plasma is reduced more than 10-fold by several mutations.

Barnes *et al.* have categorized neutralizing antibodies associated with SARS-CoV-2 into four distinct classes based on their epitope regions (Barnes *et al.*, 2020). Class 1 antibodies target the RBM located within the RBD. These antibodies bind to the "up" conformation of the RBD. Class 2 antibodies share similarities with Class 1 but exhibit greater versatility by recognizing both the "up" and "down" conformations of the spike protein. In contrast, class 3 antibodies focus on regions located outside the ACE2-binding area. They can bind to RBD regardless of whether these domains are in the "up" or "down" conformation. Class 4 antibodies target a highly conserved segment within the RBD but not directly blocking the ACE2-RBD binding. These classes of antibodies and their binding regions are shown in **Figure 13.4**. In summary, these studies helped in understanding the change in binding affinity and stability upon

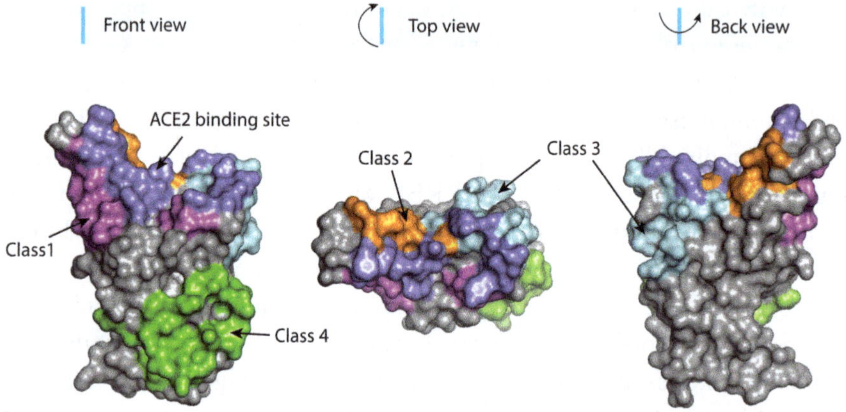

Figure 13.4 Four different classes of SARS-CoV-2 neutralizing antibody epitopes. Class 1: purple, Class 2: orange, Class 3: cyan, Class 4: green

mutation in the SARS-CoV-2 RBD–ACE2 receptor, as well as neutralization by antibodies.

13.5 Computational studies for mutational analysis of SARS-CoV-2

Numerous computational investigations have significantly enhanced our comprehension of the consequences brought about by mutations in SARS-CoV-2. These studies harness the experimentally determined information to develop databases, algorithms, machine learning models, and simulation models and predict the effects of genetic variations on the behavior of the virus, interactions, and properties, specifically virus transmissibility, infectivity, and immune escape.

13.5.1 *Computational resources developed for mutational scanning data*

There are several databases that store the information for SARS-CoV-2 sequences gathered from infected individuals. The Global Initiative on Sharing All Influenza Data (GISAID) is one such database. It fosters rapid data sharing on influenza and COVID-19 viruses, including genetic sequences and clinical details, aiding researchers in understanding virus evolution, and transmission dynamics during epidemics and pandemics. This database contains more than 15 million sequences for SARS-CoV-2 variants, including 1 million for the Alpha variant, 38,000 for the Beta variant, 78,000 for the Gamma variant, 334,000 for the Delta variant, and 465,000 for the Omicron variant, according to the data accessed on September, 14, 2023. The National Center for Biotechnology Information (NCBI), WHO, and the European Centre for Disease Prevention and Control (ECDC) also collate this information.

Researchers have developed databases using data from these repositories to analyze the SARS-CoV-2 sequences. RCoV19 from the China National Center for Bioinformation (CNCB) was one of the first databases in these directions. It integrates genomic and proteomic sequences with metadata from various sources, including GISAID, NCBI, and the China

National GeneBank (CNGB). It also encompasses scientific literature, news, and visualization tools for analyzing genome variations across the collected 2019-nCoV strains. The sequencing technologies like next-generation sequencing have led to the generation of large-scale genomic data for SARS-CoV-2, which is difficult to interpret. Consequently, scientists have developed secondary databases and resources to aggregate this wealth of sequencing data, serving as interpretable resources for further analysis and exploration. These resources could be classified into sequence- and structure-based information as shown in **Table 13.3**.

13.5.2 *Computational analysis related to SARS-CoV-2 mutations*

Computational mutagenesis at the SARS-CoV-2 spike protein interface has led researchers to understand the transmissibility and immune escape severity of COVID-19. These studies mainly delve into assessing the change in binding affinity and stability upon mutations at the interface of spike protein and ACE2 or spike protein and antibodies using bioinformatics tools (Sharma *et al.*, 2022). **Figure 13.5** shows an example where the loss of interactions in the spike protein–antibody complex (B38, PDB: 7BZ5) upon mutation decreased its binding affinity. Binding affinity refers to the strength of the interaction between molecules, such as the SARS-CoV-2 spike protein and its receptor ACE2, indicating how tightly they are bound together. Laurini *et al.* employed a combination of Molecular Mechanics/Poisson–Boltzmann Surface Area (MM/PBSA), computational mutagenesis, and interaction entropy techniques to simulate various mutation types at critical residues within ACE2 and the RBD interface in the spike protein of SARS-CoV-2 (Laurini *et al.*, 2021). Their analysis spanned hydrophobic, polar, charged, and bulky mutations, with a focus on residues driving the protein–protein interaction. Through this approach, they estimated the change in binding free energy. Notably, their findings highlighted specific variants, Q24T, T27D/K/W, D30E, H34S/T/K, E35D, Q42K, L79I/W, R357K, and R393K in ACE2, and L455D/W, F456 K/W, Q493K, N501T, and Y505W in the S-protein RBD, anticipated to enhance the binding affinity of SARS-CoV-2 spike protein and ACE2.

Table 13.3 Resources available for mutations in SARS-CoV-2

Resource name	Details	Data sources	Link	References
		Sequence-based		
Databases				
RCoV19	Genomic sequences and variations; haplotype network maps; COVID-19 literature	GISAID, NCBI, CNGB	https://ngdc.cncb.ac.cn/ncov/	Song *et al.* (2020)
SARS-CoV-2 MAT	Database of SARS-CoV-2 phylogenetic trees inferred with unrestricted public sequences	NCBI, COG-UK, CNCB	http://hgdownload.soe.ucsc.edu/goldenPath/wuhCor1/UShER_SARS-CoV-2/	McBroome *et al.* (2021)
CoV-Glue	Database of mutations, insertions, and deletions that have been observed in hCoV-19 sequences	GISAID	https://cov-glue.cvr.gla.ac.uk/	Siqueira *et al.* (2021)
Stanford University Coronavirus antiviral and resistance database (CoV-RDB)	SARS-CoV-2 variants and fold-change neutralization from antibodies	Literature, Protein Data Bank (PDB)	https://covdb.stanford.edu/susceptibility-data/table-mab-susc/	Tzou *et al.* (2022)

(Continued)

Table 13.3 (*Continued*)

Resource name	Details	Data sources	Link	References
SARS-CoV-2 Database	Consists of two databases, the SARS-CoV-2 contextual and sequence database and the SARS-CoV-2 BLAST database	INSDC (International Nucleotide Sequence Database Collaboration), Johns Hopkins University Coronavirus Resource Centre	https://covid19.sfb.uit.no/about/	https://www.covid19dataportal.org/the-european-covid-19-data-platform
Web servers				
Vcorn	Insights on COVID-19 infections and S protein mutations through correlation network analysis	WHO, NCBI	http://www.plant.osakafu-u.ac.jp/~kagiana/vcorn/sarscov2/22/	Ogata and Kitayama (2022)
SARS2Mutant	Discovers mutations and conserved regions from the SARS-CoV-2 protein sequences	GISAID	http://sars2mutant.com/	Rahimian *et al.* (2023)
Nextstrain	Analyze SARS-CoV-2 genomic sequences to aid in epidemiological understanding of pathogen spread and evolution and improve outbreak response	GISAID, open source data	https://nextstrain.org/sars-cov-2/	Hadfield *et al.* (2018)

CoV2K	An abstract model for explaining SARS-CoV-2-related concepts and interactions, focusing on viral mutations, their co-occurrence within variants, and their effects	Public Health England, COG-UK Mutation Explorer, CoVariants, ECDC, NCBI Virus, UniProtKB, IEDB	http://gmql.eu/cov2k/api/	Alfonsi *et al.* (2022)
MicroGMT	A command-line-based Python package to detect mutations from the sequence data (either for SARS-CoV-2 or other microbial genomes)	Open data source	https://github.com/qunfengdong/MicroGMT	Xing *et al.* (2020)
CovMT	Allows geographic exploration of clades and mutation fingerprints from all available isolate genomes	GISAID	https://www.cbrc.kaust.edu.sa/covmt/index.php?p=home	Alam *et al.* (2021)
COVID-19 CG	Open resource for tracking SARS-CoV-2 single-nucleotide variations (SNVs), lineages, and clades using the virus genomes	GISAID	https://covidcg.org/	Chen *et al.* (2021)

(*Continued*)

Table 13.3 (*Continued*)

Resource name	Details	Data sources	Link	References
		Structure-based		
Databases				
CoV3D	Database for coronavirus protein structures and their complexes with antibodies, receptors, and small molecules	Protein Data Bank (PDB)	https://cov3d.ibbr.umd.edu	Gowthaman *et al.* (2021)
ACovPepDB	Database of anti-coronavirus peptides	PubMed	http://i.uestc.edu.cn/ACovPepDB/	Zhang *et al.* (2022)
CoV-AbDab	Documents all published/patented antibodies and nanobodies able to bind to coronaviruses, including SARS-CoV2, SARS-CoV1, and MERS-CoV	PubMed, PDB	https://opig.stats.ox.ac.uk/webapps/covabdab/	Raybould *et al.* (2021)
Ab-CoV	Contains experimental binding affinity and neutralization profiles of coronavirus-related neutralizing antibodies	PubMed, CoV-AbDab	https://web.iitm.ac.in/bioinfo2/ab-cov/home	Rawat *et al.* (2022b)
SCoV2-MD	Resource that organizes atomistic simulations of the SARS-CoV-2 proteome	GISAID	https://submission.gpcrmd.org/covid19/	Torrens-Fontanals *et al.* (2022)

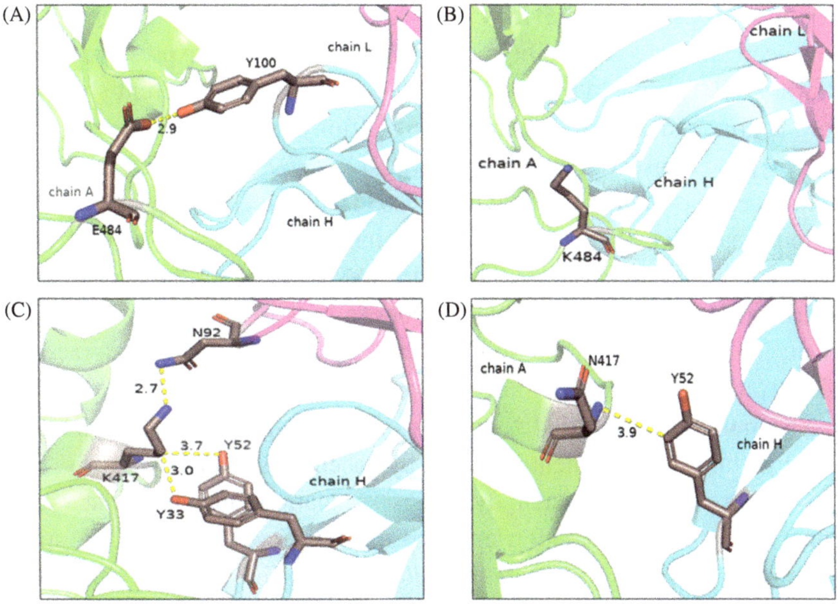

Figure 13.5 The interactions at the interface of spike–antibody complex and mutants. (A) E484, (B) E484K, (C) K417, and (D) K417N in the spike–antibody complex (PDB: 7BZ5). In the spike–protein complex (PDB: 7BZ5), E484 interacts with Y100 of the antibody heavy chain (Figure 13.5A), but the E484K mutation disrupts this interaction, reducing binding affinity by 1.83 kcal/mol (Figure 13.5B). Meanwhile, K417 forms interactions with Y33, Y52, and N92 of the antibody heavy and light chains (Figure 13.5C). However, the K417N mutation abolishes two interactions, decreasing binding affinity by 1.23 kcal/mol (Figure 13.5D). The figure was taken from Sharma *et al.* (2022)

Likewise, Gan *et al.* demonstrated that by structurally modeling the interaction between the SARS-CoV-2 spike protein and the human ACE2 receptor, numerous novel variant combinations were predicted to possess enhanced binding affinities (Gan *et al.*, 2021). The evolution of the SARS-CoV-2 S-protein seems to be influenced by protein folding and stability, along with ACE2 binding and evasion of antibodies. Investigating the impact of mutations on protein stability, Baek *et al.* employed three diverse computational methods and considered eight distinct protein structures to accommodate variations in methodology and structure (Bæk, *et al.*, 2023). Across all 24 comparisons made using these methods and structures, stability changes were observed to strongly correlate with experimental yeast expression data from previous research. This correlation suggests that enhanced protein fold stability corresponds to increased expression levels. Rawat *et al.* compared the spike-ACE2 complexes for different strains of coronavirus and assessed their binding affinity, interaction area, disease severity, and conserved residues (Rawat *et al.*, 2021). They found that the mild severity of human coronavirus NL63 is attributed to its smaller interaction area, lower hydrophobicity, and interaction energy, while SARS-COV's increased severity compared to SARS-COV-2 is due to a similar interface size but fewer flexible residues, a higher hydrophobic environment, and greater interaction energy facilitating ACE2 interaction. Moreover, changes in the conformations of the structure of spike protein due to mutations have led to changes in binding affinity and stability, which in turn affect its neutralization by vaccines and antibodies.

Insights from molecular dynamics simulations unveiled that mutations within the RBD region enhance the stability of the hACE2–spike protein complex compared to the wild-type configuration (Celik *et al.*, 2022). FoldX 5.0 (Delgado *et al.*, 2019) has also been used for calculating the change in binding affinity and stability upon mutations in SARS-CoV-2 (Delgado *et al.*, 2020). Certain inquiries have focused on the mutability of protein sites (Rawat *et al.*, 2022a) within the SARS-CoV-2 proteome, contributing to a deeper grasp of the evolutionary trajectory. They revealed that the physicochemical and structural features of the SARS-CoV-2 proteins can significantly affect the mutability of the viral proteome, and the proportion of bulky aromatic residues (F, Y, W) is significantly higher in the low-mutability sites. On the other hand, positively charged residues (R, K) occur frequently at low-mutability sites, whereas negatively charged residues (D, E) are present more in high-mutability sites (**Figure 13.6**).

Some large-scale genomic studies have provided insights into the pattern of the mutations over time (Obermeyer *et al.*, 2022; Fumagalli *et al.*, 2023). The investigation by Hossain *et al.* revealed that the most prominent alteration in the SARS-CoV-2 genome was C > T (52.67%), followed by G > T (14.59%), and A > G (11.13%) (Hossain *et al.*, 2021). Particularly, strains from India exhibited the highest mutation count (48), trailed by countries like Scotland, the USA, Netherlands, Norway, and France with up to 36 spike mutations. Noteworthy mutations like D416G,

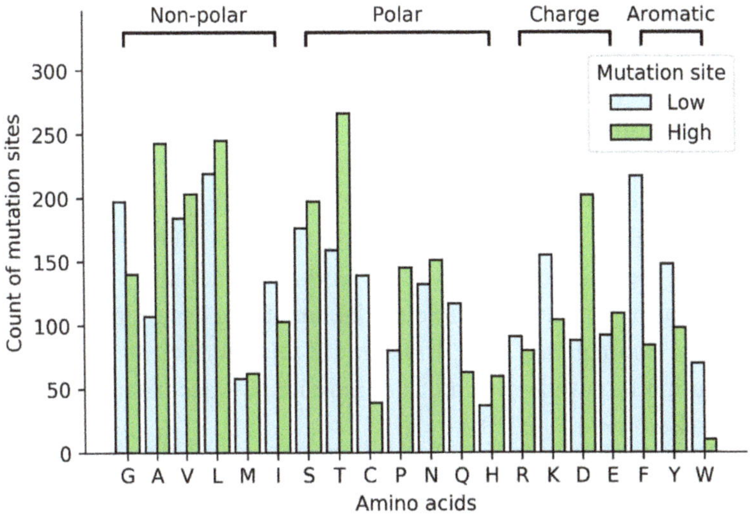

Figure 13.6 Frequency of amino acids in high and low mutability site classes. This figure was taken from Rawat *et al.* (2022a) on September, 28, 2023

F106F, P314L, C241T, L93L, A222V, A199A, V30L, and A220V took center stage as the most frequent occurrences. However, missense mutations such as D1118H, S194L, R262H, M809L, P314L, and others were found to considerably impact the structural stability of corresponding proteins. Liang (2023) examined 13 significant SARS-CoV-2 variants, revealing higher mean percent mutations in spike, ORF8, nucleocapsid, and NSP6. Omicron had more mutations in NSP6 and structural proteins, while Delta featured more in ORF7a. Omicron subvariant BA.2 had more mutations in ORF6 and BA.4 in NSP1, ORF6, and ORF7b than BA.1. Conserved proteins like NSP4, NSP13, NSP14, membrane, and ORF3a could offer better vaccine or therapeutic targets against immune evasion than mutation-prone proteins like NSP6, spike, ORF8, or nucleocapsid.

13.6 Spike mutations and immune escape

A range of mutations within the spike protein can significantly impact the effectiveness of mAbs in neutralizing the virus, enabling variants, specifically VOCs and VOIs (variants classified by WHO based on their potential impact on public health) to evade immunity and propagate within a population (Chen *et al.*, 2022). One key site of interest is the E484 residue on the RBD, which can be targeted by polyclonal antibodies in convalescent sera. Substituting amino acids K, Q, or P at this position has been shown to markedly diminish neutralization titers (Greaney *et al.*, 2021). The E484K mutation, replacing glutamic acid (E) with lysine (K) at position 484 of the RBD, is an escape mutation found in VOCs like Beta and Gamma, as well as some VOIs. This alteration facilitates evasion from convalescent plasma and multiple mAbs, including casirivimab and bamlanivimab (Baum *et al.*, 2020; Liu *et al.*, 2021). Furthermore, E484K has been identified to evade plasma from individuals vaccinated with mRNA vaccines (Baum *et al.*, 2020).

Other RBD escape mutations include K417N/T and L452R. K417N/T, present in Beta, Gamma, and Omicron VOCs, enables escape from mAbs like etesevimab (Starr *et al.*, 2020, 2021) but retains susceptibility to convalescent plasma and plasma from those immunized with BNT162b2 or mRNA-1273 (Wang *et al.*, 2021). L452R, observed in Delta and Omicron subvariants BA.4 and BA.5, reduces the neutralization efficacy of several mAbs (Li *et al.*, 2020) and convalescent plasma (Li *et al.*, 2020). Additional mutations also play a role in VOC antigenicity. For instance, N501Y, another RBD mutation across Alpha, Beta, Gamma, and Omicron VOCs, enhances ACE2 receptor affinity (Starr *et al.*, 2020), with limited impact on certain mAbs. This mutation is rarely linked to diminished neutralizing activity in convalescent plasma or vaccine sera (Wang *et al.*, 2021). P681H/R, occurring near the S1/S2 furin cleavage site, augments SARS-CoV-2 virulence through enhanced S1/S2 cleavage (Johnson *et al.*, 2021). While P681H is identified in Alpha and Omicron VOCs, P681R is characteristic of the Delta VOC. An N-terminal domain (NTD) mutation involving deletion at positions 69–70 correlates with increased viral replication (Meng *et al.*, 2021) and serves as a proxy for certain VOCs monitoring (Volz *et al.*, 2021).

Numerous investigations have demonstrated a decline in the neutralizing efficacy of vaccine serum against Omicron and its subvariants. These vaccines encompass two doses of the ChAdOx1-S vaccine (Planas *et al.*, 2022; Rössler *et al.*, 2022), two doses of mRNA vaccines (mRNA-1273 and BNT162b2 (Rössler *et al.*, 2022; Planas *et al.*, 2022;

Carreño *et al.*, 2022)), and a heterologous regimen with the ChAdOx1-S and BNT162b2 vaccines (Rössler *et al.*, 2022). Convalescent serum from individuals who had previously contracted the Alpha, Beta, or Delta VOC also exhibited diminished neutralizing activity against BA.1 (Rössler *et al.*, 2022; Schmidt *et al.*, 2022). Yet, serum samples from individuals who had experienced SARS-CoV-2 infection and subsequent vaccination, along with those who received booster shots (third dose with mRNA vaccines) (Planas *et al.*, 2021; Carreño *et al.*, 2022; Weisblum *et al.*, 2020), displayed detectable neutralizing antibodies against BA.1. Comparable neutralizing activities of vaccine serum against BA.1 and BA.2 were observed (Yu *et al.*, 2022), and booster doses exhibited similar effectiveness against symptomatic illness and hospitalization for both BA.1 and BA.2 (Sacco *et al.*, 2022). Nevertheless, one study revealed that, in comparison to BA.1, BA.2 was linked to heightened susceptibility to infection among unvaccinated, fully vaccinated, and booster-vaccinated individuals (Lyngse *et al.*, 2022). Conversely, BA.4/5 and BA.2.12.1 showcased greater evasion of neutralization compared to BA.2 when challenged with plasma from three doses of BNT162b2 (Hachmann *et al.*, 2022) and CoronaVac (Cao *et al.*, 2022).

13.7 Conclusion

Mutational studies on COVID-19 have provided profound insights into the behavioral impact and evolution of the SARS-CoV-2 virus. The emergence of this novel coronavirus and its subsequent global spread resulted in an unprecedented health crisis, leading to a pandemic with far-reaching consequences. The ability of the virus to mutate and generate new variants posed significant challenges to public health efforts, diagnostic strategies, and vaccine development. The SARS-CoV-2 virus demonstrated alarming adaptability through genetic mutations that affected its transmissibility, pathogenicity, and ability to evade immunity. The Alpha, Beta, Gamma, Delta, and Omicron variants presented distinctive sets of mutations, impacting their virulence, transmission rates, and evasion of neutralizing antibodies. These mutations within the spike protein, particularly at the interface with the ACE2 receptor and antibodies, played a pivotal role in determining the virus's behavior. Researchers employed a combination of experimental and computational approaches to investigate the effects of these mutations. Deep mutational scanning, computational modeling, molecular dynamics simulations, and binding affinity calculations aided in predicting the impact of mutations on viral behavior. Furthermore, the COVID-19 pandemic underscored the importance of global collaboration in scientific research, data sharing, and public health response. Efforts to curb the impact of the pandemic required timely information exchange, data analysis, and coordinated action across nations and scientific disciplines. In essence, the mutational studies on COVID-19 have not only provided a comprehensive understanding of the virus's behavior and evolution but have also highlighted the necessity of continuous research, collaboration, and adaptation in our fight against global health crises, considering the potential for future viral pandemic emergence.

Acknowledgments

We thank the Department of Biotechnology, IIT Madras for its computational facilities. Divya Sharma thanks the Ministry of Education, India for the PMRF fellowship.

References

Alam, I., Radovanovic, A., Incitti, R., Kamau, A. A., Alarawi, M., Azhar, E. I., & Gojobori, T. (2021). CovMT: An interactive SARS-CoV-2 mutation tracker, with a focus on critical variants. *Lancet Infectious Diseases, 21*(5), 602.

Alfonsi, T., Al Khalaf, R., Ceri, S., & Bernasconi, A. (2022). CoV2K model, a comprehensive representation of SARS-CoV-2 knowledge and data interplay. *Scientific Data, 9*(1), 260.

Andeweg, S. P., Vennema, H., Veldhuijzen, I., Smorenburg, N., Schmitz, D., Zwagemaker, F., van Gageldonk-Lafeber, A. B., Hahné, S. J. M., Reusken, C., Knol, M. J., Eggink, D., SeqNeth Molecular surveillance group., & RIVM COVID-19 Molecular epidemiology group. (2023). Elevated risk of infection with SARS-CoV-2 beta, gamma, and delta variants compared with alpha variant in vaccinated individuals. *Science Translational Medicine, 15*(684), eabn4338.

Andreano, E., Piccini, G., Licastro, D., Casalino, L., Johnson, N. V., Paciello, I., Dal Monego, S., Pantano, E., Manganaro, N., Manenti, A., Manna, R., Casa, E., Hyseni, I., Benincasa, L., Montomoli, E., Amaro, R. E., McLellan, J. S., & Rappuoli, R. (2021). SARS-CoV-2 escape from a highly neutralizing COVID-19 convalescent plasma. *Proceedings of the National Academy of Sciences of the United States of America, 118*(36).

Bæk, K. T., Mehra, R., & Kepp, K. P. (2023). Stability and expression of SARS-CoV-2 spike-protein mutations. *Molecular and Cellular Biochemistry, 478*(6), 1269–80.

Barnes, C. O., Jette, C. A., Abernathy, M. E., Dam, K. A., Esswein, S. R., Gristick, H. B., Malyutin, A. G., Sharaf, N. G., Huey-Tubman, K. E., Lee, Y. E., Robbiani, D. F., Nussenzweig, M. C., West, A. P., Jr, & Bjorkman, P. J. (2020). SARS-CoV-2 neutralizing antibody structures inform therapeutic strategies. *Nature, 588*(7839), 682–687.

Bate, N., Savva, C. G., Moody, P. C. E., Brown, E. A., Evans, S. E., Ball, J. K., Schwabe, J. W. R., Sale, J. E., & Brindle, N. P. J. (2022). In Vitro evolution predicts emerging SARS-CoV-2 mutations with high affinity for ACE2 and cross-species binding. *PLoS Pathogens, 18*(7), e1010733.

Baum, A., Fulton, B. O., Wloga, E., Copin, R., Pascal, K. E., Russo, V., Giordano, S., Lanza, K., Negron, N., Ni, M., Wei, Y., Atwal, G. S., Murphy, A. J., Stahl, N., Yancopoulos, G. D., & Kyratsous, C. A. (2020). Antibody cocktail to SARS-CoV-2 spike protein prevents rapid mutational escape seen with individual antibodies. *Science, 369*(6506), 1014–1018.

Bosch, B. J., van der Zee, R., de Haan, C. A. M., & Rottier. P. J. M. (2003). The coronavirus spike protein is a class I virus fusion protein: Structural and functional characterization of the fusion core complex. *Journal of Virology, 77*(16), 8801–8811.

Cao, Y., Yisimayi, A., Jian, F., Song, W., Xiao, T., Wang, L., Du, S., Wang, J., Li, Q., Chen, X., Yu, Y., Wang, P., Zhang, Z., Liu, P., An, R., Hao, X., Wang, Y., Wang, J., Feng, R., Sun, H., … Xie, X. S. (2022). BA.2.12.1, BA.4 and BA.5 escape antibodies elicited by omicron infection. *Nature, 608*(7923), 593–602.

Carabelli, A. M., Peacock, T. P., Thorne, L. G., Harvey, W. T., Hughes, J., COVID-19 Genomics UK Consortium, Peacock, S. J., Barclay, W. S., de Silva, T. I., Towers, G. J., & Robertson, D. L. (2023). SARS-CoV-2 variant biology: Immune escape, transmission and fitness. *Nature Reviews Microbiology, 21*(3), 162–77.

Carreño, J. M., Alshammary, H., Tcheou, J., Singh, G., Raskin, A. J., Kawabata, H., Sominsky, L. A., Clark, J. J., Adelsberg, D. C., Bielak, D. A., Gonzalez-Reiche, A. S., Dambrauskas, N., Vigdorovich, V., PSP-PARIS Study Group, Srivastava, K., Sather, D. N., Sordillo, E. M., Bajic, G., van Bakel, H., Simon, V., ... Krammer, F. (2022). Activity of convalescent and vaccine serum against SARS-CoV-2 omicron. *Nature, 602*(7898), 682–688.

Celik, I., Khan, A., Dwivany, F. M., Fatimawali,Wei, D. Q., & Tallei, T. E. (2022). Computational prediction of the effect of mutations in the receptor-binding domain on the interaction between SARS-CoV-2 and human ACE2. *Molecular Diversity, 26*(6), 3309–2334.

Chen, A. T., Altschuler, K., Zhan, S. H., Chan, Y. A., & Deverman, B. E. (2021). COVID-19 CG enables SARS-CoV-2 mutation and lineage tracking by locations and dates of interest. *eLife, 10.* https://doi.org/10.7554/eLife.63409.

Chen, J., Wang, R., Wang, M., & Wei, G. W. (2020). Mutations strengthened SARS-CoV-2 infectivity. *Journal of Molecular Biology, 432*(19), 5212–5226.

Chen, K. -W. K., Huang, D. T.-N., & Huang, L.-M. (2022). SARS-CoV-2 variants — Evolution, spike protein, and vaccines. *Biomedical Journal, 45*(4), 573–579.

Cox, M., Peacock, T. P., Harvey, W. T., Hughes, J., Wright, D. W., COVID-19 Genomics UK (COG-UK) Consortium, Willett, B. J., Thomson, E., Gupta, R. K., Peacock, S. J., Robertson, D. L., & Carabelli, A. M. (2023). SARS-CoV-2 variant evasion of monoclonal antibodies based on in vitro studies. *Nature Reviews Microbiology, 21*(2), 112–124.

Datta, P., Ukey, R., Bruiners, N., Honnen, W., Carayannopoulos, M. O., Reichman, C., Choudhary, A., Onyuka, A., Handler, D., Guerrini, V., Mishra, P. K., Dewald, H. K., Lardizabal, A., Lederer, L., Leiser, A. L., Hussain, S., Jagpal, S. K., Radbel, J., Bhowmick, T., Horton, D. B., ... Gennaro, M. L. (2021). Highly versatile antibody binding assay for the detection of SARS-CoV-2 infection and vaccination. *Journal of Immunological Methods, 499,* 113165.

Delgado B. J., Hernandez-Alias, X., Cianferoni, D., & Serrano, L. (2020). In silico mutagenesis of human ACE2 with S protein and translational efficiency explain SARS-CoV-2 infectivity in different species. *PLoS Computational Biology, 16*(12), e1008450.

Delgado, J., Radusky, L. G., Cianferoni, D., & Serrano, L. (2019). FoldX 5.0: Working with RNA, small molecules and a new graphical interface. *Bioinformatics, 35*(20), 4168–4169.

Dolgin, E. (2021). The tangled history of mRNA vaccines. *Nature, 597*(7876), 318–324.

Fan, Y., Li, X., Zhang, L., Wan, S., Zhang, L., & Zhou, F., (2022). SARS-CoV-2 omicron variant: Recent progress and future perspectives. *Signal Transduction and Targeted Therapy, 7*(1), 141.

Flores-Vega, V. R., Monroy-Molina, J. V., Jiménez-Hernández, L. E., Torres, A. G., Santos-Preciado, J. I., & Rosales-Reyes, R. (2022). SARS-CoV-2: Evolution and emergence of new viral variants. *Viruses, 14*(4).

Fowler, D. M., & Fields, S. (2014). Deep mutational scanning: A new style of protein science. *Nature Methods, 11*(8), 801–807.

Fumagalli, S. E., Padhiar, N. H., Meyer, D., Katneni, U., Bar, H., DiCuccio, M., Komar, A. A., & Kimchi-Sarfaty, C. (2023). Analysis of 3.5 million SARS-CoV-2 sequences reveals unique mutational trends with consistent nucleotide and codon frequencies. *Virology Journal, 20*(1), 31.

Gan, H. H., Twaddle, A., Marchand, B., & Gunsalus, K. C. (2021). Structural modeling of the SARS-CoV-2 spike/human ACE2 complex interface can identify high-affinity variants associated with increased transmissibility. *Journal of Molecular Biology, 433*,(15), 167051.

Garcia-Beltran, W. F., St Denis, K. J., Hoelzemer, A., Lam, E. C., Nitido, A. D., Sheehan, M. L., Berrios, C., Ofoman, O., Chang, C. C., Hauser, B. M., Feldman, J., Roederer, A. L., Gregory, D. J., Poznansky, M. C., Schmidt, A. G., Iafrate, A. J., Naranbhai, V., & Balazs, A. B. (2022). mRNA-Based COVID-19 vaccine boosters induce neutralizing immunity against SARS-CoV-2 omicron variant. *Cell, 185*(3), 457–66.e4.

Gobeil, S. M. –C., Janowska,K., McDowell, S., Mansouri, K., Parks, R., Stalls, V., Kopp, M. F., Manne, K., Li, D., Wiehe, K., Saunders, K. O., Edwards, R. J., Korber, B., Haynes, B. F., Henderson, R., & Acharya, P. (2021). Effect of natural mutations of SARS-CoV-2 on spike structure, conformation, and antigenicity. *Science, 373*(6555). https://doi.org/10.1126/science.abi6226.

Gowthaman, R., Guest, J. D., Yin, R., Adolf-Bryfogle, J., Schief, W. R., & Pierce, B. G. (2021). CoV3D: A database of high resolution coronavirus protein structures. *Nucleic Acids Research, 49*(D1), D282–D287.

Greaney, A. J., Loes, A. N., Crawford, K. H. D., Starr, T. N., Malone, k. D., Chu, H. Y., & Bloom, J. D. (2021a). Comprehensive mapping of mutations in the SARS-CoV-2 receptor-binding domain that affect recognition by polyclonal human plasma antibodies. *Cell Host & Microbe, 29*(3), 463–476.e6.

Greaney, A. J., Starr, T. N., Gilchuk, P., Zost, S. J., Binshtein, E., Loes, A. N., Hilton, S. K., Huddleston, J., Eguia, R., Crawford, K. H. D., Dingens, A. S., Nargi, R. S., Sutton, R. E., Suryadevara, N., Rothlauf, P. W., Liu, Z., Whelan, S. P. J., Carnahan, R. H., Crowe, J. E., Jr, & Bloom, J. D. (2021b). Complete mapping of mutations to the SARS-CoV-2 spike receptor-binding domain that escape antibody recognition. *Cell Host & Microbe, 29*(1), 44–57.e9.

Guy, R. K., DiPaola, R. S., Romanelli, F., & Dutch, R. E. (2020). Rapid repurposing of drugs for COVID-19. *Science, 368*(6493), 829–830.

Hachmann, N. P., Miller, J., Collier, A. R. Y., Ventura, J. D., Yu, J., Rowe, M., Bondzie, E. A., Powers, O., Surve, N., Hall, K., & Barouch, D. H. (2022). Neutralization escape by SARS-CoV-2 omicron subvariants BA.2.12.1, BA.4, and BA.5. *The New England Journal of Medicine, 387*(1), 86–88.

Hadfield, J., Megill, C., Bell, S. M., Huddleston, J., Potter, B., Callender, C., Sagulenko, P., Bedford, T., & Neher, R. A. (2018). Nextstrain: Real-time tracking of pathogen evolution. *Bioinformatics, 34*(23), 4121–4123.

Hadj Hassine, I. (2022). Covid-19 vaccines and variants of concern: A review. *Reviews in Medical Virology, 32*(4), e2313.

He, D., Chen, B., Zhao, S., & Stone, L. (2023). The immune evasion ability of delta variant is comparable to that of beta variant in South Africa. *BMC Public Health, 23*(1), 511.

Hossain, M. S., Pathan, A. Q. M. S. U., Islam, M. N., Tonmoy, M. I. Q., Rakib, M. I., Munim, M. A., Saha, O., Fariha, A., Reza, H. A., Roy, M., Bahadur, N. M., & Rahaman, M. M. (2021). Genome-wide identification and prediction of SARS-CoV-2 mutations show an abundance of variants: integrated study of bioinformatics and deep neural learning. *Informatics in Medicine Unlocked, 27*, 100798.

Hu, B., Guo, H., Zhou, P., & Shi, Z. L. (2021). Characteristics of SARS-CoV-2 and COVID-19. *Nature Reviews Microbiology, 19*(3), 141–1454.

Jacobs, J. L., Haidar, G., & Mellors, J. W. (2023). COVID-19: Challenges of viral variants. *Annual Review of Medicine, 74*, 31–53.

Johnson, B. A., Xie, X., Bailey, A. L., Kalveram, B., Lokugamage, K. G., Muruato, A., Zou, J., Zhang, X., Juelich, T., Smith, J. K., Zhang, L., Bopp, N., Schindewolf, C., Vu, M., Vanderheiden, A., Winkler, E. S., Swetnam, D., Plante, J. A., Aguilar, P., Plante, K. S., … Menachery, V. D. (2021). Loss of furin cleavage site attenuates SARS-CoV-2 pathogenesis. *Nature, 591*(7849), 293–299.

Kim, Y., Gaudreault, N. N., Meekins, D. A., Perera, K. D., Bold, D., Trujillo, J. D., Morozov, I., McDowell, C. D., Chang, K. O., & Richt, J. A. (2022). Effects of spike mutations in SARS-CoV-2 variants of concern on human or animal ACE2-mediated virus entry and neutralization. *Microbiology Spectrum, 10*(3), e0178921.

Laurini, E., Marson, D., Aulic, S., Fermeglia, A., & Pricl, S. (2021). Computational mutagenesis at the SARS-CoV-2 spike protein/angiotensin-converting enzyme 2 binding interface: Comparison with experimental evidence. *ACS Nano, 15*(4), 6929–6948.

Li, Q., Wu, J., Nie, J., Zhang, L., Hao, H., Liu, S., Zhao, C., Zhang, Q., Liu, H., Nie, L., Qin, H., Wang, M., Lu, Q., Li, X., Sun, Q., Liu, J., Zhang, L., Li, X., Huang, W., & Wang, Y. (2020). The impact of mutations in SARS-CoV-2 spike on viral infectivity and antigenicity. *Cell, 182*(5), 1284–1294.e9.

Liang, F. (2023). Quantitative mutation analysis of genes and proteins of major SARS-CoV-2 variants of concern and interest. *Viruses, 15*(5).

Liu, Z., VanBlargan, L. A., Bloyet, L. M., Rothlauf, P. W., Chen, R. E., Stumpf, S., Zhao, H., Errico, J. M., Theel, E. S., Liebeskind, M. J., Alford, B., Buchser, W. J., Ellebedy, A. H., Fremont, D. H., Diamond, M. S., & Whelan, S. P. J. (2021). Identification of SARS-CoV-2 spike mutations that attenuate monoclonal and serum antibody neutralization. *Cell Host & Microbe, 29*(3), 477–488.e4.

Low, Z. Y., Zabidi, N. Z., Yip, A. J. W., Puniyamurti, A., Chow, V. T. K., & Lal, S. K. (2022). SARS-CoV-2 non-structural proteins and their roles in host immune evasion. *Viruses, 14*(9).

Lyngse, F. P., Kirkeby, C. T., Denwood, M., Christiansen, L. E., Mølbak, K., Møller, C. H., Skov, .R.L., Krause, T. G., Rasmussen, M., Sieber, R. N., Johannesen, T. B., Lillebaek, T., Fonager, J., Fomsgaard, A., Møller, F. T., Stegger, M., Overvad, M., Spiess, K., & Mortensen, L. H. (2022). Household transmission of SARS-CoV-2 omicron variant of concern subvariants BA.1 and BA.2 in Denmark. *Nature Communications, 13*(1), 5760.

Mariano, G., Farthing, R. J., Lale-Farjat, S. L. M., & Bergeron, J. R. C. (2020). Structural characterization of SARS-CoV-2: Where we are, and where we need to be. *Frontiers in Molecular Biosciences, 7*(December), 605236.

Mbaeyi, S., Oliver, S. E., Collins, J. P., Godfrey, M., Goswami, N. D., Hadler, S. C., Jones, J., Moline, H., Moulia, D., Reddy, S., Schmit, K., Wallace, M., Chamberland, M., Campos-Outcalt, D., Morgan, R. L., Bell, B. P., Brooks, O., Kotton, C., Talbot, H. K., Lee, G., … Dooling, K. (2021). The advisory committee on immunization practices' interim recommendations for additional primary and booster doses of COVID-19 vaccines — United States, 2021. *MMWR Morbidity and Mortality Weekly Report, 70*(44), 1545–1552.

McBroome, J., Thornlow, B., Hinrichs, A. S., Kramer, A., De Maio, N., Goldman, N., Haussler, D., Corbett-Detig, R., & Turakhia, Y. (2021). A daily-updated database and tools for comprehensive SARS-CoV-2 mutation-annotated trees. *Molecular Biology and Evolution, 38*(12), 5819–5824.

McCallum, M., Czudnochowski, N., Rosen, L. E., Zepeda, S. K., Bowen, J. E., Walls, A. C., Hauser, K., Joshi, A., Stewart, C., Dillen, J. R., Powell, A. E.,

Croll, T. I., Nix, J., Virgin, H. W., Corti, D., Snell, G., & Veesler, D. (2022). Structural basis of SARS-CoV-2 omicron immune evasion and receptor engagement. *Science, 375*(6583), 864–868.

McCarthy, K. R., Rennick, L. J., Nambulli, S., Robinson-McCarthy, L. R., Bain, W. G., Haidar, G., & Paul Duprex, W. (2021). Recurrent deletions in the SARS-CoV-2 spike glycoprotein drive antibody escape. *Science, 371*(6534), 1139–1142.

Mehta, O. P., Bhandari, P., Raut, A., Kacimi, S. E. O., & Huy, N. T. (2020). Coronavirus disease (COVID-19): Comprehensive review of clinical presentation. *Frontiers in Public Health, 8*, 582932.

Meng, B., Kemp, S. A., Papa, G., Datir, R., Ferreira, I. A. T. M., Marelli, S., Harvey, W. T., Lytras, S., Mohamed, A., Gallo, G., Thakur, N., Collier, D. A., Mlcochova, P., COVID-19 Genomics UK (COG-UK) Consortium, Duncan, L. M., Carabelli, A. M., Kenyon, J. C., Lever, A. M., De Marco, A., Saliba, C., ... Gupta, R. K. (2021). Recurrent emergence of SARS-CoV-2 spike deletion H69/V70 and its role in the alpha variant B.1.1.7. *Cell Reports, 35*(13), 109292.

Mistry, P., Barmania, F., Mellet, J., Peta, K., Strydom, A., Viljoen, I. M., James, W., Gordon, S., & Pepper, M. S. (2021). SARS-CoV-2 variants, vaccines, and host immunity. *Frontiers in Immunology, 12*, 809244.

Muralidharan, N., Sakthivel, R., Velmurugan, D., & Michael Gromiha, M. (2021). Computational studies of drug repurposing and synergism of lopinavir, oseltamivir and ritonavir binding with SARS-CoV-2 protease against COVID-19. *Journal of Biomolecular Structure & Dynamics, 39*(7), 2673–2678.

Obermeyer, F., Jankowiak, M., Barkas, N., Schaffner, S. F., Pyle, J. D., Yurkovetskiy, L., Bosso, M., Park, D. J., Babadi, M., MacInnis, B. L., Luban, J., Sabeti, P. C., & Lemieux, J. E. (2022). Analysis of 6.4 million SARS-CoV-2 genomes identifies mutations associated with fitness. *Science, 376*(6599), 1327–1332.

Ogata, Y., & Kitayama, R. (2022). A database for retrieving information on SARS-CoV-2 S protein mutations based on correlation network analysis. *BMC Genomic Data, 23*(1), 34.

Ong, S. W. X., Chiew, C. J., Ang, L. W., Mak, T. M., Cui, L.,. Toh, M. P. H. S., Lim, Yi. D., Lee, P. H., Lee, T. H., Chia, P. Y., Maurer-Stroh, S., Lin, R. T. P., Leo, Y. S., Lee, V. J., Lye, D. C., & Young, B. E. (2022). Clinical and virological features of severe acute respiratory syndrome coronavirus 2 (SARS-CoV-2) variants of concern: A retrospective cohort study comparing B.1.1.7

(Alpha), B.1.351 (Beta), and B.1.617.2 (Delta). *Clinical Infectious Diseases: An Official Publication of the Infectious Diseases Society of America, 75*(1), e1128–e1136.

Onyeaka, H., Anumudu, C. K., Al-Sharify, Z. T., Egele-Godswill, E., & Mbaegbu, P. (2021). COVID-19 pandemic: A review of the global lockdown and its far-reaching effects. *Science Progress, 104*(2), 368504211019854.

Planas, D., Saunders, N., Maes, P., Guivel-Benhassine, F., Planchais, C., Buchrieser, J., Bolland, W. H. Porrot, F., Staropoli, I., Lemoine, F., Péré, H., Veyer, D., Puech, J., Rodary, J., Baele, G., Dellicour, S., Raymenants, J., Gorissen, S., Geenen, C., Vanmechelen, B., ... Schwartz, O. (2022). Considerable escape of SARS-CoV-2 omicron to antibody neutralization. *Nature, 602*(7898), 671–675.

Planas, D., Veyer, D., Baidaliuk, A., Staropoli, I., Guivel-Benhassine, F., Rajah, M. M., Planchais, C., Porrot, F., Robillard, N., Puech, J., Prot, M., Gallais, F., Gantner, P., Velay, A., Le Guen, J., Kassis-Chikhani, N., Edriss, D., Belec, L., Seve, A., Courtellemont, L., ... Schwartz, O. (2021). Reduced sensitivity of SARS-CoV-2 variant delta to antibody neutralization. *Nature, 596*(7871), 276–280.

Pondé, R. A. A. (2022). Physicochemical effect of the N501Y, E484K/Q, K417N/T, L452R and T478K mutations on the SARS-CoV-2 spike protein RBD and its influence on agent fitness and on attributes developed by emerging variants of concern. *Virology, 572*, 44–54.

Prabakaran, R., Jemimah, S., Rawat, P., Sharma, D., & Gromiha, M. M. (2021). A novel hybrid SEIQR model incorporating the effect of quarantine and lockdown regulations for COVID-19. *Scientific Reports, 11*(1), 24073.

Quinn, S. C., Jamison, A. M., & Freimuth, V. (2021). Communicating effectively about emergency use authorization and vaccines in the COVID-19 pandemic. *American Journal of Public Health, 111*(3), 355–358.

Raghu, D., Hamill, P., Banaji, A., McLaren, A., & Hsu, Y. T. (2022). Assessment of the binding interactions of SARS-CoV-2 spike glycoprotein variants. *Journal of Pharmaceutical Analysis, 12*(1), 58–64.

Rahimian, K., Arefian, E., Mahdavi, B., Mahmanzar, M., Kuehu, D. L., & Deng, Y. (2023). SARS2Mutant: SARS-CoV-2 amino-acid mutation atlas database. *NAR Genomics and Bioinformatics, 5*(2), lqad037.

Rahmani, K., Shavaleh, R., Forouhi, M., Disfani, H. F., Kamandi, M., Oskooi, R. K., Foogerdi, M., Soltani, M., Rahchamani, M., Mohaddespour, M., & Dianatinasab, M. (2022). The effectiveness of COVID-19 vaccines in reducing the incidence, hospitalization, and mortality from COVID-19: A systematic review and meta-analysis. *Frontiers in Public Health, 10*, 873596.

Raj, R. (2021). Analysis of non-structural proteins, NSPs of SARS-CoV-2 as targets for computational drug designing. *Biochemistry and Biophysics Reports, 25*, 100847.

Rawat, P., Jemimah, S., Ponnuswamy, P. K., & Gromiha, M. M. (2021a). Why are ACE2 binding coronavirus strains SARS-CoV/SARS-CoV-2 wild and NL63 mild? *Proteins, 89*(4), 389–398.

Rawat, P., Sharma, D., Pandey, M., Prabakaran, R., & Gromiha, M. M. (2022a). Understanding the mutational frequency in SARS-CoV-2 proteome using structural features. *Computers in Biology and Medicine, 147*, 105708.

Rawat, P., Sharma, D., Prabakaran, R., Ridha, F., Mohkhedkar, M., Janakiraman, V., & Gromiha, M. M. (2022b). Ab-CoV: A curated database for binding affinity and neutralization profiles of coronavirus-related antibodies. *Bioinformatics, 38(16), 4051–4052.*

Rawat, P., Sharma, D., Srivastava, A., Janakiraman, V., & Gromiha, M. M. (2021b). Exploring antibody repurposing for COVID-19: Beyond presumed roles of therapeutic antibodies. *Scientific Reports, 11(1), 10220.*

Raybould, M. I. J., Kovaltsuk, A., Marks, C., & Deane, C. M. (2021). CoV-AbDab: The coronavirus antibody database. *Bioinformatics, 37(5), 734–735.*

Rössler, A., Riepler, L., Bante, D., Laer, D. V., & Kimpel, J. (2022). SARS-CoV-2 omicron variant neutralization in serum from vaccinated and convalescent persons. *The New England Journal of Medicine, 386*(7), 698–700.

Sacco, C., Manso, M. D., Mateo-Urdiales, A., Rota, M. C., Petrone, D., Riccardo, F., Bella, A., Siddu, A., Battilomo, S., Proietti, V., Popoli, P., Menniti Ippolito, F., Palamara, A. T., Brusaferro, S., Rezza, G., Pezzotti, P., Fabiani, M., & Italian National COVID-19 Integrated Surveillance System and the Italian COVID-19 vaccines registry (2022). Effectiveness of BNT162b2 vaccine against SARS-CoV-2 infection and severe COVID-19 in children aged 5–11 years in Italy: A retrospective analysis of January-April, 2022. *The Lancet, 400*(10346), 97–103.

Saponaro, A. (2018). Isothermal titration calorimetry: A biophysical method to characterize the interaction between label-free biomolecules in solution. *Bio-Protocol, 8*(15), e2957.

Schmidt, F., Muecksch, F., Weisblum, Y., Da Silva, J., Bednarski, J., Cho, A., Wang, Z., Gaebler, C., Caskey, M., Nussenzweig, M. C., Hatziioannou, T., & Bieniasz, P. D. (2022). Plasma neutralization of the SARS-CoV-2 omicron variant. *The New England Journal of Medicine, 386*(6), 599–601.

Scovino, A. M., Dahab, E. C., Vieira, G. F., Freire-de-Lima, L., Freire-de-Lima, C. G., & Morrot, A. (2022). SARS-CoV-2's variants of concern: A brief characterization. *Frontiers in Immunology, 13*, 834098.

Shanmugam, A., Venkattappan, A., & Gromiha, M. M. (2022). Structure based drug designing approaches in SARS-CoV-2 spike inhibitor design. *Current Topics in Medicinal Chemistry, 22*(29), 2396–2409.

Sharma, D., Baas, T., Nogales, A., Martinez-Sobrido, L., & Gromiha, M. M. (2023). CoDe: A web-based tool for codon deoptimization. *Bioinformatics Advances, 3*(1), vbac102.

Sharma, D., Notarte, K. I., Fernandez, R. A., Lippi, G., Gromiha, M. M., & Henry, B. M. (2023). In silico evaluation of the impact of omicron variant of concern sublineage BA.4 and BA.5 on the sensitivity of RT-qPCR assays for SARS-CoV-2 detection using whole genome sequencing. *Journal of Medical Virology, 95*(1), e28241.

Sharma, D., Rawat, P., Janakiraman, V., & Gromiha, M. M. (2022). Elucidating important structural features for the binding affinity of spike — SARS-CoV-2 neutralizing antibody complexes. *Proteins, 90*(3), 824–834.

Singanayagam, A., Hakki, S., Dunning, J., Madon, K. J., Crone, M. A., Koycheva, A., Derqui-Fernandez, N., Barnett, J. L., Whitfield, M. G., Varro, R., Charlett, A., Kundu, R., Fenn, J., Cutajar, J., Quinn, V., Conibear, E., Barclay, W., Freemont, P. S., Taylor, G. P., Ahmad, S., … ATACCC Study Investigators (2022). Community transmission and viral load kinetics of the SARS-CoV-2 delta (B.1.617.2) variant in vaccinated and unvaccinated individuals in the UK: A prospective, longitudinal, cohort study. *The Lancet Infectious Diseases, 22*(2), 183–195.

Singh, D., & Yi, S. V. (2021). On the origin and evolution of SARS-CoV-2. *Experimental & Molecular Medicine, 53*(4), 537–547.

Siqueira, J. D., Goes, L. R., Alves, B. M., de Carvalho, P. S., Cicala, C., Arthos, J., Viola, J. P. B., de Melo, A. C., & Soares, M. A. (2021). SARS-CoV-2 genomic analyses in cancer patients reveal elevated intrahost genetic diversity. *Virus Evolution, 7*(1), veab013.

Song, S., Ma, L., Zou, D., Tian, D., Li, C., Zhu, J., Chen, M., Wang, A., Ma, Y., Li, M., Teng, X., Cui, Y., Duan, G., Zhang, M., Jin, T., Shi, C., Du, Z., Zhang, Y., Liu, C., Li, R., … Bao, Y. (2020). The global landscape of SARS-CoV-2 genomes, variants, and haplotypes in 2019nCoVR. *Genomics, Proteomics & Bioinformatics, 18*(6), 749–759.

Starr, T. N., Greaney, A. J., Addetia, A., Hannon, W. W., Choudhary, M. C., Dingens, A. S., Li, J. Z., & Bloom, J. D. (2021). Prospective mapping of viral mutations that escape antibodies used to treat COVID-19. *Science, 371*(6531), 850–854.

Starr, T. N., Greaney, A. J., Hilton, S. K., Ellis, D., Crawford, K. H. D., Dingens, A. S., Navarro, M. J. (2020). Deep mutational scanning of SARS-CoV-2 receptor binding domain reveals constraints on folding and ACE2 binding. *Cell, 182*(5), 1295–1310.e20.

Teng, S., Sobitan, A., Rhoades, R., Liu, D., & Tang, Q. (2021). Systemic effects of missense mutations on SARS-CoV-2 spike glycoprotein stability and receptor-binding affinity. *Briefings in Bioinformatics, 22*(2), 1239–1253.

Thomson, E. C., Rosen, L. E., Shepherd, J. G., Spreafico, R., da Silva Filipe, A., Wojcechowskyj, J. A., Davis, C., Piccoli, L., Pascall, D. J., Dillen, J., Lytras, S., Czudnochowski, N., Shah, R., Meury, M., Jesudason, N., De Marco, A., Li, K., Bassi, J., O'Toole, A., Pinto, D., ... Snell, G. (2021). Circulating SARS-CoV-2 spike N439K variants maintain fitness while evading antibody-mediated immunity. *Cell, 184*(5), 1171–1187.e20.

Thorne, L. G., Bouhaddou, M., Reuschl, A. K., Zuliani-Alvarez, L., Polacco, B., Pelin, A., Batra, J., Whelan, M. V. X., Hosmillo, M., Fossati, A., Ragazzini, R., Jungreis, I., Ummadi, M., Rojc, A., Turner, J., Bischof, M. L., Obernier, K., Braberg, H., Soucheray, M., Richards, A., ... Krogan, N. J. (2022). Evolution of enhanced innate immune evasion by SARS-CoV-2. *Nature, 602*(7897), 487–495.

Torrens-Fontanals, M., Peralta-García, A., Talarico, C., Guixà-González, R., Giorgino, T., & Selent. J. (2022). SCoV2-MD: A database for the dynamics of the SARS-CoV-2 proteome and variant impact predictions. *Nucleic Acids Research, 50*(D1), D858–D866.

Tortorici, M. A., & Veesler, D. (2019). Structural insights into coronavirus entry. *Advances in Virus Research, 105*, 93–116.

Toussi, S. S., Hammond, J. L., Gerstenberger, B. S., & Anderson, A. S. (2023). Therapeutics for COVID-19. *Nature Microbiology, 8*(5), 771–786.

Tzou, P. L., Tao, K., Pond, S. L. K., & Shafer, R. W. (2022). Coronavirus resistance database (CoV-RDB): SARS-CoV-2 susceptibility to monoclonal antibodies, convalescent plasma, and plasma from vaccinated persons. *PloS One, 17*(3), e0261045.

Varghese, P. M., Tsolaki, A. G., Yasmin, H., Shastri, A., Ferluga, J., Vatish, M., Madan, T., & Kishore, U. (2020). Host-pathogen interaction in COVID-19: Pathogenesis, potential therapeutics and vaccination strategies. *Immunobiology, 225*(6), 152008.

Viana, R., Moyo, S., Amoako, D. G., Tegally, H., Scheepers, C., Althaus, C., Anyaneji, U. J., Bester, P. A., Boni, M. F., Chand, M., Choga, W. T., Colquhoun, R., Davids, M., Deforche, K., Doolabh, D., du Plessis, L., Engelbrecht, S., Everatt, J., Giandhari, J., Giovanetti, M., ... de Oliveira, T. (2022). Rapid epidemic expansion of the SARS-CoV-2 omicron variant in Southern Africa. *Nature, 603*(7902), 679–686.

Volz, E., Mishra, S., Chand, M., Barrett, J. C., Johnson, R., Geidelberg, L., Hinsley, W. R., Laydon, D. J., Dabrera, G., O'Toole, Á., Amato, R., Ragonnet-Cronin, M., Harrison, I., Jackson, B., Ariani, C. V., Boyd, O., Loman, N. J., McCrone, J. T., Gonçalves, S., Jorgensen, D., ... Ferguson, N. M. (2021). Assessing transmissibility of SARS-CoV-2 lineage B.1.1.7 in England. *Nature, 593*(7858), 266–269.

Wang, P., Nair, M. S., Liu, L., Iketani, S., Luo, Y., Guo, Y., Wang, M., Yu, J., Zhang, B., Kwong, P. D., Graham, B. S., Mascola, J. R., Chang, J. Y., Yin, M. T., Sobieszczyk, M., Kyratsous, C. A., Shapiro, L., Sheng, Z., Huang, Y., & Ho, D. D. (2021). Antibody resistance of SARS-CoV-2 variants B.1.351 and B.1.1.7. *Nature, 593*(7857), 130–135.

Wang, Q., Ye, S. B., Zhou, Z. J., Li, J. Y., Lv, J. Z., Hu, B., Yuan, S., Qiu, Y., & Ge, X. Y. (2023). Key mutations on spike protein altering ACE2 receptor utilization and potentially expanding host range of emerging SARS-CoV-2 variants. *Journal of Medical Virology, 95*(1), e28116.

Wang, Y., Yang, C., Song, Y., Robert Coleman, J., Stawowczyk, M., Tafrova, J., Tasker, S., Boltz, D., Baker, R., Garcia, L., Seale, O., Kushnir, A., Wimmer, E., & Mueller, S. (2021). Scalable live-attenuated SARS-CoV-2 vaccine candidate demonstrates preclinical safety and efficacy. *Proceedings of the National Academy of Sciences of the United States of America, 118*(29), e2102775118.

Weisblum, Y., Schmidt, F., Zhang, F., DaSilva, J., Poston, D., Lorenzi, J. C., Muecksch, F., Rutkowska, M., Hoffmann, H. H., Michailidis, E., Gaebler, C., Agudelo, M., Cho, A., Wang, Z., Gazumyan, A., Cipolla, M., Luchsinger, L., Hillyer, C. D., Caskey, M., Robbiani, D. F., . . . Bieniasz, P. D. (2020). Escape from neutralizing antibodies by SARS-CoV-2 spike protein variants. *eLife*, 9, e61312

Xing, Y., Li, X., Gao, X., & Dong, Q. (2020). MicroGMT: A mutation tracker for SARS-CoV-2 and other microbial genome sequences. *Frontiers in Microbiology, 11*, 1502.

Yesudhas, D., Srivastava, A., & Gromiha, M. M. (2021). COVID-19 outbreak: History, mechanism, transmission, structural studies and therapeutics. *Infection 49*(2), 199–213.

Yu, J., Collier, A-. R. Y., Rowe, M., Mardas, F., Ventura, J. D., Wan, H., Miller, J., Powers, O., Chung, B., Siamatu, M., Hachmann, N. P., Surve, N., Nampanya, F., Chandrashekar, A., & Barouch, D. H. (2022). Neutralization of the SARS-CoV-2 omicron BA.1 and BA.2 variants. *The New England Journal of Medicine, 386*(16), 1579–1580.

Zhang, Q., Chen, X., Li, B., Lu, C., Yang, S., Long, J., Chen, H., Huang, J., & He, B. (2022). A database of anti-coronavirus peptides. *Scientific Data, 9*(1), 294.

Zhou, W., Xu, C., Wang, P., Anashkina, A. A., & Jiang, Q. (2022). Impact of Mutations in SARS-COV-2 spike on viral infectivity and Antigenicity." *Briefings in Bioinformatics* 23(1), bbab375.

Zhu, Z., Lian, X., Su, X., Wu, W., Marraro, G. A., & Zeng, Y. (2020). From SARS and MERS to COVID-19: A brief summary and comparison of severe acute respiratory infections caused by three highly pathogenic human coronaviruses. *Respiratory Research, 21*(1), 224.

https://doi.org/10.1142/9789811293269_0014

Chapter 14

Transcriptome-based analysis for understanding the effects of mutations in neurodegenerative diseases

Nela Pragathi Sneha[1], S. Akila Parvathy Dharshini[1], Y-h. Taguchi[2], and M. Michael Gromiha[1,*]

[1]*Department of Biotechnology, Bhupat and Jyoti Mehta School of Biosciences, Indian Institute of Technology Madras, Chennai, Tamil Nadu 600036, India*

[2]*Department of Physics, Chuo University, Kasuga, Bunkyo-ku, Tokyo 112-8551, Japan*

Abstract

Neurodegenerative diseases are the most common cause of death after cancer. They are characterized by gradual loss of neurons, resulting in a spectrum of symptoms ranging from motor impairments to memory loss, culminating in fatality. Neurodegenerative diseases are affected by

*Corresponding author
Tel: +91-2257-4138
Fax: +91-2257-4102
MMG: gromiha@iitm.ac.in

a range of factors including lifestyle, cellular stress, aging, and genetic variants. These variants play a significant role in understanding disease susceptibility by modifying gene regulatory elements. Investigating the role of variants provides deep insights into the disease association. Next-generation sequencing (NGS) techniques have increased the scope of understanding diseases from a genetic perspective. Covering both protein-coding and noncoding regions for identifying various disease susceptible variants using computational tools and pipelines is crucial. In this chapter, with a brief introduction to different neurodegenerative diseases and the associated variants, we explain various sequencing techniques and pipelines used to identify plausible disease-causing variants and study their effects on different functional events that occur during post-transcription. In essence, this review helps to understand various techniques and pipelines used for variant calling and predicting their effects.

14.1 Introduction

Variants are changes in the DNA sequence that can impact protein function by affecting their regulatory elements. These variants can originate from both coding and noncoding regions of the genome. Studying variants involved in regulatory activities helps us understand how genetic susceptibility influences protein function within pathways compromised by disease. Neurodegenerative diseases are majorly driven by the variants caused in the genome. Most of the neurodegenerative disorders, such as Alzheimer's disease (AD), Huntington's disease (HD), Parkinson's disease (PD), Amyotrophic Lateral Sclerosis (ALS) and frontotemporal dementia (FTD) are being steered by genetic variants. Identifying these plausible disease-causing is essential for elucidating their biological and clinical impact. Traditionally, the variants are identified using whole genome/exome sequencing to reveal all possible sequence variants on a large scale. Furthermore, RNA-seq data can be utilized to study variations in the coding region, aiding in the identification of rare variants associated with alternative splicing events. In this chapter, we discuss various neurodegenerative disorders and the variations reported in them that are crucial in disease prognosis. Further, we explain various sequencing techniques and computational pipelines to identify variants and predict their effects.

14.1.1 *Different mutations/variants in neurodegenerative disorders*

Mutations play a pivotal role in the onset and progression of neurodegenerative diseases, serving as crucial factors in the disruption of neuronal function and eventual degeneration. Understanding the intricate relationship between mutations and neurodegeneration is paramount for unraveling the underlying mechanism. AD is the most common neurodegenerative disorder, which affects the neuronal system and ultimately leads to memory loss. AD is a progressive neurodegenerative disorder that impairs short- and long-term memory due to neuronal cell death. AD patients experience a decline in cognitive thinking, memory, and organizing skills. AD is characterized by the loss of hippocampal CA1 neurons in the initial stages of the disease. Variants that are associated with the genes amyloid precursor protein (APP), presenilin 1 (PSEN1), and presenilin 2 (PSEN2) are known for their role in the onset of AD. Mutations/alterations in Apolipoprotein E (APOE4) are responsible for disease progression in the later stages of the disease.

PD is the second most common neurodegenerative disorder after AD. PD patients experience loss of dopaminergic neurons in the substantia nigra pars compacta of the brain. Synuclein Alpha (SNCA), Parkin, PTEN-induced putative kinase 1 (PINK1), DJ1, and Leucine Rich Repeat Kinase 2 (LRRK2) are PD-associated genes, and the variants in these genes cause PD (Larsen *et al.*, 2018). Heat shock protein family A member 9 (HSPA9), TNF receptor associated protein 1 (TRAP1), ras homolog family member T1 (RHOT1), and HtrA serine peptidase 2 (HTRA2) are other genes associated with PD. Variants present in these genes are known to be deleterious in PD (Larsen *et al.*, 2018).

Amyotrophic lateral sclerosis (ALS) is a motor neuron disease that affects lower and upper motor neurons along with the spinal cord and brainstem (Hardiman *et al.*, 2017). ALS patients exhibit motor symptoms along with cognitive and behavioral symptoms. Patients also experience breathing difficulties and swallowing problems (Zarei *et al.*, 2015). The currently available treatment options are for treating the symptoms, and only two FDA-approved drugs, namely riluzole and edaravone are available for ALS (Ragagnin *et al.*, 2019). Fast-fatigable motor neurons are more susceptible to degeneration in ALS patients (Mejzini *et al.*, 2019).

The most common variants of ALS are due to mutations in c9orf72, TAR DNA binding protein (TARDBP), superoxide dismutase 1 (SOD1), and fused in sarcoma (FUS) genes.

HD is an autosomal dominant neurodegenerative disorder caused due to excessive generation of critical assessment of genome (CAG) repeats in the HTT gene (Arrasate & Finkbeiner, 2012). More than 35 CAG repeats in HTT are the major cause of HD (MacDonald *et al.*, 1993). HD patients experience motor impairment along with cognitive and behavioral disturbances (De Boo *et al.*, 1997).

FTD is a syndrome caused by the loss of upper and lower motor neurons, and patients experience symptoms such as rigidity, tremors, weakness in muscles, and poor coordination (Young *et al.*, 2018). There are currently no FDA-approved drugs available for treating FTD (Tsai & Boxer, 2016), and only symptomatic treatments are available for this disease. Similar to other neurodegenerative disorders, c9orf72 is one of the prominent genes with mutations in FTD when compared to microtubule-associated protein tau (MAPT), and granulin precursor (GRN) (Tipton *et al.*, 2022). TARDBP, FUS, valosin containing protein (VCP), and charged multivesicular body protein 2B (CHMP2B) are other genes with mutations in FTD. Sequencing serves as a powerful tool for unraveling the complex landscape of mutations underlying neurodegenerative diseases, offering insights into their genetic origins, prevalence, and functional consequences. In next section we discuss about various sequencing techniques.

14.2 Introduction to sequencing technologies

14.2.1 *History of sequencing*

Sequencing techniques offer many ways for genome-wide profiling of mRNAs, transcription factors, and chromatin regulation. Sequencing was started by Sanger's dideoxy (Sanger *et al.*, 1977) and Gilbert's chemical degradation methods (Maxam & Gilbert, 1977), which are the first generation of sequencing techniques. The second-generation DNA sequencing technique known as pyrosequencing is based on luminescence emitted by the luciferase enzyme due to pyrophosphate synthesis. This method is popularly called sequencing by synthesis (SBS). Third-generation techniques introduced zero-mode waveguides that are composed of DNA polymerase.

Short-read sequencing is a cost-effective and easier method to increase genome-scale sequencing. Short-read sequencing includes semiconductor sequencing (ion torrent) that uses changes in hydrogen ion concentration (pH) to detect the incorporated base. Long-read sequencing includes nanopore technology and SMRT techniques to sequence long strands of DNA. Single molecule real time (SMRT) technique uses fluorescence, which is generated during DNA synthesis by polymerase, and these reads can have an average length of 3,000 bp. Oxford nanopore technology uses single-stranded DNA fragments, which move through the nanopore. The ionic current applied across the nanopore makes the DNA to traverse the pore, and the change in current is measured by an electronic chip inside the flow cell, which is converted to DNA sequence using a base-calling algorithm.

The single-cell sequencing technique (Kolodziejczyk *et al.*, 2015) is the most popular next-generation technique to study cell type susceptibility and cell level heterogeneity in most of diseases. The first step is to dissociate single cells from tissues. The second step is library preparation, which is a major step that uses barcodes (droplet-based microfluidics), followed by well-known tasks such as isolating RNA, cDNA synthesis, amplification, and sequencing. The latest technique in sequencing that explores the localization of cell types and their gene expression is spatial sequencing. This technique measures RNA expression using *in situ* — hybridization, sequencing, and capturing methods. The spatial sequencing technique (Ståhl *et al.*, 2016) starts with sample preparation, constructing libraries, sequencing, and downstream analysis. **Figure 14.1** provides a visual representation of the advancement in sequencing techniques over the years.

14.2.2 *Sanger sequencing*

Sanger sequencing was the most popular and widely used method for obtaining the sequence of DNA (**Figure 14.2**) before next-generation technologies were invented. It includes the basic steps such as template preparation, polymerization, and sequencing (Ansorge, 2009). Template preparation involves obtaining molecules of interest (DNA/RNA) and library preparation (attaching adapters), along with amplification. DNA fragments are glided on a glass slide (flow cell), which contains

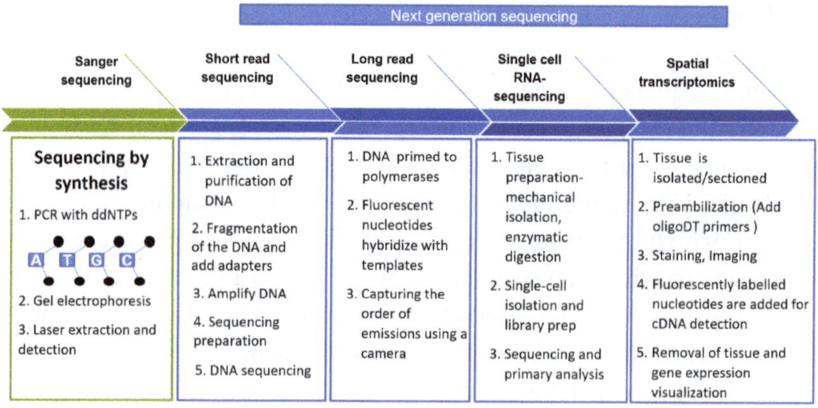

Figure 14.1 Sequencing techniques from past to present

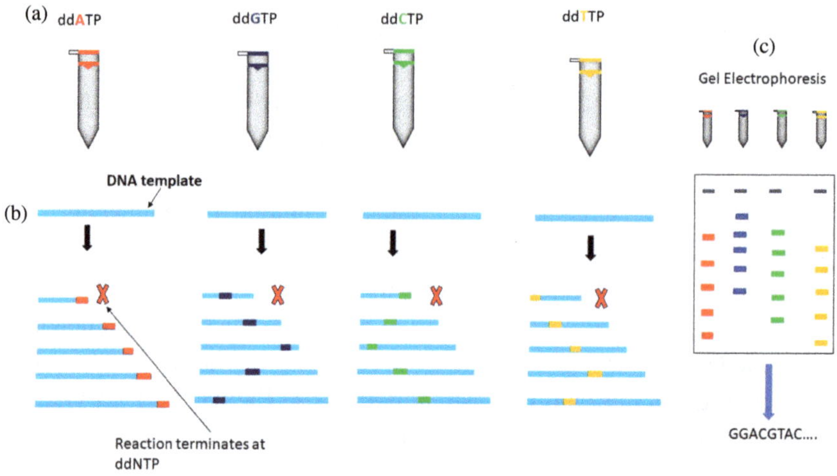

Figure 14.2 Sanger sequencing methodology. It includes (a) template preparation, (b) polymerization and chain termination, and (c) sequencing. The figure was adapted from io.com/articles/article/Overview-Sequencing-Technique

complementary oligonucleotides that can bind to the adapters. DNA polymerase and fluorescent nucleotides are introduced into the flow cell for hybridization. The amplification step is the generation of hundreds of strands/clusters. Generally, there are two types of amplification, such as bridge amplification and rolling circle amplification (Ali *et al.*, 2014).

Another step in the sequencing process involves the recording of fluorescent signals generated during hybridization. The Sanger chain termination method is the first technique to introduce sequencing. Dideoxy nucleotides are used by DNA polymerase to hybridize the DNA template, and the process ends, which is denoted as chain termination. This step is followed by polyacrylamide gel electrophoresis. The nucleotides are sequenced using these steps in this technology. The technique represented in **Figure 14.2a** shows four test tubes, where the template is divided and introduced into them. Each tube contains a template (or fragment) along with primer, DNA polymerase, dNTPs (dATP, dGTP, dTTP, dCTP), and radiolabeled dideoxyNTP. With the help of DNA polymerase, dNTPs get attached to the template inside the tube, and this process stops immediately after the dideoxyNTP (ddATP/ddCTP/ddGTP/ddTTP). The chain termination step (**Figure 14.2b**) is followed by sequencing using gel electrophoresis. The fragments that are terminated vary in their length. The length of each fragment is measured using the gel electrophoresis technique (**Figure 14.2c**). This gives the sequence of nucleotides in the fragment.

14.2.3 *Sequencing by synthesis*

The second-generation sequencing techniques include the SBS (Nyrén *et al.*, 1993) technique, which has shorter read sequences (50–300 nucleotides). SBS uses the bridge polymerase chain reaction (PCR) technique, and the reactions are run in parallel (**Figure 14.3**). The process of sequencing begins with purifying and fragmenting the DNA. The fragmented DNA is attached to adapters (**Figure 14.3a**) and added to a flow cell containing oligonucleotides for providing binding complementarity to adapters. The complimentary strand to the attached adapter is synthesized by DNA polymerase (**Figure 14.3b**). DNA folds over like a bridge to the adjacent oligonucleotide on the flow cell. This step is called bridge amplification (**Figure 14.3c**). The folded DNA strand is separated into forward and reverse strands (**Figure 14.3d**) and undergoes amplification to generate multiple strands (forward and reverse). At each round of synthesis, the camera takes a picture of the chip and records the currently added base. These steps continue until the whole DNA sequence is fully sequenced.

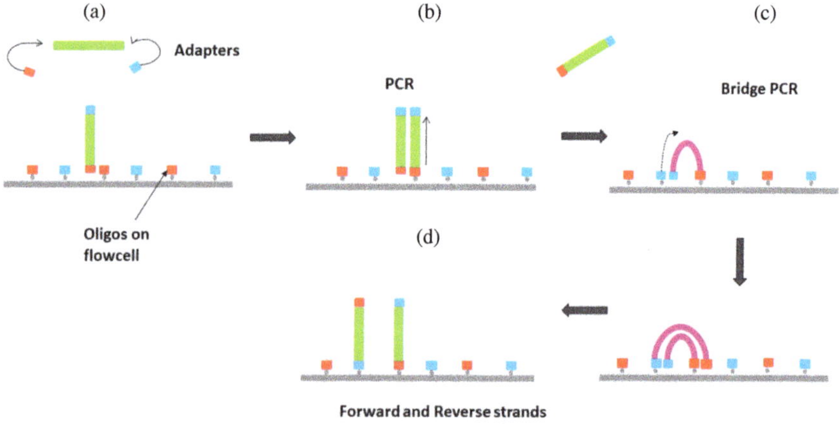

Figure 14.3 Sequencing by synthesis technique: (A) DNA fragments are bound to complimentary oligos on the flowcell, (B) PCR generates a reverse strand and the original strand is washed away, (C) Bending of reverse strand and bridge PCR initiation, and (D) formation of forward and reverse strands. The figure was adapted from https://the-dna-universe.com/2021/11/25/how-to-do-ngs-50-faster-with-our-ngs-platforms/

Short-read sequencing has their own limitations in technical details such as sample preparation, optics, and instrumentation. Long read sequencing/large fragment single-molecule sequencing is a third-generation sequencing technique used to find large fragments of DNA.

14.2.4 *Next-generation sequencing techniques*

Long-read sequencing has the capability to sequence reads ranging from 5,000 to 30,000 base pairs in length. Aiding in finding structural variants and repetitive regions in the genome. These reads are sequenced without the necessity for amplification, as in short-read sequencing. Long-read sequencing uses SMRT technique similar to circular consensus sequencing and continuous long-read sequencing (Logsdon *et al.*, 2020). The circular consensus sequencing method generates numerous observations of a base to generate high-accuracy reads from single molecules. Continuous long-read sequencing generates reads using long templates (greater than 30 kb in length) with high error rate than the circular consensus sequencing method. Long-read sequencing has its own limitations, such as a high error rate, longer processing time for large genomes, and less accuracy. Traditional sequencing techniques (bulk RNAseq) use RNA extracted from tissues made of multiple cell types. Bulk RNA-seq (Thind *et al.*, 2021) provides an average expression of genes across hundreds to millions of cells. The real picture of individual cell--to-cell heterogeneity is compromised by the bulk RNA-seq technique.

14.2.5 *Single-cell sequencing technique*

Single-cell RNA sequencing (scRNA-seq) technology is a technique to understand cell heterogeneity/subpopulation of different cells from the same tissue (**Figure 14.3**). This technique enables the understanding of RNA molecules in different cells with great resolution. Single-cell RNA-seq has three major steps: (i) tissue preparation, (ii) single-cell isolation and library preparation, and (iii) sequencing. Tissue preparation includes isolating viable cells and carefully removing the dead cells. Pipetting, washing, and straining are part of sample preparation in the single-cell RNA-seq technique. The second step, single cell isolation is carried out by different techniques such as fluorescence-activated cell sorting (FACS), magnetic activated cell sorting (MACS), microfluidics, and laser-capture microdissection. FACS is a fluorescence-based cell isolation method where fluorescent molecules are selected based on target-specific antibodies. In the laser capture microdissection method, cells are isolated using laser (infrared/ultraviolet light). Isolation of cells using reagents of water in oil droplets is done in the microfluidics technique. Library preparation involves the collection of DNA fragments derived from one cell. DNA fragments are barcoded to recognize the cell type, and specific adapter sequences are attached at 5′ and 3′ ends to get a high-quality library. The final step is sequencing by different approaches, such as SBS, Drop-seq, Smart-seq2, Split-seq, and MATQ-seq (Liu *et al.*, 2021). Even though single-cell transcriptomics has prominent advantages over traditional sequencing methods, they have some limitations. The limitations include (i) the loss of crucial information during the process of obtaining viable cells by removing cell debris, (ii) data are noisy and complex to handle, compared to bulk RNA-seq. To overcome the drawbacks of single-cell transcriptomics, spatial RNA sequencing (spRNA-seq) was introduced for better understanding of the positional variation in gene expression and to study pathological changes in tissues.

14.2.6 *Spatial-sequencing technique*

Spatial transcriptomics (Williams *et al.*, 2022) allows the simultaneous visualization of gene expression patterns and spatial organization within tissues, providing a comprehensive understanding of cellular interactions and microenvironmental dynamics. By preserving spatial information alongside transcriptomic data, spatial transcriptomics enables the identification of spatially resolved gene expression patterns, facilitating the discovery of novel cell types, biomarkers, and regulatory networks within complex biological systems. It involves three major steps: (i) staining/preparation of tissue, (ii) tissue permeabilization and library preparation, and (iii) sequencing and analysis. Preparation of tissue includes imaging of frozen tissue, which is placed on a spatial gene expression slide that has oligo capture probes. The second step involves permeabilizing the tissue, which enables the RNA to go and bind on the probes of a chip. The chip captures gene expression, and cDNA is synthesized to obtain double-stranded DNA molecules. Library preparation is a step in which cDNA-RNA hybrids are cleaved from the chip and used for library preparation. Massively parallel sequencing is carried out to sequence the libraries along with obtaining information on gene expression, tissue, and spatial information. The sequencing techniques aid in identifying disease expression patterns and disease-associated variants (**Figure 14.4**).

14.3 Computational pipelines to identify disease-associated variants

14.3.1 *Overview and importance of variant identification*

Genome wide association studies (GWAS) have proven to be important for neurodegenerative disorders in studying disease associations. To understand the phenotypic diversity among individuals, the analysis of variants on a large scale is crucial (Qin *et al.*, 2016). One of the major advantages of identifying variants is that it aids in revealing novel and functionally important variants. Variants have an effect on gene regulation indirectly/directly, and their effects on different regulatory mechanisms can be revealed through the various tools and algorithms discussed earlier. Each sequencing technique has a unique purpose to understand the

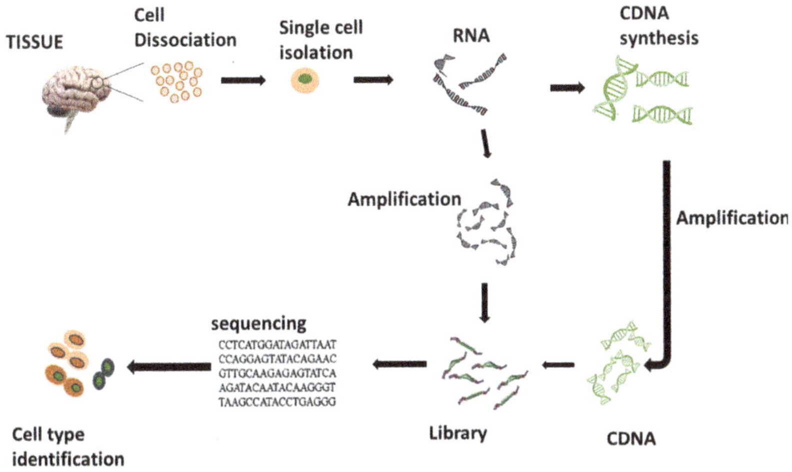

Figure 14.4 Single-cell RNA sequencing technique with steps involved in single-cell sequencing from cell isolation to cell type identification and gene expression (adapted from https://en.m.wikipedia.org/wiki/File:RNA-Seq_workflow-5.pdf)

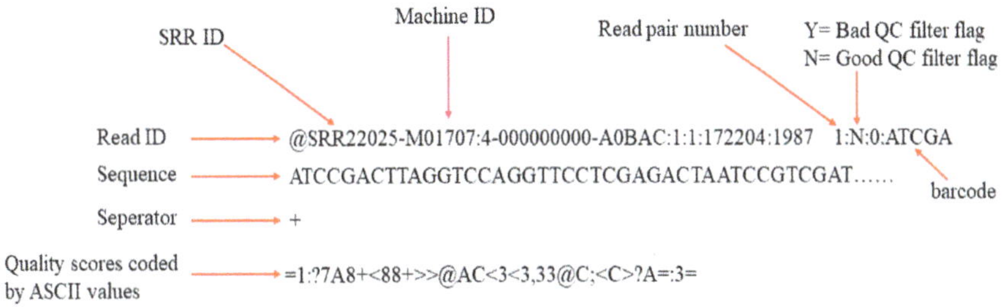

Figure 14.5 FastQ format of an Illumina read

variants. Although whole genome and whole exome techniques cover a wide range of variants, they are expensive and difficult to handle. On the other hand, RNA-seq has sequencing complexity and requires extensive computational power.

The bioinformatics approach for detecting variants from sequencing data includes three major steps: (i) read preprocessing, (ii) alignment and mapping, and (iii) variant calling. Sequencing data gives the FASTQ file (**Figure 14.5**) as output, and it is further processed for downstream analysis to identify disease-associated variants.

Variant calling is an eminent method carried out on whole genome/whole exome sequence data. Whole genome sequencing (WGS) data gives a complete overview of human genomic variants (Van El *et al.*, 2013). Whole exome sequencing (WES) data provides variant information on protein coding regions, which account for ~1.5% of the genome (Rabbani *et al.*, 2014). This analysis helps in finding variations in a large group of people with common disease/genetic conditions. Identifying single nucleotide polymorphism (SNPs), copy number variations (CNVs), insertions, deletions, inversions, and base pair substitutions are key factors in support of using variant calling procedures on whole genome data. Variant identification using the whole genome allows to predict variants in broad regions of the genome, such as noncoding RNA, intronic, exonic, regulatory, and intergenic regions. The only drawback with the WGS data is the complexity of analyzing and storing the huge data created due to the large size of the genome. Cancer and complex diseases such as neurodegenerative disorders need a concrete strategy to find disease-causing mutations/alterations of nucleotides in the genome. WGS aids in finding a wide range of variants in regulatory regions that can pinpoint the genetic etiology of complex disorders. Various genome-wide association studies (GWAS) have implied that disease-associated variants are mainly from noncoding regions of the genome. These noncoding variants can impact transcription factor binding, regulation of ncRNA expression, and modulation of chromatin/epigenetic interactions. Intergenic variants inside a pool of promoters and enhancers are important disease-associated regulators of gene expression and epigenetic modifications. Variant detection is also done using RNA-seq-based data analysis to reveal SNPs from post-transcriptional modifications that were missed from the genomic data. RNA-seq-based approaches also give information about gene fusions. Gene fusion is an event when a novel gene is generated due to the fusion of two distinct wild-type genes. Finding these events is important in cancer, as they have a prominent role in the cancer prognosis (Mertens *et al.*, 2015). Along with these findings, exploring the impact of variants on various regulatory mechanisms and transcriptional events, such as allelic expression, RNA editing, and alternative splicing is also necessary. Further, several tools are available for analyzing the effects of coding/noncoding variants using stringent cutoffs.

14.3.2 *Practices/guidelines and tools available for variant calling using whole genome sequencing*

Sequencing coverage of around 30× to 60× (coverage denotes the number of times the sequencing machine sequences the genomic region. For example, 1× is 1 cycle of sequencing) is obtained across the entire genome in WGS data (Koboldt, 2020). This coverage/depth helps in accurately identifying heterozygous variants. Some considerations need to be taken for accurate predictions in variant calling before starting the process of identifying variants. The key guidelines to be followed are (i) **PCR duplicates**: DNA fragments attached during PCR amplification in sequencing and are found during sequencing due to errors introduced by PCR polymerases and repeated during cycles of amplification. These errors can be misleading for variant calling and hence, these duplicates are either marked or removed before variant identification, (ii) **read length**: Long read sequence data is used to find structural variants that are difficult to interpret from short reads, and (iii) **coverage**: large coverage means a depth of 30× to 50× is maintained for performing variant calling using WGS data.

DNA nucleotide mutations/changes of more than 50nt at a single locus are called structural variants. The major types of structural variants identified using WGS are CNVs. Paired-end WGS offers accurate detection of structural variants using the tools DELLY (Rausch *et al.*, 2012), BreakDancer (Chen *et al.*, 2009), LUMPY (Layer *et al.*, 2014), GRIDSS (Cameron *et al.*, 2017), Manta (Chen *et al.*, 2016), SVMerge (Wong *et al.*, 2010), and Pindel (Ye *et al.*, 2009). There are tools exclusively available for the detection of CNVs, such as ExomeCNV (Sathirapongsasuti *et al.*, 2011), CoNVEX (Amarasinghe *et al.*, 2013), XHMM (Fromer *et al.*, 2012), and ExomeDepth (Plagnol *et al.*, 2012).

14.3.3 *Practices/guidelines and tools for variant calling using RNA-seq data*

Various tools are used for performing variant analysis on RNA-seq data, as discussed in Dharshini *et al.* (2019). RNA-seq data is considered for evaluating the quality of reads and to pre-process the reads with steps such as trimming and adapter contamination removal to obtain high-quality reads. Various tools are used to perform different stages of quality assessment of sequenced reads. FASTQC (https://www.bioinformatics.babraham.ac.uk/projects/fastqc/) is a compatible program for the majority of the sequencing platforms, and it generates reports and graphs to check the data quality. Other tools used for data preprocessing include NGSQCToolkit (Patel & Jain, 2012), SRAToolKit (https://hpc.nih.gov/apps/sratoolkit.html), Trimmomatic (Bolger *et al.*, 2014), and FASTX-Toolkit (Gordon *et al.*, 2014), to eliminate poor quality reads, trim adaptor sequences, and remove poor quality bases.

Reads are then mapped to the reference transcriptome/genome. There are two sources for the reference genome, namely: (i) UCSC genome browser (Mias, 2018) and (ii) Genome Reference Consortium (GRC) (Schneider & Church, 2013). There are three major types of alignments available so far, namely, reference genome-based alignment, transcriptome-based alignment, and reference-free assembly. Some of the tools used for genome-based alignment are STAR (Dobin & Gingeras, 2015), TopHat (Trapnell *et al.*, 2009), and Cufflinks (Trapnell *et al.*, 2012). Transcriptome-based alignment tools are Bowtie (Langmead & Salzberg, 2012), RSEM (Li & Dewey, 2011), Kallisto (Caldeweyher, 2021), Salmon (Bray *et al.*, 2016), eXpress (Roberts & Pachter, 2013), and sailfish (Patro *et al.*, 2014). *De novo* assembly can be performed on samples from organisms whose reference genome is unavailable. The tools available for reference-free mapping are Trinity (Grabherr *et al.*, 2011), SOAPdenovo-Trans (Xie *et al.*, 2014), and Trans-ABySS (Robertson *et al.*, 2010). Paired end reads are reliable for obtaining a better alignment rate. The percentage of mapped reads is the measure for finding the quality of an alignment or mapping. There can be multi-mapped and uniquely mapped reads during the alignment. Sequencing platforms that implement PCR steps in library preparation generate multiple reads obtained from a single template. This will result in false positives during variant calling. Removal of these PCR duplicates needs to be performed before variant calling (Ebbert *et al.*, 2016). The major tools used for the removal of PCR duplicates are Picard (Mark duplicates) (Broad Institute, 2019) and samTools — Rmdup (Li *et al.*, 2009).

GATK (McKenna *et al.*, 2010) is an important tool for variant calling used by researchers on RNA-seq data. This tool needs additional preprocessing steps before variant calling. The steps are base quality score recalibration (BQSR) and local realignment. BQSR is for adjusting the base quality scores of reads. This step will help in removing sequencer-specific biases and improve the accuracy of variant calls. Local realignment around Indels aids in reducing alignment errors and minimizing false positives in variant calling. Varscan2 (Koboldt *et al.*, 2012) is another tool for detecting SNPs/Indels from variant calling. It takes input in mpileup format, which has a sorted BAM file that is suitable to perform variant calling. Varscan considers only high-quality reads (base quality ≥20) for performing variant calling without false positives. Other variant callers used for variant identification are SOAPSNP (Li *et al.*, 2009), Atlas-SNP2 (Shen *et al.*, 2010), and Illumina CASAVA (Cao *et al.*, 2022). Variant calling step returns a VCF file (Cingolani *et al.* 2018) that contains all the information to describe variants. Annotation tools such as ANNOVAR (Wang *et al.*, 2010) and Ensembl variant effect predictor (VEP) (McLaren *et al.*, 2016) are used to annotate SNPs based on the location of the gene and the coding frame. VEP gives a detailed report on the effect of variants on regulatory regions, transcripts, and proteins, along with the annotations. The annotated variants are further considered for predicting their effect on different regulatory processes such as histone acetylation, methylation, expression, splicing, and binding of transcription factor/miRNA. Detection of gene fusion events (Nicorici *et al.*, 2014; Davidson *et al.*, 2015; Weirather *et al.*, 2015; Haas *et al.*, 2017) using RNA sequencing data is carried out using multiple tools based on different strategies/algorithms.

14.3.4 *Practices/guidelines and tools for variant calling using single-cell RNA-seq data*

Single-cell transcriptomics is pivotal for analyzing gene expression patterns at the single-cell level. Various sequencing methods are employed to generate data. The 10X method uses unique barcodes to label individual cells, enabling simultaneous sequencing of thousands in one experiment. Dropseq utilizes droplet-based microfluidics for efficient single-cell analysis. Bioparse specializes in precise and comprehensive transcriptomic profiling, particularly for rare cell types, while Smart-Seq captures full-length transcripts from single cells with high sensitivity.

These sequencing methods are integral for generating large-scale datasets, which undergo stringent quality checks and batch correction to mitigate technical artifacts. After data integration, the next step involves identifying biologically meaningful cell types through hierarchical and graph-based clustering methods. For instance, neurons can be further categorized into excitatory and inhibitory cell types, followed by the identification of their subtypes. The ultimate goal is to pinpoint these subtypes. Subsequent analyses encompass differential gene expression and co-expression network analysis. Furthermore, GWAS variants can be associated with single-nuclei data to predict the effects of susceptible variants at the single-cell resolution using the sc-linker method (Vivian Li & Li, 2021).

The sc-linker framework is instrumental in integrating scRNA-seq data, epigenomic SNP-to-gene maps, and GWAS (GWAS) summary statistics. sc-linker operates in three key steps: (i) Construct gene programs: Utilizing scRNA-seq data, sc-linker builds gene programs (continuous-valued gene sets) characterizing individual cell types, disease-dependent states, or cellular processes. These continuous values represent the likelihood that a cell belongs to a specific cell type or state or expresses a particular cellular process, (ii) Enrich gene programs with GWAS variants: ScLink enhances gene programs with GWAS variants associated with disease by determining the overlap between genes in each gene program and those containing GWAS variants, and (iii) Identify disease-critical cell types and processes: sc-linker pinpoints disease-critical cell types and processes by identifying gene programs enriched with GWAS variants, which are likely crucial for disease development.

sc-linker has been employed successfully to identify disease-critical cell types and processes in various conditions, including major depressive disorder, ulcerative colitis, and multiple sclerosis. The outcomes have provided insights into how genetic variants contribute to diseases and potential new therapeutic targets.

14.3.5 *Practices/guidelines and tools for variant calling using spatial transcriptomic data*

Spatial transcriptomics precisely determines cell spatial distribution within tissues. Visium and NanoString provide spatial context to gene expression data, allowing for gene mapping within tissue locations. For example, the human brain cortex, comprising six distinct layers, can be studied for cell presence and spatial distribution. Identifying variants affecting specific cell types can also help determine the layer where these cell types are located. Integrating single-cell transcriptomics and GWAS variants offers insights at both cellular and spatial resolution levels. Spatial eQTL mapping, a type of eQTL mapping, utilizes spatial transcriptomics data to identify eQTLs in various cell types and tissue regions. Spatial transcriptome sequencing (STS) and spRNA-seq measure gene expression in individual cells and map their locations within tissue samples. Researchers generate spatial transcriptomics data and employ statistical methods to identify eQTLs. This approach helps understand how genetic variants regulate gene expression in different cell types and tissue regions and identifies potential drug targets.

GWAS-informed spatial transcriptomics (GWAS-ST) uses GWAS summary statistics to identify cell types and tissue regions where disease-associated genetic variants are active. By integrating GWAS summary statistics with spatial transcriptomics data, GWAS-ST unveils the most affected cell types and tissue regions.

14.3.5.1 *Tools for spatial eQTL mapping and GWAS-ST*

SpatialDE is an open-source software package designed for analyzing spatial transcriptomics data, providing tools for spatial eQTL mapping, cell type identification, and result visualization. SpatialGWAS is a tool facilitating GWAS-ST by integrating GWAS summary statistics with spatial transcriptomics data to identify active cell types and tissue regions for disease-associated genetic variants. Additionally, the Spatial Transcriptomics Resource Center offers resources such as tutorials, protocols, and datasets for researchers interested in exploring these methods further.

14.4 Interpretation of pathogenic variants

Variants occurring frequently in a genome help to identify disease-associated markers/genes through an association study. GWAS is a genome-wide study performed on various diseases across genomes. GWAS steps include data (genotypic/phenotypic information) collection, quality control, imputation, and association testing. GWAS data are available in public repositories such as GWAS Central (Beck *et al.*, 2020), GRASP database (Leslie *et al.*, 2014), and the GWAS Catalog (Sollis *et al.*, 2023) as well as in the literature. Although GWAS aids in identifying novel variant-trait associations, it sometimes misses the causal variants associated with the disease. To reveal disease-associated variants and their impact, it is necessary to carry out downstream analysis to find their effects in regulatory elements. The variant effect is analyzed using different methods and tools.

Variant effects on regulatory mechanisms are histone modification, DNA methylation, chromatin interaction, transcription factor binding, miRNA binding, and posttranscriptional modifications, which can be studied using *in silico* tools to predict their effects and correlate with disease association.

14.4.1 *Variant effect on transcription factor binding*

Most of the disease-associated variants are located in noncoding regions of the genome. The variant effect on transcription factor binding can be predicted using SNP2TFBS (Logsdon *et al.*, 2020), FABIAN variant (Steinhaus *et al.*, 2022), haploreg (Ward & Kellis, 2012), rSNPBASE (Guo *et al.*, 2014), and motifbreakR (Coetzee *et al.*, 2015). These tools predict the transcription factor binding motif and then assess the variant's effect on this motif using a Position Specific Scoring Matrix (PSSM). Variants in the transcription binding motif can disrupt the binding affinity between the transcription factor and its target sequence, potentially leading to altered gene expression patterns. This alteration may result in dysregulation of downstream biological processes and contribute to the development of various phenotypic traits.

14.4.2 *Epigenetic alterations of variants*

SNPs that involve various epigenetic modifications such as histone acetylation (hQTL), DNA methylation (mQTL), gene expression alteration (eQTL), splicing pattern (sQTL), miRNA expression (miQTL), and apaQTL alternative polyadenylation changes are crucial for associating disease variants with gene regulatory mechanisms. There are databases and tools for studying these variants and their effects (Wright *et al.*, 2012). QTLBase is a repository that contains information about variants of various QTL studies published in the literature (both disease-associated and control variants).

14.4.3 *Impact of variants on miRNA binding*

Variants in the miRNA binding region may lead to miRNA-mediated gene regulation and, in turn, cause complex genetic disorders/diseases miRNASNP V3 is a repository with experimentally validated variants that alter miRNA binding. PolymiRTS (Bhattacharya *et al.*, 2014), a database of variants in experimentally valid seed regions and target sites, is used to verify the variants that disrupt the binding of miRNA. miRDB (Chen & Wang, 2020) is a tool for miRNA target prediction. These databases and tools provide insights into the experimentally validated miRNA regions and the variant information that alter the seed regions and their binding. miRNA-target gene interactions are crucial to understand the regulatory networks controlled by miRNAs. miRNAs and their targets are predicted using computational approaches and also experimentally validated to confirm their interaction. Computational methods to predict the miRNA-gene target interactions include: (i) structural accessibility of targets: the necessary free energy needed for accessing the binding site is used to rank the targets, (ii) thermodynamic properties of miRNA–mRNA complexes: Low free energy indicates that the interaction between the miRNA and mRNA is more stable, (iii) seed match: miRNA seed sequence (initial 2–8 nucleotides in the 5′ region) with complementary base pairing is used to predict the targets, and (iv) evolutionary conservation: conservation of target sites for predicting potential miRNA–mRNA interactions.

14.4.4 *Predicting disease association of variants*

Machine learning methods/tools are used to find accurate disease-causing mutations and discriminate them with neutral mutations (Kulandaisamy *et al.*, 2020; Rangaswamy *et al.*, 2020; Kulandaisamy *et al.*, 2022). These methods are specific to AD and are used to predict disease-causing mutations. Alz-disc is used to classify the disease-causing mutations in Alzheimer's based on the sequence information. It used physicochemical properties, substitution matrices, evolutionary details, and neighboring residues for predicting the pathogenic variants. VEPAD is a method that uses features such as dinucleotide properties, conservation scores, and nucleotide composition-based features to predict the function-based effects of SNPs in AD. Both coding and noncoding variants of AD were considered for this study, and this tool can predict the deleterious variants and their effects. There are several other tools (Hou & Ma, 2014; Desvignes *et al.*, 2018; Mukherjee *et al.*, 2019; Oscanoa *et al.*, 2020; Zhou *et al.*, 2023) to validate or rank the variants based on their role/functional annotation.

14.5 Case study of variant-associated genes and their impact

There are various studies on variant calling and downstream analysis of identified variants to elucidate their role in various gene-regulating mechanisms. Here, we discuss RNA-seq analysis on two brain regions: the prefrontal cortex (BA9) (Sneha *et al.*, 2022) and motor cortex (BA4) (Sneha *et al.*, 2023) of HD samples and controls. The variant calling and the subsequent steps followed for variant effects on different biological/molecular functions are shown in the **Figure 14.6**.

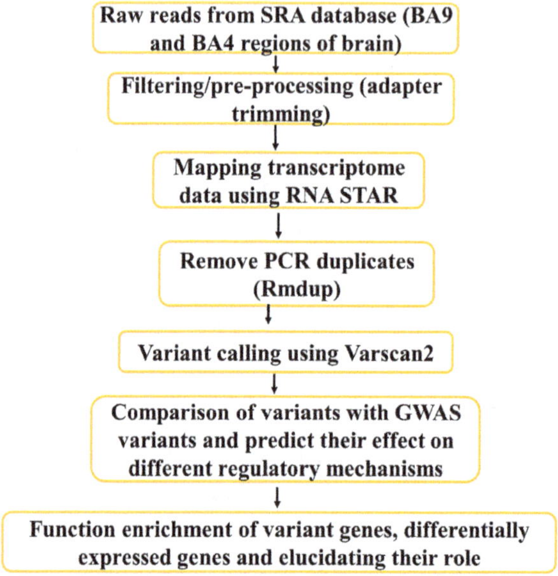

Figure 14.6 Variant analysis using RNAseq data from cortex samples of the brain (figure was taken from Sneha *et al.*, 2023)

The RNA-seq data obtained from human postmortem samples is preprocessed and subjected to alignment with the latest reference genome, and variant calling is performed using Varscan2. The identified variants are annotated and filtered based on the comparison with reported GWAS variants from various neurodegenerative disorders. The variants are filtered as novel and reported based on their association with the GWAS study. These variants and their associated genes are filtered using *in silico* tools to find their effect on transcription factors, miRNA binding, and epigenetic modifications as shown in **Figure 14.7**. Function interaction network is also constructed using information on the differential expression patterns of variant genes and transcription factors (differentially expressed in HD). Variant-associated genes and their respective effects on transcription factor binding, miRNA binding, and quantitative trait loci

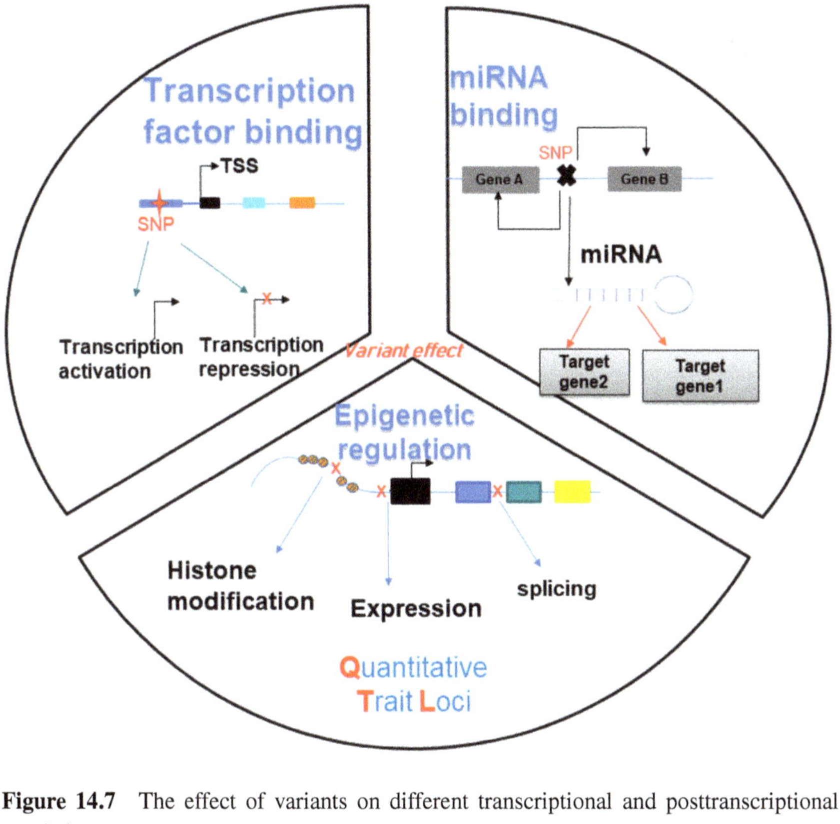

Figure 14.7 The effect of variants on different transcriptional and posttranscriptional regulations

Table 14.1 Variant effect prediction in BA9 and BA4 samples.

STUDY	CHR	REF/ ALT	RSID	GENE	TF (EXPRESSION)	miRNA (expression)	miRNA target (EXPRESSION)	QTL	GWAS
BA9	Chr 5	T/A	rs11167469	CSNK1A1 (DOWN)	Prrx2 (UP)	miR-374a (UP)	CD177, CAPN6, HOXA10, FOXD2, RGS1, SLC26A2, MS4A4A (DOWN)	eQTL	PD
BA9	Chr3	G/T	rs3796308	TMEM43 (DOWN)	FOXD1 (DOWN)	miR-30a (UP)	HOXA13, CLEC12A, MRC1, SIX1, DSP, SCAARA5, SPTLC3, RGS1 (DOWN)	eQTL, pQTL	AD, PD
BA4	Chr5	T/G	rs73781086	SEMA6A (DOWN)	EGR1 (UP)	miR-124 (UP)	GNAI3, ITGB1, PTBP1, KLF6, MTDH (DOWN)	NA	Novel
BA4	Chr5	G/A	rs2059874	SEMA6A (DOWN)	EBF1 (UP)	miR-124 (UP)	GNAI3, ITGB1, PTBP1, KLF6, MTDH (DOWN)	eQTL	AD, PD, AMD
BA4	Chr14	C/T	rs1125415	KLC1 (DOWN)	IRF1(UP)	miR-34a-5p (UP)	GAS1, PDGFR, ERBB2, SRC (DONW)	eQTL, sQTL, mQTL	AD, AMD
BA4	Chr8	G/A	rs7834598	DLC1(UP)	ARID3A(DOWN)	miR-146a-5p (DOWN)	RAC1, SOX2, BCLAF1, ERBB4 (UP)	eQTL	AMD

eQTL, expression QTL; pQTL, protein QTL; sQTL, splicing QTL; hQTL, histone modification QTL; mQTL, methylation QTL; AMD, age-related macular degeneration. Data are obtained from Sneha *et al.* (2022, 2023).

using *in silico* tools on human postmortem samples of brain regions (Sneha *et al.*, 2022, 2023) are shown in **Table 14.1**. Studies have demonstrated that the aforementioned events collectively contribute to neurodegenerative processes by disrupting gene expression networks and cellular homeostasis (Dharshini *et al.*, 2020, 2021).

14.6 Conclusion

Analyzing genetic variants and exploring their disease association are important for devising disease-modifying therapies. In this chapter, we briefly discussed various variants responsible for neurodegenerative diseases. Further, sequencing techniques have been reviewed along with their applications. We also elaborated on various tools and pipelines for variant effect prediction using different sequencing data, as well as, benefits of different sequencing techniques for their role in variant calling. A case study on RNA-seq data from HD samples provides insights on identifying variant-associated genes and their functional impacts.

Acknowledgments

We thank the Bioinformatics facility, the Department of Biotechnology, and the Indian Institute of Technology Madras for computational facilities. Nela Pragathi Sneha thanks the Ministry of Human Resource Development (MHRD), India, for providing a research fellowship.

References

Ali, M. M., Li, F., Zhang, Z., Zhang, K., Kang, D. K., Ankrum, J. A., Le, X. C., & Zhao, W. (2014). Rolling circle amplification: A versatile tool for chemical biology, materials science and medicine. *Chemical Society Reviews*, *43*(10), 3324–3341.

Amarasinghe, K. C., Li, J., & Halgamuge, S. K. (2013). CoNVEX: Copy number variation estimation in exome sequencing data using HMM. *BMC Bioinformatics*, *14*, 1–9.

Ansorge, W. J. (2009). Next-generation DNA sequencing techniques. *New Biotechnology*, *25*(4), 195–203.

Arrasate, M., & Finkbeiner, S. (2012). Protein aggregates in Huntington's disease. *Experimental Neurology*, *238*(1), 1–11.

Beck, T., Shorter, T., & Brookes, A. J. (2020). GWAS Central: A comprehensive resource for the discovery and comparison of genotype and phenotype data from genome-wide association studies. *Nucleic Acids Research*, *48*(D1), D933–940.

Bhattacharya, A., Ziebarth, J. D., & Cui, Y. (2014). PolymiRTS Database 3.0: Linking polymorphisms in microRNAs and their target sites with human diseases and biological pathways. *Nucleic Acids Research*, *42*(D1), D86–D91.

Bolger, A. M., Lohse, M., & Usadel, B. (2014). Trimmomatic: A flexible trimmer for Illumina sequence data. *Bioinformatics*, *30*(15), 2114–2120.

Bray, N. L., Pimentel, H., Melsted, P., & Pachter, L. (2016). Near-optimal probabilistic RNA-seq quantification. *Nature Biotechnology*, *34*(5), 525–527.

Broad Institute. (2019). *Picard toolkit*. Broad Institute, GitHub repository.

Caldeweyher, E. (2021). Kallisto: A command-line interface to simplify computational modelling and the generation of atomic features. *Journal of Open Source Software*, *6*(60), 3050.

Cameron, D. L., Schröder, J., Penington, J. S., Do, H., Molania, R., Dobrovic, A., Speed, T. P., & Papenfuss, A. T. (2017). GRIDSS: Sensitive and specific genomic rearrangement detection using positional de Bruijn graph assembly. *Genome Research*, *27*(12), 2050–2060.

Cao, Z., Huang, Y., Duan, R., Jin, P., Qin, Z. S., & Zhang, S. (2022). Disease category-specific annotation of variants using an ensemble learning framework. *Briefings in Bioinformatics*, *23*(1), bbab438.

Chen, K., Wallis, J. W., McLellan, M. D., Larson, D. E., Kalicki, J. M., Pohl, C. S., McGrath, S. D., Wendl, M. C., Zhang, Q., Locke, D. P., Shi, X., Fulton, R. S., Ley, T. J., Wilson, R. K., Ding, L., & Mardis, E. R. (2009). BreakDancer: An algorithm for high-resolution mapping of genomic structural variation. *Nature Methods*, *6*(9), 677–681.

Chen, X., Schulz-Trieglaff, O., Shaw, R., Barnes, B., Schlesinger, F., Källberg, M., Cox, A. J., Kruglyak, S., & Saunders, C. T. (2016). Manta: Rapid detection of structural variants and indels for germline and cancer sequencing application. *Bioinformatics*, *32*(8), 1220–1228.

Chen, Y., & Wang, X. (2020). MiRDB: An online database for prediction of functional microRNA targets. *Nucleic Acids Research*, *48*(D1), D127–D131.

Cingolani, P., Cunningham, F., Mclaren, W., & Wang, K. (2018). Variant annotations in VCF format. [cited 2023 July 31] VCFannotationformat_v1. 0. pdf. (n.d.).

Coetzee, S. G., Coetzee, G. A., & Hazelett, D. J. (2015). MotifbreakR: An R/ Bioconductor package for predicting variant effects at transcription factor binding sites. *Bioinformatics*, *31*(23), 3847–3849.

Davidson, N. M., Majewski, I. J., & Oshlack, A. (2015). JAFFA: High sensitivity transcriptome-focused fusion gene detection. *Genome Medicine*, *7*(1), 1–12.

De Boo, G. M., Tibben, A., Lanser, J. B. K., Jennekens-Schinkel, A., Hermans, J., Maat-Kievit, A., & Roos, R. A. C. (1997). Early cognitive and motor symptoms in identified carriers of the gene for Huntington disease. *Archives of Neurology*, *54*(11), 1353–1357.

Desvignes, J. P., Bartoli, M., Delague, V., Krahn, M., Miltgen, M., Béroud, C., & Salgado, D. (2018). VarAFT: A variant annotation and filtration system for human next generation sequencing data. *Nucleic Acids Research*, *46*(W1), W545–W553.

Dharshini, S. A. P., Jemimah, S., Taguchi, Y. H., & Gromiha, M. M. (2021). Exploring common therapeutic targets for neurodegenerative disorders using transcriptome study. *Frontiers in Genetics, 12*, 639160.

Dharshini, S. A. P., Taguchi, Y. H., & Gromiha, M. M. (2019). Investigating the energy crisis in Alzheimer disease using transcriptome study. *Scientific Reports, 9*(1), 18509.

Dharshini, S. A. P., Taguchi, Y. H., & Gromiha, M. M. (2020). Identifying suitable tools for variant detection and differential gene expression using RNA-seq data. *Genomics, 112*(3), 2166–2172.

Dobin, A., & Gingeras, T. R. (2015). Mapping RNA-seq reads with STAR. *Current Protocols in Bioinformatics*, *51*(1), 11–14.

Ebbert, M. T. W., Wadsworth, M. E., Staley, L. A., Hoyt, K. L., Pickett, B., Miller, J., Duce, J., Kauwe, J. S. K., & Ridge, P. G. (2016). Evaluating the necessity of PCR duplicate removal from next-generation sequencing data and a comparison of approaches. *BMC Bioinformatics*, *17*, 491–500.

Fromer, M., Moran, J. L., Chambert, K., Banks, E., Bergen, S. E., Ruderfer, D. M., Handsaker, R. E., McCarroll, S. A., O'Donovan, M. C., Owen, M. J., Kirov, G., Sullivan, P. F., Hultman, C. M., Sklar, P., & Purcell, S. M. (2012). Discovery and statistical genotyping of copy-number variation from whole-exome sequencing depth. *American Journal of Human Genetics*, *91*(4), 597–607.

Gordon, A., Hannon, G. J., & Gordon. (2014). FASTX-Toolkit. In [Online] http://hannonlab.cshl.edu/fastx_toolkit accessed on 09 July 2023.

Grabherr, M. G., Haas, B. J., Yassour, M., Levin, J. Z., Thompson, D. A., Amit, I., Adiconis, X., Fan, L., Raychowdhury, R., Zeng, Q., Chen, Z., Mauceli, E., Hacohen, N., Gnirke, A., Rhind, N., Di Palma, F., Birren, B. W., Nusbaum, C.,

Lindblad-Toh, K., … Regev, A. (2011). Trinity: reconstructing a full-length transcriptome without a genome from RNA-Seq data HHS public access. *Nature Biotechnology, 29*(7), 644.

Guo, L., Du, Y., Chang, S., Zhang, K., & Wang, J. (2014). RSNPBase: A database for curated regulatory SNPs. *Nucleic Acids Research, 42*(D1), D1033–1039.

Haas, B., Dobin, A., Stransky, N., Li, B., Yang, X., Tickle, T., Bankapur, A., Ganote, C., Doak, T., & Pochet, N. (2017). STAR-Fusion: Fast and accurate fusion transcript detection from RNA-Seq. *BioRxiv*, 12095.

Hardiman, O., Al-Chalabi, A., Chio, A., Corr, E. M., Logroscino, G., Robberecht, W., Shaw, P. J., Simmons, Z., & Van Den Berg, L. H. (2017). Amyotrophic lateral sclerosis. *Nature Reviews Disease Primers, 3*, 1–19.

Hou, J. P., & Ma, J. (2014). DawnRank: Discovering personalized driver genes in cancer. *Genome Medicine, 6*(7), 1–16.

Koboldt, D. C. (2020). Best practices for variant calling in clinical sequencing. *Genome Medicine, 12*(1), 1–13.

Koboldt, D. C., Zhang, Q., Larson, D. E., Shen, D., McLellan, M. D., Lin, L., Miller, C. A., Mardis, E. R., Ding, L., & Wilson, R. K. (2012). VarScan 2: Somatic mutation and copy number alteration discovery in cancer by exome sequencing. *Genome Research, 22*(3), 568–576.

Kolodziejczyk, A. A., Kim, J. K., Svensson, V., Marioni, J. C., & Teichmann, S. A. (2015). The Technology and biology of single-Cell RNA sequencing. *Molecular Cell, 58*(4), 610–640.

Kulandaisamy, A., Parvathy Dharshini, S. A., & Gromiha, M. M. (2022). Alz-Disc: A tool to discriminate disease-causing and neutral mutations in Alzheimer's Disease. *Combinatorial Chemistry & High Throughput Screening, 26*(4), 769–777.

Kulandaisamy, A., Zaucha, J., Sakthivel, R., Frishman, D. and Gromiha, M.M. (2020) Pred-MutHTP: Prediction of disease-causing and neutral mutations in human transmembrane proteins. *Human Mutation.* 41, 581–590.

Langmead, B., & Salzberg, S. L. (2012). Fast gapped-read alignment with Bowtie 2. *Nature Methods, 9*(4), 357–359.

Larsen, S. B., Hanss, Z., & Krüger, R. (2018). The genetic architecture of mitochondrial dysfunction in Parkinson's disease. *Cell and Tissue Research, 373*, 21–37.

Layer, R. M., Chiang, C., Quinlan, A. R., & Hall, I. M. (2014). LUMPY: A probabilistic framework for structural variant discovery. *Genome Biology, 15*(6), 1–19.

Leslie, R., O'Donnell, C. J., & Johnson, A. D. (2014). GRASP: Analysis of genotype-phenotype results from 1390 genome-wide association studies and corresponding open access database. *Bioinformatics*, *30*(12), i185–194.

Li, B., & Dewey, C. N. (2011). RSEM: Accurate transcript quantification from RNA-Seq data with or without a reference genome. *BMC Bioinformatics*, *12*, 1–16.

Li, H., Handsaker, B., Wysoker, A., Fennell, T., Ruan, J., Homer, N., Marth, G., Abecasis, G., & Durbin, R. (2009). The sequence alignment/Map format and SAMtools. *Bioinformatics*, *25*(16), 2078–2079.

Li, R., Li, Y., Fang, X., Yang, H., Wang, J., Kristiansen, K., & Wang, J. (2009). SNP detection for massively parallel whole-genome resequencing. *Genome Research*, *19*(6), 1124–1132.

Life Technologies. (2023, July 7). Application note: Transcriptome sequencing using the Ion Proton System [Internet]. http://tools.thermofisher.com/content/sfs/brochures/Transcriptome-Seq-App-Note.pdf. (n.d.).

Liu, J., Xu, T., Jin, Y., Huang, B., & Zhang, Y. (2021). Progress and clinical application of single-cell transcriptional sequencing technology in cancer research. *Frontiers in Oncology*, *10*, 593085.

Logsdon, G. A., Vollger, M. R., & Eichler, E. E. (2020). Long-read human genome sequencing and its applications. *Nature Reviews Genetics*, *21*(10), 597–614.

MacDonald, M. E., Ambrose, C. M., Duyao, M. P., Myers, R. H., Lin, C., Srinidhi, L., Barnes, G., Taylor, S. A., James, M., Groot, N., MacFarlane, H., Jenkins, B., Anderson, M. A., Wexler, N. S., Gusella, J. F., Bates, G. P., Baxendale, S., Hummerich, H., Kirby, S., . . . Harper, P. S. (1993). A novel gene containing a trinucleotide repeat that is expanded and unstable on Huntington's disease chromosomes. *Cell*, *72*(6), 971–983.

Maxam, A. M., & Gilbert, W. (1977). A new method for sequencing DNA. *Proceedings of the National Academy of Sciences of the United States of America*, *74*(2), 560–564.

McKenna, A., Hanna, M., Banks, E., Sivachenko, A., Cibulskis, K., Kernytsky, A., Garimella, K., Altshuler, D., Gabriel, S., Daly, M., & DePristo, M. A. (2010). The genome analysis toolkit: A MapReduce framework for analyzing next-generation DNA sequencing data. *Genome Research*, *20*(9), 1297–1303.

McLaren, W., Gil, L., Hunt, S. E., Riat, H. S., Ritchie, G. R. S., Thormann, A., Flicek, P., & Cunningham, F. (2016). The Ensembl variant effect predictor. *Genome Biology*, *17*(1), 1–14.

Mejzini, R., Flynn, L. L., Pitout, I. L., Fletcher, S., Wilton, S. D., & Akkari, P. A. (2019). ALS Genetics, mechanisms, and therapeutics: Where are we now? *Frontiers in Neuroscience*, *13*, 1310.

Mertens, F., Johansson, B., Fioretos, T., & Mitelman, F. (2015). The emerging complexity of gene fusions in cancer. *Nature Reviews Cancer*, *15*(6), 371–381.

Mias, G., & Mias, G. (2018). Databases: E-utilities and UCSC genome browser. *Mathematica for Bioinformatics: A Wolfram Language Approach to Omics*, 133–170.

Mukherjee, S., Perumal, T. M., Daily, K., Sieberts, S. K., Omberg, L., Preuss, C., Carter, G. W., Mangravite, L. M., & Logsdon, B. A. (2019). Identifying and ranking potential driver genes of Alzheimer's disease using multiview evidence aggregation. *Bioinformatics*, *35*(14), i568–576.

Nicorici, D., Satalan, M., Edgren, H., Kangaspeska, S., Murumagi, A., Kallioniemi, O., Virtanen, S., & Kilkku, O. (2014). FusionCatcher — a tool for finding somatic fusion genes in paired-end RNA-sequencing data. *bioRxiv, 011650.*

Nyrén, P., Pettersson, B., & Uhlén, M. (1993). Solid phase DNA minisequencing by an enzymatic luminometric inorganic pyrophosphate detection assay. *Analytical Biochemistry*, *208*(1), 171–175.

Oscanoa, J., Sivapalan, L., Gadaleta, E., Dayem Ullah, A. Z., Lemoine, N. R., & Chelala, C. (2020). SNPnexus: A web server for functional annotation of human genome sequence variation (2020 update). *Nucleic Acids Research*, *48*(W1), W185–W192.

Patel, R. K., & Jain, M. (2012). NGS QC toolkit: A toolkit for quality control of next generation sequencing data. *PLoS One*, *7*(2), e30619.

Patro, R., Mount, S. M., & Kingsford, C. (2014). Sailfish enables alignment-free isoform quantification from RNA-seq reads using lightweight algorithms. *Nature Biotechnology*, *32*(5), 462–464.

Plagnol, V., Curtis, J., Epstein, M., Mok, K. Y., Stebbings, E., Grigoriadou, S., Wood, N. W., Hambleton, S., Burns, S. O., Thrasher, A. J., Kumararatne, D., Doffinger, R., & Nejentsev, S. (2012). A robust model for read count data in exome sequencing experiments and implications for copy number variant calling. *Bioinformatics*, *28*(21), 2747–2754.

Qin, L., Wang, J., Tian, X., Yu, H., Truong, C., Mitchell, J. J., Wierenga, K. J., Craigen, W. J., Zhang, V. W., & Wong, L. J. C. (2016). Detection and quantification of mosaic mutations in disease genes by next-generation sequencing. *Journal of Molecular Diagnostics*, *18*(3), 446–453.

Rabbani, B., Tekin, M., & Mahdieh, N. (2014). The promise of whole-exome sequencing in medical genetics. *Journal of Human Genetics, 59*(1), 5–15.

Ragagnin, A. M. G., Shadfar, S., Vidal, M., Jamali, M. S., & Atkin, J. D. (2019). Motor neuron susceptibility in ALS/FTD. *Frontiers in Neuroscience, 13*, 532.

Rangaswamy, U., Dharshini, S. A. P., Yesudhas, D., & Gromiha, M. M. (2020). VEPAD — Predicting the effect of variants associated with Alzheimer's disease using machine learning. *Computers in Biology and Medicine, 124*, 103933.

Rausch, T., Zichner, T., Schlattl, A., Stütz, A. M., Benes, V., & Korbel, J. O. (2012). DELLY: Structural variant discovery by integrated paired-end and split-read analysis. *Bioinformatics, 28*(18), i333–i339.

Roberts, A., & Pachter, L. (2013). Streaming fragment assignment for real-time analysis of sequencing experiments. *Nature Methods, 10*(1), 71–73.

Robertson, G., Schein, J., Chiu, R., Corbett, R., Field, M., Jackman, S. D., Mungall, K., Lee, S., Okada, H. M., Qian, J. Q., Griffith, M., Raymond, A., Thiessen, N., Cezard, T., Butterfield, Y. S., Newsome, R., Chan, S. K., She, R., Varhol, R., ... Birol, I. (2010). De novo assembly and analysis of RNA-seq data. *Nature Methods, 7*(11), 909–912.

Sanger, F., Nicklen, S., & Coulson, A. R. (1977). DNA sequencing with chain-terminating inhibitors. *Proceedings of the National Academy of Sciences of the United States of America, 74*(12), 5463–5467.

Sathirapongsasuti, J. F., Lee, H., Horst, B. A. J., Brunner, G., Cochran, A. J., Binder, S., Quackenbush, J., & Nelson, S. F. (2011). Exome sequencing-based copy-number variation and loss of heterozygosity detection: ExomeCNV. *Bioinformatics, 27*(19), 2648–2654.

Schneider, V., & Church, D. (2013). *Genome reference consortium*. The NCBI Handbook.

Shen, Y., Wan, Z., Coarfa, C., Drabek, R., Chen, L., Ostrowski, E. A., Liu, Y., Weinstock, G. M., Wheeler, D. A., Gibbs, R. A., & Yu, F. (2010). A SNP discovery method to assess variant allele probability from next-generation resequencing data. *Genome Research, 20*(2), 273–280.

Sneha, N. P., Dharshini, S. A. P., Taguchi, Y. H., & Gromiha, M. M. (2022). Integrative meta-analysis of Huntington's disease transcriptome landscape. *Genes, 13*(12), 2385.

Sneha, N. P., Dharshini, S. A. P., Taguchi, Y. H., & Gromiha, M. M. (2023). Investigating neuron degeneration in Huntington's disease using RNA-Seq based transcriptome study. *Genes, 14*(9), 1801.

Sollis, E., Mosaku, A., Abid, A., Buniello, A., Cerezo, M., Gil, L., Groza, T., Güneş, O., Hall, P., Hayhurst, J., Ibrahim, A., Ji, Y., John, S., Lewis, E., Macarthur, J. A. L., Mcmahon, A., Osumi-Sutherland, D., Panoutsopoulou, K., Pendlington, Z., ... Harris, L. W. (2023). The NHGRI-EBI GWAS Catalog: knowledgebase and deposition resource. *Nucleic Acids Research*, *51*(1 D) D9777–D985.

Ståhl, P. L., Salmén, F., Vickovic, S., Lundmark, A., Navarro, J. F., Magnusson, J., Giacomello, S., Asp, M., Westholm, J. O., Huss, M., Mollbrink, A., Linnarsson, S., Codeluppi, S., Borg, Å., Pontén, F., Costea, P. I., Sahlén, P., Mulder, J., Bergmann, O., … Frisén, J. (2016). Visualization and analysis of gene expression in tissue sections by spatial transcriptomics. *Science, 353*(6294), 78–82.

Steinhaus, R., Robinson, P. N., & Seelow, D. (2022). FABIAN-variant: predicting the effects of DNA variants on transcription factor binding. *Nucleic Acids Research*, *50*(W1), W322–329.

SRA Toolkit. (2023, August 01). Nation of National Center for Biotechnology.

The Genome Reference consortium . https://www.ncbi.nlm.nih.gov/grc [accessed on July 26].

Thind, A. S., Monga, I., Thakur, P. K., Kumari, P., Dindhoria, K., Krzak, M., Ranson, M., & Ashford, B. (2021). Demystifying emerging bulk RNA-Seq applications: The application and utility of bioinformatic methodology. *Briefings in Bioinformatics*, *22*(6), bbab259.

Tipton, P. W., Deutschlaender, A. B., Savica, R., Heckman, M. G., Brushaber, D. E., Dickerson, B. C., Gavrilova, R. H., Geschwind, D. H., Ghoshal, N., Graff-Radford, J., Graff-Radford, N. R., Grossman, M., Hsiung, G. Y. R., Huey, E. D., Irwin, D. J., Jones, D. T., Knopman, D. S., Mcginnis, S. M., Rademakers, R., ... Wszolek, Z. K. (2022). Differences in motor features of C9orf72, MAPT, or GRN variant carriers with familial frontotemporal lobar degeneration. *Neurology*, *99*(11), e1154–e116.

Trapnell, C., Pachter, L., & Salzberg, S. L. (2009). TopHat: Discovering splice junctions with RNA-Seq. *Bioinformatics*, *25*(9), 1105–1111.

Trapnell, C., Roberts, A., Goff, L., Pertea, G., Kim, D., Kelley, D. R., Pimentel, H., Salzberg, S. L., Rinn, J. L., & Pachter, L. (2012). Differential gene and transcript expression analysis of RNA-seq experiments with TopHat and Cufflinks. *Nature Protocols*, *7*(3), 562–578.

Tsai, R. M., & Boxer, A. L. (2016). Therapy and clinical trials in frontotemporal dementia: Past, present, and future. *Journal of Neurochemistry, 138*, 211–221.

Van El, C. G., Cornel, M. C., Borry, P., Hastings, R. J., Fellmann, F., Hodgson, S. V., Howard, H. C., Cambon-Thomsen, A., Knoppers, B. M., Meijers-Heijboer, H., Scheffer, H., Tranebjaerg, L., Dondorp, W., & De Wert, G. M. W. R. (2013). Whole-genome sequencing in health care. *European Journal of Human Genetics*, *21*(6), 580–584.

Vivian Li, W., & Li, Y. (2021). scLink: Inferring sparse gene co-expression networks from single-cell expression data. *Genomics, Proteomics and Bioinformatics*, *19*(3), 475–492.

Wang, K., Li, M., & Hakonarson, H. (2010). ANNOVAR: Functional annotation of genetic variants from high-throughput sequencing data. *Nucleic Acids Research*, *38*(16), e164–e164.

Ward, L. D., & Kellis, M. (2012). HaploReg: A resource for exploring chromatin states, conservation, and regulatory motif alterations within sets of genetically linked variants. *Nucleic Acids Research*, *40*(D1), D930–D934.

Weirather, J. L., Afshar, P. T., Clark, T. A., Tseng, E., Powers, L. S., Underwood, J. G., Zabner, J., Korlach, J., Wong, W. H., & Au, K. F. (2015). Characterization of fusion genes and the significantly expressed fusion isoforms in breast cancer by hybrid sequencing. *Nucleic Acids Research*, *43*(18), e116–e116.

Williams, C. G., Lee, H. J., Asatsuma, T., Vento-Tormo, R., & Haque, A. (2022). An introduction to spatial transcriptomics for biomedical research. *Genome Medicine, 14*(1), 1–18.

Wong, K., Keane, T. M., Stalker, J., & Adams, D. J. (2010). Enhanced structural variant and breakpoint detection using SVMerge by integration of multiple detection methods and local assembly. *Genome Biology*, *11*(12), 1–9.

Wright, F. A., Shabalin, A. A., & Rusyn, I. (2012). Computational tools for discovery and interpretation of expression quantitative trait loci. *Pharmacogenomics, 13*(3), 343–352.

Xie, Y., Wu, G., Tang, J., Luo, R., Patterson, J., Liu, S., Huang, W., He, G., Gu, S., Li, S., Zhou, X., Lam, T. W., Li, Y., Xu, X., Wong, G. K. S., & Wang, J. (2014). SOAPdenovo-Trans: De novo transcriptome assembly with short RNA-Seq reads. *Bioinformatics*, *30*(12), 1660–1666.

Ye, K., Schulz, M. H., Long, Q., Apweiler, R., & Ning, Z. (2009). Pindel: A pattern growth approach to detect break points of large deletions and medium sized insertions from paired-end short reads. *Bioinformatics*, *25*(21), 2865–2871.

Young, J. J., Lavakumar, M., Tampi, D., Balachandran, S., & Tampi, R. R. (2018). Frontotemporal dementia: Latest evidence and clinical implications. *Therapeutic Advances in Psychopharmacology*, *8*(1), 33–48.

Zarei, S., Carr, K., Reiley, L., Diaz, K., Guerra, O., Altamirano, P. F., Pagani, W., Lodin, D., Orozco, G., & Chinea, A. (2015). A comprehensive review of amyotrophic lateral sclerosis. *Surgical Neurology International* (Vol. 6,171).

Zhou, H., Arapoglou, T., Li, X., Li, Z., Zheng, X., Moore, J., Asok, A., Kumar, S., Blue, E. E., Buyske, S., Cox, N., Felsenfeld, A., Gerstein, M., Kenny, E., Li, B., Matise, T., Philippakis, A., Rehm, H. L., Sofia, H. J., ... Lin, X. (2023). FAVOR: functional annotation of variants online resource and annotator for variation across the human genome. *Nucleic Acids Research*, *51*(D1). D1300–1311.

https://doi.org/10.1142/9789811293269_bmatter

Appendix A

List of databases for understanding the effects of mutations in proteins

Database	Link
Protein aggregation	
WALTZ-DB 2.0	http://waltzdb.switchlab.org/
AmyLoad	*http://compreclin.iiar.pwr.edu.pl/amyload/*
AmyPro	*https://amypro.net/*
PDB_Amyloid	https://pitgroup.org/amyloid/
A3D	http://biocomp.chem.uw.edu.pl/A3D2/yeast
AL-Base	http://albase.bumc.bu.edu/aldb
AmyloGraph	https://amylograph.com/
AmyloBase	http://150.217.63.173/biochimica/bioinfo/amylobase/pages/view.html
SNPeffect	https://snpeffect.switchlab.org/
CPAD 2.0	https://web.iitm.ac.in/bioinfo2/cpad2/
Protein stability	
ProThermDB	https://web.iitm.ac.in/bioinfo2/prothermdb/.
ThermoMutDB	http://biosig.unimelb.edu.au/thermomutdb/
MPTherm	https://www.iitm.ac.in/bioinfo/mptherm/

(*Continued*)

(*Continued*)

Database	Link
Protein folding rates	
ACPro	https://www.ats.amherst.edu/protein
PFDB	*http://lee.kias.re.kr/~bala/PFDB*
K-Pro	https://folding.biofold.org/k-pro
Intrinsically disordered proteins and regions	
MFIB	http://mfib.enzim.ttk.mta.hu/
DIBS	http://dibs.enzim.ttk.mta.hu/
ELM	http://elm.eu.org
DisProt	https://disprot.org
Mobi-DB	https://mobidb.org
IDEAL	https://www.ideal-db.org
FuzDB	https://fuzdb.org
Binding affinity of protein-protein complexes	
PROXiMATE	https://www.iitm.ac.in/bioinfo/PROXiMATE/
SKEMPI 2.0	https://life.bsc.es/pid/skempi2
MPAD	https://web.iitm.ac.in/bioinfo2/mpad/
Binding affinity of protein-nucleic acid complexes	
dbAMEPNI	*http://zhulab.ahu.edu.cn/dbAMEPNI*
ProNAB	https://web.iitm.ac.in/bioinfo2/pronab/.
PDBbind	http://www.pdbbind.org.cn/
PNATDB	http://chemyang.ccnu.edu.cn/ccb/database/PNAT/
ProNIT	*http://dna00.bio.kyutech.ac.jp/pronit/*
PDBbind	http://www.pdbbind.org.cn/
Binding affinity of protein-carbohydrate complexes	
ProCaff	https://web.iitm.ac.in/bioinfo2/procaff/
ProCarbDB[#]	*http://www.procarbdb.science/procarb*
PDBBind	http://www.pdbbind.org.cn/
BindingDB	https://www.bindingdb.org/bind/index.jsp

(*Continued*)

Database	Link
Cancer hotspots	
HotSpotAnnotations	https://bio.tools/hotspotannotations
Cancer Hotspots	https://www.cancerhotspots.org/
The TP53 Database	https://tp53.isb-cgc.org/
3D Hotspots	https://www.3dhotspots.org/
Cancer mutations	
COSMIC	https://cancer.sanger.ac.uk/cosmic
TCGA	https://www.cancer.gov/ccg/research/genome-sequencing/tcga
ICGC	https://dcc.icgc.org/
CBioPortal	http://cbioportal.org
OncoKB	http://oncokb.org/
Disease-causing mutations	
OMIM	https://www.omim.org/
SwissVar/Uniport	https://www.uniprot.org/
ClinVar	https://www.ncbi.nlm.nih.gov/clinvar/
HuVarBase	https://www.iitm.ac.in/bioinfo/huvarbase/
MaveDB	https://www.mavedb.org/.
HGMD	http://www.hgmd.org.
gnomAD	https://gnomad.broadinstitute.org/
MutHTP	http://www.iitm.ac.in/bioinfo/MutHTP/
HumSavar	http://www.uniprot.org/docs/humsavar
TMSNPdb	http://lmc.uab.es/tmsnp/tmsnpdb
Neutral Mutations	
DbSNP	*http://www.ncbi.nlm.nih.gov/SNP*
1000 Genomes	https://www.internationalgenome.org/home
dbCPM	*http://bioinfo.ahu.edu.cn:8080/dbCPM*

(*Continued*)

(*Continued*)

Database	Link
SARS-CoV-2	
RCoV19	https://ngdc.cncb.ac.cn/ncov/
SARS-CoV-2 MAT	http://hgdownload.soe.ucsc.edu/goldenPath/wuhCor1/UShER_SARS-CoV-2/
CoV-Glue	https://cov-glue.cvr.gla.ac.uk/
CoV-RDB	https://covdb.stanford.edu/susceptibility-data/table-mab-susc/
SARS-CoV-2 Database	https://covid19.sfb.uit.no/about/
CoV3D	https://cov3d.ibbr.umd.edu
ACovPepDB	http://i.uestc.edu.cn/ACovPepDB/
CoV-AbDab	https://opig.stats.ox.ac.uk/webapps/covabdab/
Ab-CoV	https://web.iitm.ac.in/bioinfo2/ab-cov/home
SCoV2-MD	https://submission.gpcrmd.org/covid19/

Databases, which are not accessible on 24th April 2024 are shown in italics

https://doi.org/10.1142/9789811293269_bmatter

Appendix B

List of tools for understanding the effects of mutations in proteins

Tools	Link
Aggregation propensity and aggregation prone regions in a protein	
AGGRESCAN	http://bioinf.uab.es/aggrescan/
Pafig	*http://www.mobioinfor.cn/pafig/*
WALTZ	https://waltz.switchlab.org/
AbAmyloid	http://iclab.life.nctu.edu.tw/abamyloid
FoldAmyloid	http://bioinfo.protres.ru/fold-amyloid/
iAMY-SCM	http://camt.pythonanywhere.com/iAMY-SCM
APPNN	http://cran.r-project.org/web/packages/appnn/index.html
Amylogram	http://www.smorfland.uni.wroc.pl/shiny/AmyloGram/; http://github.com/michbur/AmyloGramAnalysis
ANuPP	https://web.iitm.ac.in/bioinfo2/ANuPP/
TANGO	http://tango.crg.es/
SecStr	http://biophysics.biol.uoa.gr/SecStr/
NetCSSP	http://cssp2.sookmyung.ac.kr/

(*Continued*)

(*Continued*)

Tools	Link
BetaSerpentine	https://bioinfo.crbm.cnrs.fr/index.php?route=tools&tool=25
BETASCAN	http://betascan.csail.mit.edu
AmyloidMutants	http://amyloid.csail.mit.edu/
PASTA 2	http://old.protein.bio.unipd.it/pasta2/
GAP	http://www.iitm.ac.in/bioinfo/GAP/
FISH Amyloid	http://comprec-lin.iiar.pwr.edu.pl/fishInput/
AgMata	https://bitbucket.org/bio2byte/agmata
3D PROFILE	https://services.mbi.ucla.edu/zipperdb/submit
CORDAX	https://cordax.switchlab.org/
PATH	https://github.com/KubaWojciechowski/PATH
AMYLPRED2	http://aias.biol.uoa.gr/AMYLPRED2/
MetAmyl	http://metamyl.genouest.org/
AGGRESCAN3D 2.0	http://biocomp.chem.uw.edu.pl/A3D2/ https://bitbucket.org/lcbio/aggrescan3d
Change in aggregation rate upon mutation	
AggreRATE-Disc	https://www.iitm.ac.in/bioinfo/aggrerate-disc/
AggreRATE-Pred	https://www.iitm.ac.in/bioinfo/aggrerate-pred/
AbsoluRATE-Pred	https://web.iitm.ac.in/bioinfo2/absolurate-pred/
Protein stability change (ΔΔG) upon mutation	
DDGUN	https://folding.biofold.org/ddgun/
FoldX	https://foldxsuite.crg.eu/
CUPSAT	https://cupsat.brenda-enzymes.org/
SDM2/SDM	*http://marid.bioc.cam.ac.uk/sdm2*
POPMuSiC	https://soft.dezyme.com/
Rosetta	https://www.rosettacommons.org/software
SAAFEC-SEQ	http://compbio.clemson.edu/SAAFEC-SEQ/
EASE-MM	https://sparks-lab.org/server/ease-mm/

(*Continued*)

Tools	Link
INPS	https://inps.biocomp.unibo.it/inpsSuite/default/index
PON-tstab	http://structure.bmc.lu.se/PON-Tstab/
MAESTROweb	https://biwww.che.sbg.ac.at/maestro/web
STRUM	http://zhanglab.ccmb.med.umich.edu/STRUM/
mCSM	http://biosig.unimelb.edu.au/mcsm/
NeEMO	http://protein.bio.unipd.it/neemo/
iStable2.0	*http://ncblab.nchu.edu.tw/iStable2*
DeepDDG	http://protein.org.cn/ddg.html
DynaMut	http://biosig.unimelb.edu.au/dynamut/
DUET	http://biosig.unimelb.edu.au/duet/stability
iStable	http://predictor.nchu.edu.tw/iStable/
Protein folding rates and change in folding rates upon mutations	
FOLD-RATE Q	http://bioinformatics.tcu.edu.tw/FOLDRATE20r/foldrate20.htm
SFoldRate	http://gila.bioe.uic.edu/lab/tools/foldingrate/fr0.html
Pred-PFR	http://www.csbio.sjtu.edu.cn/bioinf/FoldingRate/
FoldRate	http://www.csbio.sjtu.edu.cn/bioinf/FoldRate/
Folding RaCe	http://www.iitm.ac.in/bioinfo/proteinfolding/foldingrace.html
FOLD-RATE	https://www.iitm.ac.in/bioinfo/fold-rate/
K-Fold	https://folding.biofold.org/k-fold/
FREEDOM	http://bioinformatics.tcu.edu.tw/FREEDOMr/freedom.htm
FORA	http://bioinformatics.tcu.edu.tw/FORAr/fora.htm

(*Continued*)

(*Continued*)

Tools	Link
Disordered regions in proteins	
IUPred	https://iupred.elte.hu/
IUPRed2A	http://iupred2a.elte.hu
IUPred3	https://iupred3.elte.hu
ANCHOR2	https://iupred2a.elte.hu/
POODLE-I	http://cblab.my-pharm.ac.jp/poodle/poodle.html
SPOT-Disorder2	https://sparkslab.org/server/ spot-disorder2/
IDP-Seq2Seq	*http://bliulab.net/IDP-Seq2Seq*
RFPR-IDP	*http://bliulab.net/RFPR-IDP/server*
DeepIDP-2L	*http://bliulab.net/DeepIDP-2L*
Trans-DFL	*http://bliulab.net/TransDFL*
IDPFusion	*http://bliulab.net/IDP-Fusion/*
fIDPnn	http://biomine.cs.vcu.edu/servers/flDPnn
CLIP	http://biomine.cs.vcu.edu/servers/CLIP
DisoMine	https://www.bio2byte.be/b2btools/disomine
DEPICTER	http://biomine.cs.vcu.edu/servers/DEPICTER
DEPICTER2	http://biomine.cs.vcu.edu/servers/DEPICTER2
Tools for extracting features in proteins, nucleic acids and complexes	
AAindex	https://www.genome.jp/aaindex/
DSSP	https://www3.cmbi.umcn.nl/xssp/
Naccess	http://www.bioinf.manchester.ac.uk/naccess/
HBPLUS	https://www.ebi.ac.uk/thornton-srv/software/HBPLUS/
AL2CO	http://prodata.swmed.edu/al2co/al2co.php
AACon	https://www.compbio.dundee.ac.uk/aacon/
Consurf	https://consurf.tau.ac.il/
DynaMut	https://biosig.lab.uq.edu.au/dynamut/

(*Continued*)

Tools	Link
ENDES	http://sparks.informatics.iupui.edu
NAPS	http://bioinf.iiit.ac.in/NAPS/
PSAIA	http://complex.zesoi.fer.hr/tools/PSAIA.html
Curves+	http://gbio-pbil.ibcp.fr/cgi/Curves_plus/
w3DNA 2.0	http://web.x3dna.org/
Arpeggio	http://structure.bioc.cam.ac.uk/arpeggio/
PDBsum1	https://www.ebi.ac.uk/thornton-srv/software/PDBsum1/
DNAproDB	https://dnaprodb.usc.edu/
ProDFace	http://structbioinfo.iitj.ac.in/resources/bioinfo/pd_interface/
FoldX	http://foldx.embl.de/
RNAmap2D	https://genesilico.pl/software/stand-alone/rnamap2d
3V	http://3vee.molmovdb.org/
Prince	http://www.facweb.iitkgp.ac.in/~rbahadur/prince/home.html
Change in binding affinity upon mutation in protein-protein complexes	
BeAtMuSiC	http://babylone.ulb.ac.be/beatmusic/index.php
BindProf	http://zhanglab.ccmb.med.umich.edu/BindProf/
Mutabind	http://www.ncbi.nlm.nih.gov/projects/mutabind/
BindProfX	https://zhanglab.ccmb.med.umich.edu/BindProfX/
iSEE	https://github.com/haddocking/iSee
mCSM-PPI2	https://biosig.lab.uq.edu.au/mcsm_ppi2/
TopNetTree	https://codeocean.com/capsule/2202829/tree/v1
GeoPPI	https://github.com/Liuxg16/GeoPPI
SAAMBE-3D	http://compbio.clemson.edu/saambe_webserver/

(*Continued*)

(*Continued*)

Tools	Link
ProAffiMuSeq	https://web.iitm.ac.in/bioinfo2/proaffimuseq/
PANDA	*https://pandaaffinity.pythonanywhere.com/*
SAAMBE-SEQ	http://compbio.clemson.edu/saambe_webserver/indexSEQ.php
Hotspot residues in protein-DNA complexes	
PrPDH	http://bioinfo.ahu.edu.cn:8080/PrPDH
inpPDH	http://bioinfo.ahu.edu.cn/inpPDH
sxPDH	https://github.com/xialab-ahu/sxPDH
PreHots	http://dmb.tongji.edu.cn/tools/PreHots/
WTL-PDH	https://github.com/chase2555/WTL-PDH
HISNAPI	http://agroda.gzu.edu.cn:9999/ccb/server/HISNAPI/
Hotspot residues in protein-RNA complexes	
HotSPRing	http://www.csb.iitkgp.ernet.in/applications/HotSPRing/main.
PrabHot	http://denglab.org/PrabHot/.
XGBPRH	https://github.com/SupermanVip/XGBPRH
Change in binding affinity in protein-DNA complexes	
SAMPDI	http://compbio.clemson.edu/SAMPDI-3D/
PremPDI	https://lilab.jysw.suda.edu.cn/research/PremPDI/
PEMPNI	http://liulab.hzau.edu.cn/PEMPNI
SAMPDI-3D	http://compbio.clemson.edu/SAMPDI-3D/
Change in binding affinity in protein-RNA complexes	
mCSM–NA	http://structure.bioc.cam.ac.uk/mcsm_na
PEMPNI	http://liulab.hzau.edu.cn/PEMPNI
PremPRI	https://lilab.jysw.suda.edu.cn/ research/PremPRI/

(*Continued*)

Tools	Link
Binding affinity and change in binding affinity in protein-carbohydrate complexes	
PCA-Pred	https://web.iitm.ac.in/bioinfo2/pcapred/
PCA-MutPred	https://web.iitm.ac.in/bioinfo2/pcamutpred
CSM-carbohydrate	http://biosig.unimelb.edu.au/csm_carbohydrate/
Variant effects/disease-causing mutations	
SIFT	https://sift.bii.a-star.edu.sg/index.html
PROVEAN	http://provean.jcvi.org/index.php
MutationAssessor	http://mutationassessor.org/r3/
LIST_S2	https://list-s2.msl.ubc.ca/
COSMIS	https://github.com/CapraLab/cosmis
EVmutation	http://evmutation.org/
GEMME	www.lcqb.upmc.fr/GEMME/
DeepSequence	https://github.com/debbiemarkslab/DeepSequence
EVE	https://evemodel.org/
DeMaSk	https://demask.princeton.edu/
ESM-1v	https://github.com/facebookresearch/esm/tree/main/examples/variant-prediction
SeqDesign	https://github.com/mjuraska/seqDesign
EvoRator	*https://evorator.tau.ac.il/*
Envision	https://envision.gs.washington.edu/
Sequence UNET	https://sequence-unet.readthedocs.io/en/latest/
nn4dms	https://github.com/gitter-lab/nn4dms
Condel	http://bbglab.irbbarcelona.org/fannsdb/home
SNAP	http://www.rostlab.org/services/SNAP
FATHMM	http://fathmm.biocompute.org.uk

(*Continued*)

(*Continued*)

Tools	Link
PolyPhen-2	http://genetics.bwh.harvard.edu/pph2/
PredictSNP	http://loschmidt.chemi.muni.cz/predictsnp
PON-P2	http://structure.bmc.lu.se/PON-P2/
AlphaMissense	https://github.com/google-deepmind/alphamissense
Pred-MutHTP	https://www.iitm.ac.in/bioinfo/PredMutHTP/
TMSNP	http://lmc.uab.es/tmsnp/
BorodaTM	https://www.iitm.ac.in/bioinfo/MutHTP/boroda.php
mCSM-membrane	http://biosig.unimelb.edu.au/mcsm_membrane/
MutTMPredictor	http://csbio.njust.edu.cn/bioinf/muttmpredictor/
Cancer hotspot residues	
HotMAPS	https://github.com/KarchinLab/HotMAPS
Mutation3D	http://mutation3d.org
QuartPAC	http://bioconductor.jp/packages/3.1/bioc/html/QuartPAC.html
OncodriveCLUST	http://bg.upf.edu/oncodriveclust
HotSpot3d	https://github.com/ding-lab/hotspot3d
CanProSite	https://web.iitm.ac.in/bioinfo2/CanProSite/
MutBLESS	https://web.iitm.ac.in/bioinfo2/MutBLESS/index.html
Cancer mutations	
CanDrA	http://bioinformatics.mdanderson.org/main/CanDrA
CScape	http://CScape.biocompute.org.uk/
CHASMplus	http://karchinlab.github.io/CHASMplus
CADD	https://cadd.gs.washington.edu/
MVP	https://github.com/ShenLab/missense

(*Continued*)

Tools	Link
GBM Driver	https://web.iitm.ac.in/bioinfo2/GBMDriver/index.html
OncodriveFM	http://www.intogen.org/oncodrivefml
IntOGen-mutations	http://www.intogen.org/mutations/
MetaSVM	https://sites.google.com/site/sunghwanshome/
SARS-CoV-2	
Vcorn	http://www.plant.osakafu-u.ac.jp/~kagiana/vcorn/sarscov2/22/
SARS2Mutant	http://sars2mutant.com/
Nexstrain	https://nextstrain.org/sars-cov-2/
CoV2K	http://gmql.eu/cov2k/api/
MicroGMT	https://github.com/qunfengdong/MicroGMT
CovMT	https://www.cbrc.kaust.edu.sa/covmt/index.php?p=home
COVID-19 CG	https://covidcg.org/

Tools not available on 24th April are shown in italics

https://doi.org/10.1142/9789811293269_bmatter

Index

www.ingramcontent.com/pod-product-compliance
Lightning Source LLC
Jackson TN
JSHW010846180125
77306JS00002B/125
* 9 7 8 9 8 1 1 2 9 3 2 5 2 *